Springer Monographs in Mathematics

G.F. Roach

An Introduction to Echo Analysis

Scattering Theory and Wave Propagation

Emeritus Professor G.F. Roach, BSc; MSc; PhD; DSc; ScD; FRSE; FRAS; C.Math; FIMA; FRSA
Department of Mathematics, University of Strathclyde,
26 Richmond Street, Glasgow, G1 1XH, UK

Mathematics Subject Classification (2000): 35A05, 35P25, 35P05

British Library Cataloguing in Publication Data
A catalogue record for this book is available from the British Library

Library of Congress Control Number: 2007922045

ISBN-13: 978-1-84628-851-7   e-ISBN-13: 978-1-84628-852-4
DOI: 10.1007/978-1-84628-852-4

Printed on acid-free paper

9 8 7 6 5 4 3 2 1

Springer Science+Business Media
springer.com

Preface

One of the more promising non-destructive means of diagnosing the properties of quite complicated materials is to use wave energy as a probe. An analysis of this technique requires a detailed understanding of first how signals evolve in the medium of interest in the absence of inhomogeneities and, second, the nature of the scattered or echo field when the original signal is perturbed by inhomogeneities which might exist in the medium. The overall aim of the analysis is to calculate relationships between an unperturbed signal waveform and an associated echo waveform and indicate how these relationships can be used to characterise inhomogeneities in the medium.

An initial aim of this monograph is to give a largely self-contained, introductory account of acoustic wave propagation and scattering in the presence of time independent perturbations. Later chapters of the book will indicate how the approach adopted here for dealing with acoustic problems can be extended to cater for similar problems in electromagnetism and elasticity.

In this monograph we gather together the principal mathematical topics which are used when dealing with wave propagation and scattering problems involving time independent perturbations. In so doing we will provide a unified and reasonably self-contained introduction to an active research area which has been developing over recent years. We will also indicate how the material can be used to develop constructive methods of solution. The overall intention is to present the material so that is just as persuasive to the theoretician as to the applied scientist who may not have the same mathematical background. This book is meant to be a guide which indicates the technical requirements when investigating wave scattering problems. Throughout the emphasis will be on concepts and results rather than on the fine detail of proof. The proofs of results which are simply stated in the text, many of which are lengthy and very much of a technical nature, can be found in the references cited.

Many of the results described in this book represent the works of a large number of authors and an attempt has been made to provide a reasonably comprehensive Bibliography. However, particular mention must be made of the pioneering works of Ikebe, Lax and Phillips and of Wilcox. The influence of the works of these authors has been considerable and is gratefully acknowledged. In particular, a profound debt of gratitude is owed to Rolf Leis and Calvin Wilcox who have been such an inspiration over the years.

I would also like to express my gratitude to the many colleagues with whom I have had such useful discussions. In particular, I would thank Christos Athanasiadis, Aldo Belleni-Morante, Wilson Lamb and Ioannis Stratis who have read various parts of the manuscript and offered so many suggestions.

A special word of thanks must also go to Mary Sergeant for the patient way in which she tackled the seemingly endless task of typing and proof corrections.

Finally, I want to express my appreciation to Karen Borthwick and Helen Desmond of Springer Verlag for their efficient and friendly handling of the publication of this book.

GFR
Glasgow
2005

Contents

1

Introduction and Outline of Contents

The study of wave phenomena in a material can often yield significant information about the characteristics of that material. In this monograph we shall be interested in how this can be achieved in a number of areas in the applied sciences. This study has been motivated by problems arising in such areas as, for example, radar, sonar, non-destructive testing and ultrasonic medical diagnosis. In all these areas a powerful diagnostic is the dynamical response of the media to a given excitation.

The fundamental problem with which we shall be concerned is, in its simplest terms, the following one.

A system of interest consists of a medium, a source of energy, that is, a "transmitter", and a detector , that is, a "receiver". The transmitter emits a signal which is detected at the receiver, possibly after becoming perturbed, that is, scattered, by some inhomogeneity in the medium. We are interested in the manner in which the signal evolves throughout the given medium and in the form it assumes, with or without being scattered, at the receiver.

We take as a starting point the assumption that all media involved consist of a continuum of interacting, infinitesimal elements. Consequently, a disturbance in some small region of a medium will induce an associated disturbance in neighbouring regions with the result that some sort of disturbance eventually spreads throughout the medium. The progress or evolution of such disturbances we shall call **propagation**. In this book we will be particularly interested in those cases when the disturbance is a wave. Typical examples of this phenomenon include, for instance, waves on water where the medium is the layer of water close to the surface, the interaction forces are fluid pressure and gravity and the resulting waveform is periodic. Again, acoustic waves in gases, liquids and solids are supported by an elastic interaction and exhibit a variety of waveforms which can be, for example, sinusoidal, periodic, transient pulse or arbitrary. In principle any type of waveform can be set in motion in a given system provided suitable initial or source conditions are imposed. These various processes can be conveniently expressed, symbolically, in the form

$$u(x, t) = U(t - s)u(x, s)$$

where $u(x, t)$ is a quantity which describes the **state** of the system at the point x and time t whilst $U(t - s)$ is a quantity which characterises the evolution, in time, of the system from an initial state $u(x, s)$ to a state $u(x, t)$. Furthermore, if $V(t - s)$ characterises the evolution of some other system then we can expect that a comparison, in some suitable sense, of $U(t - s)$ and $V(t - s)$ could provide information regarding wave phenomena in the two systems. In the following chapters we shall discuss means of determining the form of U and V, both in abstract and in certain specific cases of practical interest, and indicate how they might be used in the development of constructive methods for determining the wave processes in the two systems.

In the earlier parts of the book attention will be confined to an investigation of the properties of the **scalar wave equation** as it appears in acoustics. Later material will be devoted to an examination of wave equations appearing in studies of the elastic field and the electromagnetic field.

Our main interest here will be in those physical phenomena whose evolution can be described in terms of propagating waves. A simple example of such phenomena is a physical quantity, y, that is defined in the form

$$y(x, t) = f(x - ct), \quad (x, t) \in \mathbf{R} \times \mathbf{R} \tag{1.1}$$

where c is a real constant. We notice that y has the same value at all those x and t for which $(x - ct)$ has same values. For example, the function f defined by

$$f(x - ct) = \exp(-(x - ct))$$

has the value one when $(x, t) = (0, 0)$ and also whenever x and t have values which ensure that $(x - ct) = 0$. Thus (1.1) represents a wave which moves with constant velocity c along the *x–axis*, without changing shape. If, further, f is assumed to be sufficiently differentiable then on differentiating (1.1) twice with respect to x and t we obtain

$$\left\{\frac{\partial^2}{\partial t^2} - c^2 \frac{\partial^2}{\partial x^2}\right\} y(x,t) = 0, \quad (x, t) \in \mathbf{R} \times \mathbf{R} \tag{1.2}$$

Similarly, we notice that a physical quantity, w, defined by

$$w(x, t) = g(x + ct), \quad (x, t) \in \mathbf{R} \times \mathbf{R}$$

where c is a real constant, represents a wave moving with constant velocity c along the x-axis without changing shape, but moving in the opposite direction to the wave $y(x, t)$ defined in (1.1). Furthermore, we see that $w(x, t)$ also satisfies an equation of the form (1.2). The equation (1.2) is referred to as the **classical**

wave equation. The prefix "classical" will only be used in order to avoid possible confusion with other wave equations which might come into consideration.

Because the wave equation is linear the compound wave

$$u(x, t) = y(x, t) + w(x, t) = f(x - ct) + g(x + ct) \tag{1.3}$$

where f, g are arbitrary functions, is also a solution of the wave equation. This is the celebrated d'Alembert solution of the wave equation. In a specific problem the functions f and g are determined in terms of the imposed initial conditions which solutions of (1.2) are required to satisfy [3],[4].

We would emphasise at this stage that not all solutions of the wave equation yield propagating waves. For example, if the wave equation can be solved using a separation of variables technique then so-called **stationary wave** solutions can be obtained. They are called stationary waves since they have certain features such as nodes and anti-nodes which retain their position permanently with respect to time. Such solutions can be related to the **bound states** appearing in quantum mechanics and to the **trapped wave** phenomenon of classical wave theory.

In practical situations experimental measurements are mainly made far from the inhomogeneity, that is, in the so-called far field of the scatterer.

In the particular case of acoustic wave scattering processes, we shall see later that in the absence of any scatterers, the acoustic field, $u_0(x, t)$, has the asymptotic form

$$u_0(x,t) = \frac{s(x \cdot \theta_0 - t + |x_0|)}{|x_0|} + O\left(\frac{1}{|x_0|^2}\right), \quad |x_0| \to \infty$$

The quantity s is defined by

$$s(\tau) = \frac{1}{4\pi} \int_{|y-x_0| \le \delta_0} f(\theta_0.(y - x_0) - \tau, y)\, dy, \quad \tau \in \mathbf{R}$$

where f is a source function which characterises the signal emitted by a transmitter placed at a point x_0 and θ_0 is a unit vector defined by $x_0 = -|x_0|\,\theta_0$. The quantity s is referred to as the **signal waveform.**

When the signal from the transmitter is scattered by some inhomogeneity in the medium then the scattered field, $u_s(x, t)$, can be shown to have the asymptotic form

$$u_s(x,t) = \frac{e(|x| - t, \theta, \theta_0)}{|x|} + O\left(\frac{1}{|x|^2}\right), \quad |x| \to \infty$$

where $e(\tau, \theta, \theta_0)$ is called the **echo waveform**. It depends on θ_0, the direction of incidence of the signal, and θ, the direction of observation.

The principal aim in practical wave scattering problems is to calculate the relationship between the echo and signal waveforms and to highlight their dependency on the source function and the scatterers. We shall see, in the following chapters, that many of the results obtained when developing an associated scattering theory lead to efficient, constructive means of performing these calculations for a wide range of physically relevant problems.

Scattering means different things to different people. Broadly speaking it can be thought of as the interaction of an evolutionary process with a non-homogeneous medium which might also have a non-linear response. In the following chapters a mathematical framework will be developed which is particularly well suited to the study of scattering processes from both a theoretical and a constructive point of view. The intention throughout will be to present the material so that it is just as persuasive to mathematicians interested in the spectral analysis of initial boundary value problems for partial differential equations as to the applied scientist who, understandably, might require more quantitative results from their mathematical model. Consequently, certain sections and chapters can be safely passed over by those already familiar with the material.

Scattering phenomena arise as a consequence of some perturbation of a given system and they are analysed by developing an associated **scattering theory**. These scattering theories are concerned with the comparison of two systems, one being regarded as a perturbation of the other, and with the study of the manner in which the two systems might eventually become equal in some sense. As we shall see, this will lead quite automatically to a discussion of certain entities called **wave operators**. The availability of an appropriate scattering theory would, therefore, seem to offer good prospects for providing a sound basis from which to develop robust approximation methods for obtaining solutions to some given, but difficult, problem in terms of a more readily obtainable solution to an associated problem.

There are two main formalisms available when developing a scattering theory, the time dependent and the time independent. The time dependent formalism deals with a time dependent **wave function**, that is, with a solution of the wave equation which describes the evolution of the wave as it actually occurs. In the distant past the wave is virtually unaffected by the presence of any scatterer. Hence, the corresponding wave function might be expected to behave asymptotically when $t \to -\infty$ like the wave function for a truly unperturbed or free problem. Similarly, in the distant future, when again the wave can be expected to be virtually unaffected by any scatterer, the corresponding wave function might be expected to behave asymptotically when $t \to +\infty$ like the wave function for some other free problem. It turns out, as we shall see, that it is possible to relate the essentially free wave function which obtains in the distant past before any scattering takes place to the essentially free wave function which obtains in the distant future well after any scattering has occurred. This connection is effected by means of a certain operator, denoted by S, called the **scattering operator**.

Since it is mainly the asymptotically free motions which are observed in practice the single operator S contains all the information of experimental interest.

Thus, if we know how to compute the scattering operator, S, then the scattering problem is solved. Working with these asymptotic values is experimentally sensible and, in addition, it avoids any explicit discussion of the, often quite complicated, nature of the wave function near the scatterer.

In a time-dependent scattering theory the time evolution of the states of a system and their asymptotic behaviour for large times are of dominant interest. This gives rise to the so-called **time domain analysis** of a given problem.

The time-independent formalism arises as a result of trying to expand the actual wave function in terms of so-called stationary scattering states. These states are obtained as a consequence of separating out, in some way, the time dependence in the problem and then studying the solutions of the resulting spatial equations. This approach gives rise to the so-called **frequency domain analysis** of a given problem in which a central interest is the asymptotic behaviour of solutions at large distances. These stationary states turn out to be eigenfunctions, often in some generalised sense, of an associated spatial operator. A principal usefulness of this formalism is that it provides a powerful means for dealing with the actual computation of the scattering operator and for establishing its properties.

We have already remarked that not all states of a system need, necessarily, lead to scattering events. However, those states of a system that do lead to scattering events and which, as $t \to \pm\infty$, are **asymptotically equal** (AE), in some sense, to scattering states of some simpler system are said to satisfy an **asymptotic condition.** A precise description of this condition is a principal study of time-dependent scattering theory. As we shall see, once we have introduced the appropriate mathematical structures, it can be formulated abstractly in terms of a pair of one-parameter groups of linear operators $\{U(t)\}$ and $\{V(t)\}$ with $t \in \mathbf{R}$, the relation between these groups usually being given in terms of their infinitesimal generators. We remark that one of the reasons for the name "time-independent scattering theory" is the fact that in this theory these groups can be replaced by the resolvent of related spatial operators. The importance of the time-independent theory lies in the fact that the actual calculation of various expressions can often be more readily carried out than for similar expressions occurring in the time-dependent theory.

In the quantum mechanics of the scattering of elementary particles by a potential the wave packets describing scattered particles can be shown to be AE, for large time, to the corresponding wave packets for free particles. The correspondence between these two systems, one describing the scattered particles the other describing the free particles, is effected by means of so-called Møller operators [3]. Wilcox [5] has developed analogous concepts for wave propagation problems associated with the equations of classical physics. Specifically, Wilcox showed that waves propagating in an inhomogeneous medium are AE, for large time, to corresponding waves propagating in an homogeneous medium. The correspondence between the two wave systems is given by an analogue of the Møller operators of quantum scattering which we shall simply refer to as **wave operators** (WO). Since wave propagation problems in homogeneous media can often be solved explicitly, a knowledge of an appropriate WO would then provide information concerning the asymptotic behaviour of solutions to wave propagation problems in inhomogeneous media. We remark that the Wilcox approach leads to the comparison of unitary groups whilst an alternative approach developed by Lax and Phillips [2] compares so-called incoming and outgoing subspaces.

Mathematically, many scattering problems can be conveniently modelled in terms of **initial boundary value problems (IBVP)** for the wave equation in either scalar or vector form.

For much of this monograph we shall be concerned with so-called **direct problems** for which the signal and target characteristics are known and the aim is to predict the scattered field. Nevertheless, we would remark that of prime practical interest are the much more difficult **inverse problems** in which the initiating signal and the scattered field are known and the target characteristics have to be determined. However, before we can comfortably start to discuss inverse problems a proper understanding of direct problems is required. Therefore, to fix ideas here we shall confine attention, initially, to direct problems and to IBVPs that have the following typical form.

Determine a quantity $u(x, t)$ which satisfies

$$\{\partial_t^2 + L(x, t)\}\, u(x, t) = f(x, t), \quad (x, t) \in Q \tag{1.4}$$

$$u(x, s) = \varphi_s(x), \quad u_t(x, s) = \psi_s(x), \quad x \in \Omega(s) \tag{1.5}$$

$$u(x, t) \in (bc)(t), \quad (x, t) \in \partial\Omega(s) \times \mathbf{R} \tag{1.6}$$

where

$$L(x, t) = -c^2\Delta + V(x, t) \tag{1.7}$$

with Δ denoting the usual Laplacian in $\mathbf{R}^n$, and

$$Q \subset \{(x, t) \in \mathbf{R}^n \times \mathbf{R}\}$$

$$\Omega(t) := \{x \in \mathbf{R}^n : (x, t) \in Q\}$$

$$B(t) := \{x \in \mathbf{R}^n : (x, t) \notin Q\}$$

The region Q is open in $\mathbf{R}^n \times \mathbf{R}$ and $\Omega(t)$ denotes the exterior, at time t, of a scattering target $B(t)$. For each value of t the domain $\Omega(t)$ is open in $\mathbf{R}^n$ and is assumed to have a smooth boundary $\partial\Omega(t)$. The lateral surface of Q, denoted ∂Q, is defined by

$$\partial Q := \bigcup_{t \in I} \partial\Omega(t) \tag{1.8}$$

where $I := \{t \in R : 0 \leq t < T\}$. The quantities f, c, φ_s, ψ_s and V are given data functions and $s \in R$ denotes a fixed initial time. The notation in (1.6) indicates that the solution, $u(x, t)$, is required to satisfy certain conditions, denoted (bc), imposed on the boundary which, in general, might depend on t.

Problems of the form (1.4) to (1.6) divide quite naturally into the following two broad classes.

P1: (i) $V(x, t) \equiv 0$

(ii) $(bc)(t)$ imposed

P2: (i) $V(x, t) \neq 0$

(ii) $(bc)(t)$ not imposed.

Problems of the form P1 are referred to as **target scattering** problems whilst problems of the form P2 describe **potential scattering**.

The first thing that has to be done when analysing any such problems is to declare the type of solution that is actually being sought; that is, a **solution concept** has to be introduced. To this end we often make use of $C^m(\mathbf{R}^n \times \mathbf{R}, \mathbf{R})$ which denotes the collection of m-times continuously differentiable functions of $x \in \mathbf{R}^n$ and $t \in [0,T)$, $T > 0$ which have values in $\mathbf{R}$.

A local classical solution of (1.4) to (1.6) is an element

$$u \in C(\mathbf{R}^n \times [0,T), \mathbf{R}) \cap C^2(\mathbf{R}^n \times (0,T), \mathbf{R}) \tag{1.9}$$

which for some $T > 0$, satisfies (1.4) to (1.6).

A local classical solution is called a **global classical solution** if we can take $T = \infty$.

A ***p*-periodic classical solution** is a global classical solution which is p-periodic in $t \in R$.

Since we shall be concerned with IBVP that model real-life situations we shall require that the problem is **well-posed**, that is,

- a solution exists and is unique for a large class of initial data,
- the solution depends continuously on the given data.

Even when it is possible in principle, the actual determination of a classical solution is often a very difficult task in practice. The situation can be eased considerably by realising, that is, by interpreting or seeking solutions of, the IBVP (1.4) to (1.6) in some more convenient collection of functions than that indicated in (1.9). For example, consider the case of potential scattering as indicated in P2 above but with V independent of t. We might choose to realise the IBVP (1.4)–(1.6) in the collection $L_2(\mathbf{R}^n \times \mathbf{R}) \equiv L_2(\mathbf{R}^n \times \mathbf{R}, \mathbb{C})$ the familiar space of (equivalence classes of) square integrable functions with domain $\mathbf{R}^n \times \mathbf{R}$ and with values in $\mathbb{C}$. With the understanding that we would then have to work with distributional rather than classical derivatives we can obtain a realisation of the IBVP (1.4)–(1.6) in the collection $L_2(\mathbf{R}^n \times \mathbf{R}) \equiv L_2(\mathbf{R}^n \times \mathbf{R}, \mathbb{C})$ by introducing the spatial mapping or operator A defined by (see Chapter 2 for more analytical details)

$$\begin{aligned} \mathrm{A}\colon L_2(\mathbf{R}^n, \mathbb{C}) &\to L_2(\mathbf{R}^n, \mathbb{C}) \equiv L_2(\mathbf{R}^n) \\ Av &= -\Delta v + V, \quad v \in D(A) \end{aligned} \tag{1.10}$$

where $D(A)$ denotes the **domain of definition** of A which in this instance is taken to be

$$D(A) = \{v \in L_2(\mathbf{R}^n)\colon -\Delta v + V \in L_2(\mathbf{R}^n),\ v \in (bc)\}$$

The classical IBVP (1.4)–(1.6) can now be replaced by the following abstract Initial Boundary Problem (IVP)

$$\{\partial_t^2 + \mathrm{A}\}u(x, t) = f(x, t), \quad (x, t) \in \mathbf{R}^n \times \mathbf{R} \tag{1.11}$$

$$u(x, s) = \varphi_s(x), \quad u_t(x, s) = \psi_s(x), \quad x \in \mathbf{R}^n \tag{1.12}$$

the boundary conditions (1.6) having been accommodated in the definition of $D(A)$.

If we now repeat this strategy for all the problems which could come under the heading of P1 and P2 above, then we arrive at the following hierarchy of IVPs.

$$\{\partial_t^2 + A_j(t)\}u_j(x, t) = f_j(x, t), \quad (x, t) \in \mathbf{R}^n \times \mathbf{R} \tag{1.13}$$

$$u_j(x, s) = \varphi_j(x, s), \quad \frac{\partial}{\partial t}u_j(n, s) = \psi_j(x, s) \tag{1.14}$$

where $j = 0, 1, 2, 3, 4$, $s \in \mathbf{R}$ is a fixed initial time and φ_j, ψ_j, f_j are given data functions. The operators $A_j(t)\colon H \to H =: L_2(\mathbf{R}^n)$, $j = 0, 1, 2, 3, 4$ which appear in the above have the following specific forms.

$$\begin{aligned} A_0u_0 &= -c^2\Delta u_0 =: L_0u_0, \quad u_0 \in D(A_0) \\ D(A_0) &= \{u_0 \in H\colon L_0u_0 \in H\} \end{aligned} \tag{1.15}$$

$$\begin{aligned} A_1u_1 &= \{-c^2\Delta + V\}u_1 =: L_1u_1, \quad u_1 \in D(A_1) \\ D(A_1) &= \{u_1 \in H\colon L_1u_1 \in H\} \end{aligned} \tag{1.16}$$

$$\begin{aligned} A_2u_2 &= -c^2\Delta u_2 =: L_2u_2, \quad u_2 \in D(A_2) \\ D(A_2) &= \{u_2 \in H\colon L_2u_2 \in H,\ u_2 \in (bc)\} \end{aligned} \tag{1.17}$$

$$\begin{aligned} A_3(t)u_3 &= \{-c^2\Delta + V(x, t)\}u_3 =: L_3(x, t)u_3, \quad u_3 \in D(A_3(t)) \\ D(A_3(t)) &= \{u_3 \in H\colon L_3(\cdot, t)u_3 \in H\} \end{aligned} \tag{1.18}$$

$$\begin{aligned} A_4(t)u_4 &= -c^2\Delta u_2 =: L_4u_4, \quad u_4 \in D(A_4(t)) \\ D(A_4(t)) &= \{u_4 \in H\colon L_4u_4 \in H,\ u_4 \in (bc)(t)\} \end{aligned} \tag{1.19}$$

We shall see later that as a consequence of the **Duhamel Principle** we can, without any loss of generality, confine attention to problems involving a homogeneous form of the equation (1.13).

We remark that A_0 generates a **free problem** (FP) whilst A_1 and A_2 each generate a **perturbed problem** (PP) with respect to the FP generated by A_0. These operators are clearly independent of t and as such give rise to so-called **autonomous problems** (AP). The operators $A_3(t)$ and $A_4(t)$, being dependent on time, generate so-called **non-autonomous problems** (NAPs) which are PPs with respect to both the FP generated by A_0 and also the problems generated by A_1 and A_2. We also notice that each of these operators generate an IVP. The boundary conditions required when discussing target scattering centred on either A_2 or $A_4(t)$ are accommodated in the definition of the domain of the operators.

An investigation of NAPs is complicated and technically more demanding than that for APs. Such problems will not be discussed here. However, an overview of the main features of such problems together with a set of references for further reading can be found in the Commentary.

Our principal interests in this monograph are centred on IVPs that have the generic form (1.11), (1.12). These encompass a wide range of physically significant problems in such fields as acoustics, electromagnetics, elasticity and their various combinations. Throughout we shall have four main aims.

- To provide results concerning the existence and uniqueness of solution to such IVPs.

  This we have to do in order that we can be in a position meaningfully to compare systems governed by such IVPs. The required results will be obtained using semigroup theory.

- To provide means of interpreting the various solution forms that might be obtained.

  This will involve a detailed investigation of the spectral properties of such spatial operators as A in (1.11) and the development of associated generalised eigenfunction expansion theorems.

- To provide practically realistic mechanisms for comparing systems governed by such IVPs.

  This will involve a detailed study of the relevant wave operators and the scattering operator.

- To develop a constructive scattering theory which will yield, in particular, computable expressions for the echo wave form.

  This aspect could often be suggested by results from the previous aims.

We shall first examine these aspects in an abstract setting and then illustrate their use by working in detail some particular problems in acoustics, electromagnetics and elasticity.

Since this book is meant to be an introductory text, for some if not all readers, it is felt that it should be as self-contained as possible. The intention is first to give, at a leisurely pace, a reasonably comprehensive overview of the various concepts, techniques and strategies associated with the development of constructive scattering theories.

To give some indication of the various strategies that we shall adopt we shall fix ideas by confining attention, for the moment, to the IVP (1.11), (1.12) with the source term f set to zero for the reason explained above and with A replaced by A_0 to characterise the FP and with A replaced by A_I for the PP, the initial conditions in each case being appropriately subscripted. We now proceed in an entirely formal manner. First, we will make what appears to be an outrageous assumption; we assume that in (1.11) the operator A can be treated as a constant. By doing this we effectively suppress the x dependence in (1.11) and (1.12). This will imply that the partial differential equation (1.11) involving the unknown quantity (number), $u(x, t)$, can be replaced by an ordinary differential equation for the unknown quantity (function) $u(\cdot, t)$. This procedure will be made more precise in the following chapters. If we now solve this ordinary differential equation for $u(\cdot, t)$ and replace the x dependence we then find that the FP (with $s = 0$ for convenience) has a solution, denoted by $u_0(x, t)$, which can be written in the form

$$u_0(x,t) = \left(\cos\left(tA_0^{1/2}\right)\right)\varphi_0(x) + A_0^{-1/2}\left(\sin\left(tA_0^{1/2}\right)\right)\psi_0(x) \tag{1.20}$$

If we now assume $\varphi_0 \in L_2(\mathbf{R}^n)$ and $\psi_0 \in D\left(A_0^{-1/2}\right)$ and define

$$h_0(x) := \varphi_0(x) + iA_0^{-1/2}\psi_0(x) \tag{1.21}$$

then

$$u_0(x, t) = \operatorname{Re}\{v_0(x, t)\} \tag{1.22}$$

where

$$v_0(x,t) = \exp\left(-itA_0^{1/2}\right)h_0(x) =: U_0(t)h_0(x) \tag{1.23}$$

is the complex-valued solution of the FP (1.11), (1.12). The quantity $U_0(t)$ is called an **evolution operator** which maps $h_0(x)$, the initial state of the system, into $v_0(x, t)$, the state of the system at some other time t.

For PPs we replace A by A_1, where with a slight abuse of notation, A_1 will be understood to represent one or other of the operators appearing in (1.16) to (1.19). On repeating the same procedure as for the FP we obtain a solution, $u_1(x, t)$, for the PP in the form

$$u_1(x,t) = \left(\cos\left(tA_1^{1/2}\right)\right)\varphi_1(x) + A_1^{-1/2}\left(\sin\left(tA_1^{1/2}\right)\right)\psi_1(x) \tag{1.24}$$

If $\varphi_1 \in L_2(\mathbf{R}^n)$ and $\psi_1 \in D\left(A_1^{-1/2}\right)$ and if we define

$$h_1(x) := \varphi_1(x) + iA_1^{-1/2}\psi_1(x) \tag{1.25}$$

then we can re-write (1.24) in the form

$$u_1(x, t) = \mathrm{Re}\{v_1(x, t)\} \tag{1.26}$$

where

$$v_1(x,t) = \exp\left(-itA_1^{1/2}\right)h_1(x) =: U_1(t)h_1(x) \tag{1.27}$$

The quantity $U_1(t)$ is an **evolution operator** which maps $h_1(x)$, the initial state of the perturbed system, into $v_1(x, t)$, the state of the perturbed system at some later time t.

In order to develop a scattering theory we must first show that the FP and the PP have solutions in some convenient collection of functions and, moreover, that they can be meaningfully written in the form (1.23) and (1.27), respectively. Once this has been done then we will need a means of comparing these solutions, that is, we will need some suitable formula for measuring their distance apart as either $t \to \pm\infty$ or $x \to \infty$. In general, such a formula is called a **norm**. A familiar example of this is the formula for measuring the distance, in $\mathbf{R}^2$, between the origin, O, and a point P with coordinates (x, y). The length of the line OP, denoted $\| OP \|$, is well known to be

$$\| OP \|^2 := x^2 + y^2 \tag{1.28}$$

In (1.28) we are working with *numbers*. However, we shall see (Chapter 3) that the notion of a norm can be generalised so that we can measure the separation of two *functions* rather than the separation of the *numerical values* of the functions. Being able to do this will have the advantage that we will be able to settle questions of existence and uniqueness of physically significant solutions of the FP and the PP more readily than by working with numerical values. With all this in mind we shall assume here, for the sake of illustration, that the solutions we require belong to some suitable collection of functions, denoted by X, say. We will then write the *number* $v(x, t)$ in the form

$$v(x, t) = v(\cdot, t)(x) = v(t)\mathrm{x} \tag{1.29}$$

and refer to v as an X-valued function of t, that is, for each value of t we have $v(t) \in X$. We then compare the solutions of the FP and the PP by considering expressions of the form

$$\| v_1(t) - v_0(t) \| \tag{1.30}$$

where $\| \cdot \|$ denotes some suitably chosen norm defined, on X, in terms of spatial coordinates. We then find that

$$\begin{aligned} \| v_1(t) \to v_0(t) \| &= \| U_1(t)h_1 - U_0(t)h_0 \| \\ &= \| U_0^*(t)U_1(t)h_1 - h_0 \| \\ &=: \| \Omega(t)h_1 - h_0 \| \end{aligned} \tag{1.31}$$

where $*$ denotes that $U_0(t)$ has a modified form (it will be something like an inverse) as a result of moving it from one quantity to another. The time dependence in the above comparison can be removed by taking the limit as $t \to \pm\infty$. We then obtain

$$\lim_{t\to\pm\infty} \|v_1(t) - v_0(t)\| = \|\Omega_\pm h_1 - h_0\| \tag{1.32}$$

where

$$\Omega_\pm := \lim_{t\to\pm\infty} \Omega(t) = \lim_{t\to\pm\infty} U_0^*(t)U_1(t) \tag{1.33}$$

If it can be shown that the two solutions $v_1(t)$ and $v_0(t)$ exist and that the various steps leading to (1.32) can be justified then it will still remain to show that the limits in (1.33), which define the so-called **wave operators** (WOs), $\Omega_\pm$, actually exist.

When all the above has been achieved we see that if the initial data for the FP and PP are related according to

$$h_0 = \Omega_\pm h_1 \tag{1.34}$$

then the limit in (1.32) is zero thus indicating that the PP is asymptotically free as $t \to \pm\infty$. That is, solutions of the PP with initial data (state) h_1 are time asymptotically equal to solutions of an FP with initial data (state) h_0 which is given by (1.34).

Consequently, if solutions of the two systems are known to exist in the form (1.23) and (1.27) then, keeping (1.34) in mind, we would expect there to exist elements $h_\pm$ such that

$$v_1(t) \sim U_0(t)h_\pm \quad \text{as} \quad t \to \pm\infty \tag{1.35}$$

where $\sim$ will be used to denote **asymptotic equality** (AE) and the $\pm$ are used to indicate, the possibly different, limits as $t \to \pm\infty$. We would emphasise that it is not automatic that both the limits implied by (1.35) should exist. Indeed a solution such as v_1 could be asymptotically free as $t \to +\infty$ but not as $t \to -\infty$. If we combine (1.27) and (1.35) then we have

$$U_0^*(t)U_1(t)h_1 =: \Omega(t)h_1 \sim h_\pm \tag{1.36}$$

Thus, we can conclude that

$$\Omega_\pm : h_1 \to h_\pm \tag{1.37}$$

The two initial conditions $h_\pm$ for the FP which yield the initial condition for the PP are related. This is illustrated by noticing that the above discussion implies

$$h_+ = \Omega_+ h_1 = \Omega_+ \Omega_-^* h_- =: Sh_- \tag{1.38}$$

where S is called the **scattering operator** (SO) for the problem.

Although the above discussion has been abstract and almost entirely formal, nevertheless it has been sufficient to indicate that the following strategy could be profitably adopted when investigating wave phenomena in specific physical problems.

- Represent in some suitable collection of functions, the given physical problems, FP and PP, as operator equations which are characterised by self-adjoint operators A_0 and A_1 respectively.
- Obtain existence and uniqueness results for solution to the FP and PP.
- Determine the spectral properties of A_0 and A_1. This will enable appropriate (generalised) eigenfunction expansion theorems to be obtained by means of which the various solution forms indicated above can be interpreted in a constructive manner.
- Investigate the AE of solutions to the FP and PP and determine conditions which ensure that the WOs, $\Omega_\pm$, exist.
- Obtain far-field approximations for solutions to the FP and PP in terms of the SO.
- Determine the structure of WOs and SO in terms of generalised eigenfunctions.
- Investigate whether or not *all* solutions of the perturbed system are asymptotically free as $t \to \pm\infty$. This is the so-called **asymptotic completeness** property. It is closely related to the existence of such elements as $h_\pm$ which were introduced in (1.35).
- Provide computable representations of solutions to the mathematical problems which have been introduced.
- Provide computable relations between the signal and echo wave forms.
- Develop robust approximation methods.

The above strategy only has credibility if all the steps leading up to it can be justified, that is, made more precise mathematically. To be able to do this there are two immediate requirements. First, we must introduce into the discussion elements of linear operator theory. This will provide a natural framework within which the various physical problems can be given a sound mathematical formulation. Second, we must be able to provide an interpretation of such quantities as $A^{1/2}$ and $\cos(tA^{1/2})$ which have appeared above. We shall see that we will be able to do this by using results from the spectral theory of operators and the theory of semigroups of operators. Broadly speaking, we shall use semigroup theory to obtain existence and uniqueness results and spectral theory to develop constructive methods. Furthermore, with this additional mathematical structure available we shall be able to give a precise meaning to what is actually meant by wave scattering. So far, our understanding of this has been entirely intuitive.

A Commentary on the material presented in this monograph will be found in the final chapter. This will provide more historical background, additional references and a guide to further reading and more specialised texts. The remaining chapters of the book are arranged as follows.

In Chapter 2 we provide some motivation for the remaining chapters. This we do by first recalling a number of elementary concepts and results from the classical analysis of waves on a string. Of particular importance for later analysis is the Duhamel Principle introduced at this stage. The ideas of scattering processes are illustrated and a scattering matrix introduced. Jost type solutions are introduced as a means of discussing the asymptotic behaviour of solutions at large distances from the scatterer.

We introduce in Chapters 3 to 5 a number of concepts and results from functional analysis and operator theory which are used regularly in the subsequent chapters. The presentation is mainly restricted to giving basic definitions, formulating the more important theorems and illustrating results by examples. More advanced topics in analysis will be introduced as required, thus emphasising their role when discussing wave phenomena.

In Chapter 6 we use the technical material introduced in the previous chapters to outline a strategy for discussing wave propagation and scattering. The discussion is strongly directed towards the development of constructive methods.

In Chapter 7 we set up a mechanism for discussing echo wave forms. This chapter relies very much on the materials and strategies introduced in Chapter 6. For ease of presentation and for the sake of clarity the development is made in terms of an acoustic problem. Detailed workings for more general problems will be deferred to the specific cases studied in later chapters.

The remaining chapters treat a number of problems that arise quite frequently in practice. These fall into three groups, acoustic, electromagnetic and elastic problems. Although coupled fields problems are of considerable interest in practice they will not be discussed in this monograph.

Chapters 8 and 9 deal with acoustic waves in nonhomogeneous media; specifically with acoustic waves in spatially stratified media and with acoustic waves in spatially periodically stratified media. In Chapter 8 the notion of trapped waves is introduced and appropriate generalised eigenfunction expansion theorems are discussed. Chapter 9 introduces the elements of Floquet theory and its bearing on the scattering problems of interest here.

The discussion of acoustic problems ends, in Chapter 10, with some remarks on acoustic inverse problems. In this chapter a promising method for treating the inverse problem is outlined.

Electromagnetic wave problems and elastic wave problems are considered in Chapter 11. The development follows the pattern adopted when discussing acoustic problems.

Chapter 12 indicates possibilities for further reading by providing some additional remarks and references for the material in the previous chapters. The chapter ends with a number of appendices which recall some frequently used topics from analysis.

References and Further Reading

[1] T. Ikebe: Eigenfunction expansions associated with the Schrodinger Operators and their application to scattering, *Theory. Arch. Rat. Mech. Anal.* 5, 1960, 2–33.

[2] P.D. Lax and R.S. Phillips: *Scattering Theory*, Academic Press, NewYork, 1967.

[3] R. Leis: *Initial Boundary Value Problems of Mathematical Physics*, John Wiley & Sons, Chichester 1986.

[4] G.F. Roach: *An Introduction to Linear and Nonlinear Scattering Theory*, Pitman Monographs and Surveys in Pure and Applied Mathematics, Vol. 78, Longman, Essex, 1995.

[5] C.H. Wilcox: *Scattering Theory for the d'Alembert Equation in Exterior Domains*, Lecture Notes in Mathematics, No. 442, Springer, Berlin, 1975.

2

Some One-Dimensional Examples

2.1 Introduction

Many of the strategies and techniques which are used in solving quite demanding scattering problems have been inspired by the methods used and results obtained when investigating wave motions on strings. In this connection we gather together in this chapter, for convenience and completeness, some of these details. The material is essentially well known and, consequently, is presented here in a largely formal manner. Full details can be found in the references cited in the Commentary.

Wave scattering phenomena involve three ingredients, an incoming or incident wave, an interaction zone in which the incident wave is perturbed in some manner and an outgoing wave which arises as a consequence of the perturbation. The waves in the interaction zone have, almost always, a very complicated structure. We shall see that this particular difficulty can be avoided, to a large extent, if we concentrate on the *consequence*s of the interaction rather than on the interaction itself. This we shall do by developing relationships between the incoming and outgoing processes, that is, we shall construct a scattering theory. To do this we first need to know how waves propagate in the absence of perturbations, that is, we need to study the FP. When details of the solutions to FP are well understood we can then turn to an investigation of the more demanding PP which can embrace such features as boundary conditions, forcing terms, variable coefficients and so on.

In the following sections we work through a number of specific problems associated with waves on an infinite string.

2.2 Free Problems

It is well known that the small amplitude transverse wave motion of a string is governed by an equation of the form [2] [7]

$$\{\partial_t^2 - c^2\partial_x^2\}u(x,t) = f(x,t), \quad (x,t) \in \mathbf{R} \times \mathbf{R} \tag{2.1}$$

where f characterises a force applied to the string, $u(x, t)$ denotes the transverse displacement of the string at a point x at time t and c represents the velocity of a wave which might have been generated in the string. Throughout, we use the notation ∂_t^n to denote the nth partial derivative with respect to the variable t and similarly for other variables. Also, the subscript notation will be used to denote differentiation of dependent variables.

We shall see later that it is quite sufficient for most of our purposes to study only the homogeneous form of (2.1). Specifically, we shall see in Section 2.8 that if we can solve the homogeneous equation then we will also be able to solve the inhomogeneous equation by using Green's function techniques and Duhamel's Principle. Consequently, in this chapter, unless otherwise stated, we take as a prototype equation

$$\{\partial_t^2 - c^2\partial_x^2\}u(x,t) = 0, \quad (x,t) \in \mathbf{R} \times \mathbf{R} \tag{2.2}$$

Solutions, $u(x, t)$, of this equation will, in general, be required to satisfy appropriate initial conditions to control the variations with respect to the time variable t and, similarly, certain boundary conditions to control the variations in the displacements with respect to the space variable x. However, in most cases of practical interest at all those points x that are a long way away from any boundary the effect of the boundary will be minimal since it could take quite a time, depending on the wave velocity, c, for the boundary influences to have any substantial effect at these points x. Thus for large values of t the wave motion is largely unaffected by the boundaries, that is the waves are (virtually) free of the boundary influence.

In this chapter we shall take as our prototype free problem (FP) the Initial Value problem (IVP)

$$\{\partial_t^2 - c^2\partial_x^2\}u(x,t) = 0, \quad (x,t) \in \mathbf{R} \times \mathbf{R} \tag{2.3}$$

$$u(x,0) = \varphi(x), \quad u_t(x,0) = \psi(x), \quad x \in \mathbf{R} \tag{2.4}$$

Associated with this FP are a variety of perturbed problems (PP) which could involve, for example, either forcing terms or variable coefficients or boundary conditions or combinations of these. Consequently, in any study of waves and their echoes there are three things that have to be done before anything else.

- Determine the general form of solutions to the equation governing the wave motion.
- Investigate initial value problems IVPs associated with the defining equation governing the wave motion and develop, in the absence of any perturbation, constructive methods of solution. This is taken as the underlying FP.
- Investigate PPs associated with the above FP and develop for them constructive methods of solution.

Once these three matters have been satisfactorily addressed then we will be well placed actually to compare solutions of the FP and the PPs and so develop a scattering theory. By means of such a scattering theory, which sometimes might appear to have a very abstract structure, we will see that we will be able to analyse, in an efficient and thoroughly constructive manner, the echo signals arising from perturbations of an otherwise free system. Furthermore, we shall see in later chapters that many of these one-dimensional techniques and strategies can be extended to cater for much more complicated systems than those dealing with waves on strings.

2.3 Solutions of the Wave Equation

In this section we obtain the general form of solutions of the one-dimensional wave equation

$$\{\partial_t^2 - c^2\partial_x^2\}u(x, t) = 0, \quad (x, t) \in \mathbf{R} \times \mathbf{R} \tag{2.5}$$

To this end introduce new variables, the so-called **characteristic coordinates**,

$$\xi = x - ct \quad \text{and} \quad \eta = x + ct \tag{2.6}$$

We remark that the lines $\xi =$ constant and $\eta =$ constant are called **characteristic lines** for (2.5) [9]. Transforming (2.5) to the new variables ξ, η we have

$$2x = \eta + \xi \quad \text{and} \quad 2ct = \eta - \xi \tag{2.7}$$

and we will write

$$u(x, t) = v(\xi, \eta) \tag{2.8}$$

Consequently, using the chain rule we obtain

$$\begin{aligned} 2cv_\xi &= cu_x - u_t \\ 4c^2 v_{\xi\eta} &= c^2 u_{xx} + cu_{xt} - cu_{tx} - u_{tt} = c^2 u_{xx} - u_{tt} \end{aligned} \tag{2.9}$$

Thus the wave equation (2.5) transforms under (2.6) into

$$v_{\xi\eta}(\xi, \eta) = 0 \tag{2.10}$$

The equation (2.10) has a general solution of the form

$$v(\xi, \eta) = f(\xi) + g(\eta) \tag{2.11}$$

where f, g are arbitrary, but sufficiently differentiable functions which are determined in terms of imposed conditions to be satisfied by a solution of interest.

Returning to the original variables we have

$$u(x, t) = f(x - ct) + g(x + ct) \tag{2.12}$$

In order that (2.12) should be a solution of (2.5), in the classical sense, then the functions f and g must be twice continuously differentiable functions of their arguments. We shall see later that we can relax this requirement.

In (2.12) the function f characterises a wave travelling to the right, unchanging in shape and moving with a velocity $c > 0$. To see this consider the f-wave when it is at position x_0 at time $t = 0$. Then in this case the wave has a shape (profile) given by $f(x_0)$. At some future time $t \neq 0$ the wave will have reached a point $x = x_0 + ct$. Consequently,

$$f(x - ct) = f(x_0)$$

which indicates that the shape of the wave is the same at the point (x, t) as it is at the point (x_0, t). Clearly, since $x > x_0$ we see that $f(x - ct)$ represents a wave travelling to the right with velocity c and which is unchanging in shape.

Similarly, $g(x + ct)$ represents a wave travelling to the left with velocity c and which is unchanging in shape.

We notice that since

$$\{c\partial_x + \partial_t\}f(x - ct) = 0, \quad \{c\partial_x - \partial_t\}g(x + ct) = 0 \tag{2.13}$$

then both $f(x - ct)$ and $g(x + ct)$ individually satisfy the wave equation

$$\{\partial_t^2 - c^2\partial_x^2\}w(x, t) = 0 \tag{2.14}$$

It will be convenient at this stage to introduce some notation. This we can do quite conveniently by considering the following particular solution of (2.14)

$$w(x, t) = a\cos(kx - \omega t - \varepsilon), \quad a > 0, \quad \omega > 0 \tag{2.15}$$

This is a **harmonic wave** defined in terms of the quantities

k = wave (propagation) number
ω = angular frequency
a = amplitude
ε = phase angle.

A number of perhaps more familiar wave features can be defined in terms of these quantities; specifically,

$c = \omega/k =$ **wave velocity**. The wave travelling in the positive direction if $k > 0$.
$\lambda = 2\pi/k =$ **wave length**
$f = \omega/2\pi =$ **frequency** of the (harmonic) oscillation
$T = f^{-1} = 2\pi/\omega =$ **period** of the oscillation
$\theta(x, t) = kx - \omega t - \varepsilon =$ the **phase** of the wave.

It will often be convenient to define the corresponding **complex harmonic wave**

$$\psi(x, t) = a\exp\{i\theta(x, t)\} = C\exp\{i(kx - \omega t)\} \tag{2.16}$$

where $w(x, t) = \mathrm{Re}(\psi(x, t))$ with $a = |C|$ and $\varepsilon = -\arg C$.

If a depends on either x or t then w is referred to as an **amplitude modulated wave**.

If $\theta(x, t)$ is nonlinear in either x or t then w is referred to as a **phase modulated wave**.

A wave of the form

$$w(x, t) = e^{-pt}\cos(kx - \omega t - \varepsilon), \quad p > 0 \tag{2.17}$$

is a **damped harmonic wave**.

A wave of the form

$$w(x, t) = e^{-qx}\cos(kx - \omega t - \varepsilon), \quad p > 0 \tag{2.18}$$

is an **attenuated harmonic wave**.

Solutions of (2.14) that have the specific form

$$w(x, t) = X(x)T(t) \tag{2.19}$$

are known as **separable solutions**. A typical example of such a solution is

$$w(x, t) = \sin(\pi x)\cos(\pi ct) \tag{2.20}$$

Direct substitution of (2.20) into (2.14) readily shows that (2.20) is indeed a solution of the wave equation. A general feature of waves such as (2.20) is that they are constant in time, that is they are **stationary or non-propagating waves**. To see this notice that the **nodes** of (2.20), that is, those points x at which $w(x, t) = 0$ and the **antinodes** of (2.20), that is those points x at which $w_x(x, t) = 0$ maintain permanent positions, $x = \ldots -1, 0, 1, 2 \ldots$ and $x = \ldots -\frac{1}{2}, \frac{1}{2}, \frac{3}{2} \ldots$ respectively, for all time t. We notice that (2.20) can be written in the form

$$w(x, t) = \sin(\pi x)\cos(\pi ct) = \frac{1}{2}\sin(\pi(x - ct)) + \frac{1}{2}\sin(\pi(x + ct)) \tag{2.21}$$

Thus the **stationary wave** (2.20) is seen to be a superposition of two travelling waves, travelling in opposite directions with velocity c.

We have seen that solutions of (2.14) have the general form

$$w(x, t) = f(x - ct) + g(x + ct) \tag{2.22}$$

We will often be interested in solutions (2.22) that have a particular time dependence. This can mean that we might look for solutions that have, for example, the following specific and separated form

$$w(x, t) = X(x)\exp\{i\omega t\} \tag{2.23}$$

Substituting (2.23) into (2.14) we find that

$$\{\partial_x^2 + k^2\}X(x) = 0, \quad k = \omega/c \tag{2.24}$$

This equation has the solution

$$X(x) = Ae^{ikx} + Be^{-ikx} \tag{2.25}$$

and we thus obtain, using (2.23) and (2.25)

$$w(x, t) = A\exp\{i(kx - \omega t)\} + B\exp\{-i(kx + \omega t)\} \tag{2.26}$$

Thus, comparing (2.22) and (2.26) we arrive at the following sign convention.

For waves with a time dependence $\exp\{-i\omega t\}$, then
$\exp\{ikx\}$ characterises a wave travelling to the right (increasing x)
$\exp\{-ikx\}$ characterises a wave travelling to the left (decreasing x).

2.4 Solutions of Initial Value Problems

In this section we study IVP of the form (2.3), (2.4) with $|x| < \infty$ and $t > 0$.

We have seen that the general solution of (2.3) has the form

$$u(x, t) = f(x - ct) + g(x + ct) \tag{2.27}$$

Substituting the initial conditions (2.4) into (2.27) we obtain

$$\varphi(x) = f(x) + g(x) \tag{2.28}$$

$$\psi(x) = -cf'(x) + cg'(x) \tag{2.29}$$

We notice that since f and g are assumed, at this stage, to be twice continuously differentiable it follows that the initial conditions must be such that φ is twice continuously differentiable and ψ is once continuously differentiable.

Integrate (2.29) and obtain

$$\frac{1}{c}\int_{x_0}^{x}\psi(s)ds = -f(x)+g(x) \tag{2.30}$$

where x_0 is an arbitrary constant.

From (2.28) and (2.30) we obtain

$$f(x)=\frac{1}{2}\varphi(x)-\frac{1}{2c}\int_{x_0}^{x}\psi(s)ds \tag{2.31}$$

$$g(x)=\frac{1}{2}\varphi(x)+\frac{1}{2c}\int_{x_0}^{x}\psi(s)ds \tag{2.32}$$

Substituting (2.31), (2.32) into (2.27) we obtain

$$u(x,t)=\frac{1}{2}\{\varphi(x-ct)+\varphi(x+ct)\}+\frac{1}{2c}\int_{x-ct}^{x+ct}\psi(s)ds \tag{2.33}$$

This is the celebrated **d'Alembert solution** of the one-dimensional wave equation. The interval $[x_1 - ct_1, x_1 + ct_1]$, $t > 0$ of the x-axis is the **domain of dependence** of the point (x_1, t_1). The reason for this name is that (2.33) indicates that $u(x_1, t_1)$ depends only on the values of φ taken at the ends of this interval and the values of ψ at all points of this interval. The region, D, for which $t > 0$, $x - ct \leq x_1$ and $x + ct \geq x_1$ is known as the **domain of influence** of the point x_1. This is because the value of φ at x_1 influences the solution $u(x, t)$ on the boundary of D whilst the value of ψ at x_1 influences $u(x, t)$ throughout D.

It will be instructive to consider the d'Alembert solution (2.33) in the following two particular cases.

Example 2.1. Assume

- Initial velocity of the string is everywhere zero.
- Initial displacement of the string is only nonzero in the interval (x, t).

In this case (2.33) reduces to

$$u(x,t)=\frac{1}{2}\{\varphi(x-ct)+\varphi(x+ct)\} \tag{2.34}$$

Thus we see that the forward wave $\varphi(x - ct)$ and the backward wave $\varphi(x + ct)$ each travel with a velocity c and have initial amplitude $\frac{1}{2}\varphi(x)$, that is with amplitude one half of the original (initial) amplitude $u(x, t)$. To find the solution, $u(x, t)$, at some other time we displace the graph $\varphi(x)$ by an amount ct in opposite directions. For instance, suppose the initial displacement was a triangle with base (α, β). Then as we have seen, this initial waveform will be the sum of two triangular waveforms each having half the original amplitude. At any other time, t, the displacement waveform will again be the sum of two triangular waveforms, each of half the original amplitude, one travelling to the right, the other to the left. The displacement waveform is obtained by summing the ordinates of the two displaced graphs. This can turn out to be a very complicated process. Furthermore, to write down the formula for $u(x, t)$ at any point (x, t) can often be very difficult. In the relatively simple case of waves on strings an indication of the complicated nature of the displacement waveform can be obtained by graphical means. To see this consider the specific case when the initial displacement is the triangular waveform

$$\varphi(x) = u(x,0) = \begin{cases} 0; & x \leq 0 \\ \dfrac{2u_0(x-\alpha)}{\beta-\alpha}; & \alpha \leq x \leq \dfrac{1}{2}(\alpha+\beta) \\ \dfrac{2u_0(\beta-x)}{\beta-\alpha}; & \dfrac{1}{2}(\alpha+\beta) \leq x \leq \beta \\ 0; & x \geq \beta \end{cases} \tag{2.35}$$

Displacing this waveform in the manner mentioned above and plotting the results on a graph of u against x yields a series of graphs which are snapshots of the displacement waveform at a fixed time. Carrying this through for time steps of duration

$$\frac{m(\beta-\alpha)}{10c}, \quad m = 0, 1, 2, \ldots$$

we will readily notice the following behaviour:

For $t = 0$ to $t = \dfrac{2(\beta-\alpha)}{5c}$

In this interval the backward and forward waves interact and produce a very complicated graph for the displacement waveform.

For $t \geq \dfrac{3(\beta-\alpha)}{5c}$

The backward and forward waves would seem to have "passed through" each other and exhibit no evidence of the "interaction" which was seen earlier. However, simple as the graph would now appear to be we do know that these two waves will have interacted and will "contain" information to that effect. We would like to obtain this information without going through all the complexities which would arise when investigating the interaction zone. This is a principal goal of scattering theory. We shall demonstrate how this can be achieved in the chapters that follow.

Example 2.2. Assume

- Initial velocity of the string is only nonzero in the interval (α, β).
- Initial displacement of the string is everywhere zero.

In this case (2.33) reduces to

$$\frac{1}{2c}\int_{x-ct}^{x+ct} \psi(s)ds \tag{2.36}$$

We argue in much the same way as in Example 2.1. The difference now is that ψ rather than φ is prescribed. As an illustration we shall consider the case when $\varphi(x)$ has the triangular waveform given in (2.35). Once again we confine attention to a graphical method and plot "snapshots" of the displacement $u(x, t)$ at various fixed times. We quickly see that there are five regions of interest,

1. $x + ct \leq \alpha$
2. $a \leq x + ct \leq \beta$
3. $x - ct \leq \alpha$ and $x + ct \geq \beta$
4. $\alpha \leq x - ct \leq \beta$
5. $x - ct \geq \beta$

In Region 1 we have $x + ct \leq \alpha$ and hence the entire range of integration in (2.36) is outside the interval (α, β). Consequently, $u(x, t) = 0$ in Region 1.

In a similar manner $u(x, t) = 0$ in Region 5.

In Region 2 we have $\alpha \leq x + ct \leq \beta$. Consequently

$$x - ct = x + ct - 2ct \leq \beta - 2ct < \alpha \tag{2.37}$$

The last inequality will only hold for

$$t > (\beta - \alpha)/2c \tag{2.38}$$

Therefore, in Region 2,

$$u(x,t) = \frac{1}{2c}\int_{\alpha}^{x+ct} \psi(s)ds \tag{2.39}$$

Similarly, in Region 4, together with (2.38) we obtain

$$u(x,t) = \frac{1}{2c}\int_{x-ct}^{\beta} \psi(s)ds \tag{2.40}$$

Finally, in Region 3 since $x - ct \leq \alpha$ and $x + ct \geq \beta$ it follows that

$$u(x,t) = \frac{1}{2c}\int_{\alpha}^{\beta} \psi(s)ds \tag{2.41}$$

and we conclude, since α and β, are constants, that $u(x, t)$ is a constant in Region 3.

In the case when

$$t < \frac{\beta - \alpha}{2c} \tag{2.42}$$

then arguing as before we find that in Region 3 we now have

$$\frac{1}{2c}\int_{x-ct}^{x+ct} \psi(s)ds \tag{2.43}$$

which is not a constant.

These two examples illustrate quite clearly how complicated the wave structure can be. The situation can, of course, be expected to be even worse if any form of perturbation, such as a boundary condition for instance, is involved. In the following two sections we introduce some methods that will ease these difficulties. Whilst these methods may seem, at first sight, to be a little heavy-weight for use when discussing waves on strings nevertheless they do ease considerably the difficulties we have so far mentioned and more importantly they offer good prospects for dealing with more complicated problems than waves on strings.

Finally, in this section we return to (2.33) the d'Alembert solution form. We notice that if we introduce

$$u_+(x,t) := \begin{cases} \frac{1}{2}\varphi(x-ct) + \frac{1}{2c}\int_{x-ct}^{\infty} \psi(s)ds, & x > 0 \\ \frac{1}{2}\varphi(x+ct) + \frac{1}{2c}\int_{-\infty}^{x+ct} \psi(s)ds, & x < 0 \end{cases} \tag{2.44}$$

$$u_-(x,t) := \begin{cases} \frac{1}{2}\varphi(x+ct) - \frac{1}{2c}\int_{x+ct}^{\infty} \psi(s)ds, & x > 0 \\ \frac{1}{2}\varphi(x-ct) - \frac{1}{2c}\int_{-\infty}^{x-ct} \psi(s)ds, & x < 0 \end{cases} \tag{2.45}$$

then (2.33) can be written in the alternative form

$$u(x, t) = u_+(x, t) + u_-(x, t) \tag{2.46}$$

This form will be useful when we try to learn more about the behaviour of the solution, $u(x, t)$, for large t. We notice that $u_+(x, t)$ is a function of $(x - ct)$ for $x > 0$ and a function of $(x + ct)$ for $x < 0$. Consequently, we can write more compactly

$$u_+(x, t) = f_+(|x| - ct), \quad x \in \mathbf{R} \tag{2.47}$$

Similarly, we have

$$u_-(x,t) = f_-(|x| + ct), \quad x \in \mathbf{R} \tag{2.48}$$

We also notice that in the region $x > 0$ there are two "waves", $u_+(x, t)$ travelling to the right (increasing x) and $u_-(x, t)$ travelling to the left (decreasing x). Thus, with respect to the origin $x = 0$ the wave $u_+(x, t)$ is **outgoing** whilst the wave $u_-(x, t)$ is **incoming**.

Similarly, in the region $x < 0$ the wave $u_+(x, t)$ is **outgoing** (increasing negative x) with respect to the origin whilst $u_-(x, t)$ is **incoming** (decreasing negative x).

The concepts of incoming and outgoing waves are of crucial importance in scattering theory. They will be discussed in more detail in Chapter 6 and the Commentary.

2.5 Integral Transform Methods

In this and the following section we introduce some alternative methods of constructing solutions to FP for wave problems on strings. These methods have the virtue that they generalise quite readily when we need to deal with more complicated and demanding problems than waves on strings. Furthermore, we shall see that they also provide an efficient means for developing robust constructive methods for solving quite difficult problems.

An explicit method for constructing solutions to IVP for the wave equation is provided by the Plancherel theory of the Fourier transform [6], [8], [10]. Specifically we have the following basic formulae in $\mathbf{R}^n$.

$$(Ff)(p) =: \hat{f}(p) = \lim_{M\to\infty} \frac{1}{(2\pi)^{n/2}} \int_{|x|\leq M} \exp(-ixp) f(x)dx \tag{2.49}$$

$$f(x) = (F^* \hat{f}) = \lim_{M\to\infty} \frac{1}{(2\pi)^{n/2}} \int_{|p|\leq M} \exp(ixp) \hat{f}(p)dp \tag{2.50}$$

where $x = (x_1, x_2, \ldots, x_n)$, $p = (p_1, p_2, \ldots, p_n)$ and $x \cdot p = \Sigma_{j=1}^n x_j p_j$. Here F^* denotes the inverse of the transform F. We would emphasise that the integrals in (2.49), (2.50) are improper integrals and care must be taken when interpreting the limits in (2.49), (2.50). We return to these points in detail in Chapter 6. With this understanding we shall refer to (2.49), (2.50) as a **Fourier inversion theorem**.

This inversion theorem can be used to provide a representation, a so-called **spectral representation**, of differential expressions with constant coefficients. Such a representation will often reduce the complexities and inherent difficulties of a given problem. This is a consequence of the relation

$$(F(D_j f))(p) = ip_j(Ff)(p) \tag{2.51}$$

where $D_j = \partial/\partial x_j$, $j = 1, 2, \ldots, n$. For example if we write

$$A := -\Delta = -\sum_{j=1}^{n} \partial^2/\partial x^2$$

and if Φ is a "sufficiently nice" function then using (2.51) we can obtain the representation

$$(\Phi(A)f)(x) = \lim_{M\to\infty} \frac{1}{(2\pi)^{n/2}} \int_{|p|\le M} \exp(ixp)\Phi(|p|^2)\hat{f}(p)dp \tag{2.52}$$

In later chapters we shall refer to the three results (2.49), (2.50), (2.52) collectively either as a **(generalised) eigenfunction expansion theorem** or as a **spectral representation theorem** (with respect to A).

To illustrate the use of the above Fourier transforms we consider again the following IVP governing waves on a string

$$\{\partial_t^2 - c^2\partial_x^2\}u(x, t) = 0, \quad (x, t) \in \mathbf{R} \times \mathbf{R} \tag{2.53}$$

$$u(x, 0) = \varphi(x), \quad u_t(x, 0) = \psi(x), \quad x \in \mathbf{R} \tag{2.54}$$

We now only need consider the case when $n = 1$ and then the inversion theorem (2.49), (2.50) can be conveniently written in the form

$$(Ff)(p) =: \hat{f}(p) = \frac{1}{(2\pi)^{1/2}} \int_{\mathbf{R}} \exp(-ixp)f(x)dx = \int_{\mathbf{R}} \overline{w(x, p)}f(x)dx \tag{2.55}$$

$$f(x) = (F^*\hat{f}) = \frac{1}{(2\pi)^{1/2}} \int_{\mathbf{R}} \exp(ixp)\hat{f}(p)dp = \int_{\mathbf{R}} w(x, p)\hat{f}(p)dp \tag{2.56}$$

where it is understood that the improper integrals appearing in (2.55), (2.56) are interpreted as limits as indicated above. We notice that the Fourier kernel

$$w(x, p) = \frac{1}{(2\pi)^{1/2}} \exp(ixp) \tag{2.57}$$

satisfies

$$\{\partial_x^2 + p^2\}w(x, p) = 0 \tag{2.58}$$

If we take the Fourier transform, (2.55), of the IVP for the partial differential equation (2.53) we obtain the following IVP for an ordinary differential equation

$$\{d_t^2 + c^2p^2\}\,\hat{u}(p,t) = 0 \tag{2.59}$$

$$\hat{u}(p,0) = \hat{\varphi}(p), \quad \hat{u}_t\,(p,0) = \hat{\psi}(p) \tag{2.60}$$

This IVP is easier to solve than that for the partial differential equation (2.53). Indeed, we see immediately that the solution is

$$\hat{u}(p,t) = (\cos(pct))\hat{\varphi}(p) + \frac{1}{pc}(\sin(pct))\hat{\psi}(p) \tag{2.61}$$

If we now apply the inverse Fourier transform, (2.56) to (2.61), then we obtain the required solution of the IVP (2.53), (2.54) in the form

$$\begin{aligned} u(x,t) &= \int_R w(x,p)\left\{(\cos(pct))\hat{\varphi}(p) + \frac{1}{pc}(\sin(pct))\hat{\psi}(p)\right\}dp \\ &= \{F * \hat{u}(\cdot,t)\}(x) \end{aligned} \tag{2.62}$$

To see how this form relates to that obtained earlier we first expand $\cos(pct)$ in the form

$$\cos(pct) = \frac{1}{2}\{e^{ipct} + e^{-ipct}\}$$

and use the result [3], [4, vol II]

$$(F(f(x-L)))(p) = e^{-ipL}\hat{f}(p) \tag{2.63}$$

It is then a straightforward matter to show that

$$F*(\hat{\varphi}(p)\cos(pct)) = \frac{1}{2}\{\varphi(x+ct) + \varphi(x-ct)\} \tag{2.64}$$

Similarly

$$\begin{aligned} F*\left(\frac{1}{pc}\hat{\varphi}(p)\sin(pct)\right) &= \frac{1}{2i\sqrt{2\pi}}\int_{\mathbf{R}}\hat{\psi}(p)\frac{1}{pc}\{e^{ip(x+ct)} - e^{ip(x-ct)}\}dp \\ &= \frac{1}{2c\sqrt{2\pi}}\int_{\mathbf{R}}\hat{\psi}(p)\left\{\int_{x-ct}^{x+ct} e^{ips}ds\right\}dp \\ &= \frac{1}{2c}\int_{x-ct}^{x+ct}\psi(s)ds \end{aligned} \tag{2.65}$$

Combining (2.62), (2.64) and (2.65) we obtain

$$u(x,t) = \frac{1}{2}\{\varphi(x+ct) + \varphi(x-ct)\} + \frac{1}{2c}\int_{x-ct}^{x+ct}\psi(s)ds \tag{2.66}$$

which is the familiar d'Alembert solution obtained earlier. We remark again that the Fourier transform of the given IVP for a *partial* differential equation yields, in this instance, an IVP for an *ordinary* differential equation. Whilst the ordinary differential equation is more readily solved than the partial differential equation there will remain the matter of inversion of the Fourier transform. Thus, three questions will always have to be addressed if we choose to adopt the integral transform approach.

- First, what is the most appropriate integral transform for use in reducing the given partial differential equation to an equivalent ordinary differential equation?
- Second, is there an inversion theorem of the form (2.49) available for use in dealing with the given IVP?
- Third, is there available a (spectral) representation theorem of the form (2.49), (2.50), (2.52) for use in dealing with the given IVP?

We emphasise that in dealing with our present FP we have been very lucky because if we use the Fourier integral transform then the Fourier Plancherel theory is available quite independently of any scattering requirements and we can answer the last two questions above in the affirmative. However, for a perturbation of this FP and indeed for more general problems than waves on a string we must always **prove** the availability of a representation theorem of the form (2.52). We return to this matter in more detail in Chapter 6 and subsequent chapters.

Finally, in this section we remark that we could have obtained (2.62) and hence (2.66) in another way. It turns out that this other approach will offer potentially powerful means of addressing a wide range of physically realistic problems.

Essentially, the method rests on how the partial differential equation for the problem of interest is cast into the form of an equivalent ordinary differential equation. Again for the purposes of illustration we consider the IVP (2.53), (2.54). We start by setting $A = -c^2\partial_x^2$ and then make what seems to be an outrageous assumption namely that for our immediate purposes A can be treated as a constant! This being done we arrive at the following IVP for an ordinary differential equation

$$\{d_t^2 + A\}u(x, t) = 0, \quad u(x, 0) = \varphi(x), \quad u_t(x, 0) = \psi(x) \tag{2.67}$$

This IVP has a solution which can be written in the form

$$u(x,t) = (\cos(tA^{1/2}))\varphi(x) + A^{-1/2}(\sin(tA^{1/2}))\psi(x) \tag{2.68}$$

It now remains to interpret such quantities as $\cos tA^{1/2}$.

From the standard theory of Fourier transforms [6]

$$(F(Af))(p) = (F(-c^2\partial_x^2 f))(p) = c^2p^2\hat{f}(p) \tag{2.69}$$

It then follows, because of the particularly simple form of A that we are using here, that for a "sufficiently nice" function Φ we have

$$(F(\Phi(A)f))(p) = \Phi(c^2p^2)\hat{f}(p) \tag{2.70}$$

Consequently, combining (2.70) and (2.68) we obtain

$$u(x,t) = \int_R w(x,p)\left\{(\cos(pct))\hat{\varphi}(p) + \frac{1}{pc}(\sin(pct))\hat{\psi}(p)\right\}dp \tag{2.71}$$

which is the same as (2.62) obtained by other means. We shall see that in this particular method the "outrageous assumption" can be justified, thus making the approach mathematically respectable.

Finally, we notice that the Fourier kernel $w(x, p)$, given by (2.57), satisfies

$$(A + c^2p^2)w(x, p) = 0 \tag{2.72}$$

Consequently, $w(x, p)$ would appear to be, in some sense, an **eigenfunction** of A with **eigenvalue** $(-c^2p^2)$. (See Chapter 4 for more details on this aspect.)

2.6 On the Reduction to a First Order System

An alternative method frequently used when discussing wave motions governed by an IVP of the form (2.53), (2.54) is to replace the IVP for the partial differential equation by an equivalent problem for a first order system. This approach has a number of advantages. Existence and uniqueness results can be readily obtained and, furthermore, energy considerations can be included quite automatically. We shall illustrate this approach here in an entirely formal manner. Precise analytical details will be provided in later chapters.

The initial value problem (2.53), (2.54) can be written in the form

$$\begin{bmatrix} u \\ u_t \end{bmatrix}(x,t) - \begin{bmatrix} 0 & -I \\ A & 0 \end{bmatrix}\begin{bmatrix} u \\ u_t \end{bmatrix}(x,t) = \begin{bmatrix} 0 \\ 0 \end{bmatrix} \tag{2.73}$$

$$\begin{bmatrix} u \\ u_t \end{bmatrix}(x,0) = \begin{bmatrix} \varphi \\ \psi \end{bmatrix}(x) \tag{2.74}$$

where, as before $A = -c^2\partial_x^2$.

These equations can be conveniently written in the form

$$\{\partial_t - iM\}\Psi(x, t) = 0, \quad \Psi(x, 0) = \Psi_0(x) \tag{2.75}$$

where

$$\Psi(x,t)=\begin{bmatrix} u \\ u_t \end{bmatrix}(x,t), \quad \Psi_0(x)=\begin{bmatrix} \varphi \\ \psi \end{bmatrix}(x) \tag{2.76}$$

$$-iM=\begin{bmatrix} 0 & -I \\ A & 0 \end{bmatrix} \tag{2.77}$$

If we again make the "outrageous" assumption that A is a constant then it will follow that M is a constant and hence (2.75) can be reformulated as an IVP for an ordinary differential equation of the form

$$\{d_t - iM\}\Psi(t) = 0, \quad \Psi(0) = \Psi_0 \tag{2.78}$$

where we have used the notation

$$\Psi(x, t) = \Psi(\cdot, t)(x) =: \Psi(t)(x) \tag{2.79}$$

The solution of (2.78) can be obtained, by using an integrating factor technique, in the form

$$\Psi(t) = \exp(itM)\Psi(0) \tag{2.80}$$

Writing the exponential term in a series form we obtain

$$\begin{aligned} e^{itM} &= \sum_{n=0}^{\infty} \frac{(itM)^n}{n!} = \left\{ \sum_{n=\text{even}} + \sum_{n=\text{odd}} \right\} \frac{(itM)^n}{n!} \\ &= \left\{ I - \frac{t^2M^2}{2!} + \frac{t^4M^4}{4!} - \ldots \right\} + i\left\{ tM - \frac{t^3M^3}{3!} + \frac{t^5M^5}{5!} - \ldots \right\} \end{aligned}$$

Now, using,

$$M = i\begin{bmatrix} 0 & -I \\ A & 0 \end{bmatrix}, \quad M^2 = A\begin{bmatrix} I & 0 \\ 0 & I \end{bmatrix} \tag{2.81}$$

and recalling the series expansions for $\sin x$ and $\cos x$ we obtain

$$\begin{aligned} \exp(it\,M) &= \left\{ I - \frac{t^2A}{2!} + \frac{t^4A^2}{4!} - \ldots \right\} \begin{bmatrix} I & 0 \\ 0 & I \end{bmatrix} \\ &\quad - A^{-1/2}\left\{ tA^{1/2} - \frac{t^3A^{3/2}}{3!} + \frac{t^5A^{5/2}}{5!} - \ldots \right\} \begin{bmatrix} 0 & -I \\ A & 0 \end{bmatrix} = \cos(tA^{1/2}) \begin{bmatrix} I & 0 \\ 0 & I \end{bmatrix} \\ &\quad - A^{-1/2}\sin(tA^{1/2}) \begin{bmatrix} 0 & -I \\ A & 0 \end{bmatrix} = \begin{bmatrix} \cos(tA^{1/2}) & A^{1/2}\sin(tA^{1/2}) \\ -A^{1/2}\sin(tA^{1/2}) & \cos(tA^{1/2}) \end{bmatrix} \end{aligned} \tag{2.82}$$

If we substitute this expression into (2.79) it is clear that the first component of the solution (2.80) yields the same solution of the given IVP as obtained earlier.

This approach can be given a rigorous mathematical development as we shall see. We shall make considerable use of it in this book since, on the one hand, it provides a relatively easy means of settling questions of existence and uniqueness and, on the other, it offers good prospects for developing constructive methods.

So far we have only been discussing IVP for the one-dimensional wave equation, that is, the FP for waves on a string. In the next few sections we turn our attention to some PPs associated with this FP and indicate how the various methods discussed so far are either inadequate or need to be modified in certain ways.

2.7 Waves on Sectionally Homogeneous Strings

In our investigations, so far, of waves on strings we have considered the string to have uniform density throughout. This generated what we came to call the Free Problem (FP) associated with the classical wave equation. Associated with this FP there is a whole hierarchy of Perturbed Problems (PPs). Perhaps the simplest PP, in this case, would arise when we investigate waves on a string that has a piecewise uniform density. In this section we shall consider two particular cases.

2.7.1 A Two-Part String

Consider two semi-infinite strings Ω_1 and Ω_2 of (linear) density ρ_1 and ρ_2 respectively that are joined at the point $x = r$ and stretched at tension T with Ω_1 occupying the region $x < r$ and Ω_2 the region $x > r$. As the two strings have different (linear) densities then it follows that their associated wave speeds c_1, c_2 are also different.

We shall see that this problem is a one-dimensional version of the more general interface problems. In this latter problem a (given) incident wave travels in a homogeneous medium which terminates at an interface with another, different, homogeneous medium in which the wave can also travel. Examples of such problems are given, for instance, by electromagnetic waves travelling in air meeting the surface of a dielectric and by acoustic waves travelling in air meeting an obstacle. In this class of problems the interface "scatters" the given incident wave and gives rise to reflected and transmitted waves. When all these waves combine the resulting wave fields are readily seen to be quite different to those occurring in the related FP. Such problems are examples of so-called **target scattering** problems. We illustrate some of the features of such problems by considering the following one-dimensional problem.

The governing wave equation is

$$\{\partial_t^2 - c^2(x)\partial_x^2\}u(x, t) = 0, \quad (x, t) \in \mathbf{R} \times \mathbf{R} \tag{2.83}$$

$$c(x) = \begin{cases} c_1, & x < r, \quad x \in \mathbf{R} \\ c_2, & x > r, \quad x \in \mathbf{R} \end{cases} \tag{2.84}$$

$$u(x, 0) = \varphi(x), \quad u_t(x, 0) = \psi(x) \tag{2.85}$$

At the interface we shall require continuity of the displacement and of the transverse forces. This leads to boundary conditions of the form

$$u(r^-, t) = u(r^+, t), \quad u_x(r^-, t) = u_x(r^+, t) \tag{2.86}$$

We remark that we assume here, unless otherwise stated, that $r > 0$. This is not just to increase the complexity of the presentation. It is simply because in many later illustrations we shall find it a convenient means of keeping track of the target (i.e. interface).

Let $f(x - c_1t)$ denote a given incident wave in Ω_1. We assume that Ω_2 is initially at rest so that $u(x, 0) = 0$ and $u_t(x, 0) = 0$ for $x > r$.

The wave field, $u(x, t)$ which must satisfy (2.83), (2.84), (2.85) has the general form

$$u(x, t) = \begin{cases} f(x - c_1t) + g(x + c_1t), & x < r \\ h(x - c_2t) + H(x + c_2t), & x > r \end{cases} \tag{2.87}$$

However, since Ω_2 is initially at rest we must have

$$h'(\zeta) = 0 \quad \text{and} \quad H'(\zeta) = 0 \quad \text{for} \quad \zeta > 0 \tag{2.88}$$

Hence $H(x + c_2t)$ is a constant for $t > 0$ and therefore may be discarded. The appropriate wavefield is thus

$$u(x, t) = \begin{cases} f(x - c_1t) + g(x + c_1t), & x < r, \quad t > 0 \\ h(x - c_2t), & x > r, \quad t > 0 \end{cases} \tag{2.89}$$

We also notice that

$$\begin{aligned} f(\zeta) = h(\zeta) = 0, & \quad \zeta > r \\ g(\zeta) = 0, & \quad \zeta < r \end{aligned} \tag{2.90}$$

At the interface certain boundary conditions will always have to be satisfied by solutions of (2.83). The most immediate conditions are, as we have already mentioned, continuity of displacement:

$$u(r^-, t) = u(r^+, t) \tag{2.91}$$

continuity of transverse force:

$$u_x(r^-, t) = u_x(r^+, t) \tag{2.92}$$

where

$$u(r^-,t) = \lim_{\substack{x\to r\\ x<r}} u(x,t), \quad u(r^+,t) = \lim_{\substack{x\to r\\ x>r}} u(x,t) \tag{2.93}$$

and similarly for the derivatives.

Substitute (2.89) into (2.91), (2.92) to obtain

$$f(r - c_1t) + g(r + c_1t) = h(r - c_2t) \tag{2.94}$$

$$f'(r - c_1t) + g'(r + c_1t) = h'(r - c_2t) \tag{2.95}$$

where the primes denote differentiation with respect to the argument.

Integrate (2.95) to obtain

$$\frac{f(r-c_1t)}{-c_1} + \frac{g(r+c_1t)}{c_1} = \frac{h(r-c_2t)}{-c_2} \tag{2.96}$$

where the integration constant has been set to zero in order to ensure (2.90) is satisfied.

Solving (2.94), (2.96) for the unknowns g and h we obtain

$$g(x+c_1t) = \left(\frac{c_2-c_1}{c_1+c_2}\right) f(2r - x - c_1t), \quad x<r \tag{2.97}$$

$$h(x-c_2t) = \left(\frac{2c_2}{c_1+c_2}\right) f\left(r\left(1-\frac{c_1}{c_2}\right) + \frac{c_1}{c_2}x - c_{1t}\right), \quad x>r \tag{2.98}$$

It is worth noticing a number of interesting features of the solutions represented by (2.97) and (2.98). For convenience of presentation and without any loss of generality, at this stage, we shall assume that $r = 0$.

(i) When $c_2 = 0$ there is no transmitted wave, the reflected wave has the form

$$g(x + c_1t) = -f(-x - c_1t)$$

and the required solution is

$$u(x, t) = f(x - c_1t) - f(-x - c_1t) \tag{2.99}$$

Thus as expected the reflected wave travels in the opposite direction to the incident wave. However although the incident and reflected waves have the same shape they are seen to have opposite signs.

Finally we notice from (2.99) that $u(0, t) = 0$ for all t. Thus, the solution (2.99) describes waves on a semi-infinite string with a fixed point at $x = 0$.

(ii) When $c_1 = c_2$ there is no reflected wave; again as would be expected.

(iii) When $c_2 > c_1$ then the incident and reflected waves are seen to travel in opposite directions with the same profile but no change in sign.

Thus the reflected and incident waves are in phase at the junction (interface) provided $c_2 > c_1$ and are otherwise totally out of phase.

We will be able to obtain more detailed information about the wave field once we have introduced the notions of eigenfunction expansions and Green's functions.

Finally, in this section, we consider the case when the incident wave is a simple harmonic wave of angular frequency ω. In this case we will have an incident wave of the form

$$\begin{aligned} u_i(x, t) &= f(x - ct) = \exp\{ik(x - ct)\} = \exp\{i(kx - \omega t)\} \\ &= \exp\{ikx\}\exp\{-i\omega t\} \end{aligned} \tag{2.100}$$

where $\omega = kc$.

We notice that the incident wave separates into the product of two components, one only dependent on x, the other only dependent on t. It is natural to expect that this will be the case for the complete wave field. Consequently, for the nonhomogeneous string problem we are considering we can expect the complete wave field to be separable and to have the form

$$u(x,t) = \begin{cases} e^{-i\omega_1 t} v_1(x), & \omega_1 = c_1 k_1, \quad x < 0 \\ e^{-i\omega_2 t} v_2(x), & \omega_2 = c_2 k_2, \quad x > 0 \end{cases} \tag{2.101}$$

Therefore, bearing in mind the sign convention introduced just after (2.26) the space-dependent component of the wave field, $v(x)$, can be written in the form

$$v(x) = \begin{cases} e^{ik_1 x} + Re^{-ik_1 x}, & x < 0 \\ Te^{ik_2 x}, & x > 0 \end{cases} \tag{2.102}$$

If we now apply the boundary conditions (2.91), (2.92) which must hold for all $t > 0$ then we readily find

$$R = \frac{k_2 - k_1}{k_1 + k_2}, \quad T = \frac{2k_2}{k_1 + k_2}\exp\{i(\omega_2 - \omega_1)t\} \tag{2.103}$$

Here R and T are known as the **reflection** and **transmission coefficients** respectively. In the case when we are only interested in solutions that have the same frequency then these coefficients assume the simpler form

$$R = \frac{c_2 - c_1}{c_1 + c_2}, \quad T = \frac{2c_2}{c_1 + c_2}$$

Combining (2.101), (2.102) and (2.103) we see that we recover the solutions (2.97), (2.98) in the case when $r = 0$.

2.7.2 A Three-Part String

We shall assume in this section that a portion of a homogeneous string is subjected to an elastic restoring force $E(x)$ per unit length of the string. Newton's laws of motion then indicate that the equation governing wave motion on the string is

$$\{\partial_t^2 - c^2\partial_x^2\}u(x, t) - c^2\mu^2(x)u(x, t) = 0 \tag{2.104}$$

where $\mu^2(x) = E(x)/T_s$ and T_s denotes the string tension.

Equation (2.104) is of a form which is typical when investigating **potential scattering** problems. Here $c^2\mu^2(x)$ can be viewed as the potential term. We shall only be interested here in the case when $E(x)$ is localised. That is, $E(x)$ will be assumed either to vanish outside a finite region of the string or to decay exponentially away from some fixed reference point.

A wave incident on the elastic region (the "potential") will be partly reflected and partly transmitted. However, even if $\mu(x)$ has a constant value the solution of (2.104) is not as easy to obtain as the solutions (2.83). To see this consider the case when $\mu(x)$ has the constant value μ_0 and we seek solutions of (2.104) that have the one angular frequency ω. When this is the situation we assume a solution of the form

$$u(x, t) = w(x, \omega)e^{-i\omega t} \tag{2.105}$$

and, by direct substitution into (2.104), we find that $w(x, \omega)$ must satisfy

$$\left\{d_x^2 + \left(\frac{\omega}{v}\right)^2\right\}w(x, \omega) = 0 \tag{2.106}$$

where

$$\frac{1}{v^2} = \frac{1}{c^2} - \left(\frac{\mu_0}{\omega}\right)^2 \tag{2.107}$$

We then obtain the string displacement $u(x, t)$, by (2.105) in the form

$$u(x, t) = a\exp\left\{-i\omega\left(t - \frac{x}{v}\right)\right\} + b\exp\left\{-i\omega\left(t + \frac{x}{v}\right)\right\} \tag{2.108}$$

We see that, in (2.108), v is the phase velocity of the wave. Furthermore, we notice that the wave motion represented by (2.108) is dispersive since by (2.107) the phase velocity, v, of the wave is frequency dependent. It follows that distortionless propagation of the wave as described by f and g in the general solution of the classical wave equation, is no longer possible.

We also notice that there is a "cut off" frequency associated with the wave motions generated in this system. According to (2.106), (2.107) frequencies that are less than $\mu_0 c$ lead to an imaginary propagation constant. These low frequency disturbances do not propagate as waves, they merely move the string up and down in phase. Thus it is possible that localised wave motion might be excited on a nonhomogeneous string.

To illustrate the scattering of an incident wave by the elastic region (potential) consider a string with a segment that has a constant elastic restoring force so that we have

$$\mu(x) = \begin{cases} 0, & |x| > r \\ \mu_0, & |x| < r \end{cases} \tag{2.109}$$

If a wave of frequency ω and unit magnitude is incident on this region then the resulting spatial part of the wavefield may be written, as in the previous section, in the form

$$v(x) = \begin{cases} e^{ikx} + Re^{-ikx}, & x < -r \\ Te^{ik_2 x}, & x > +r \\ Ae^{i\alpha x} + Be^{-i\alpha x}, & |x| < r \end{cases} \tag{2.110}$$

where $k = \omega/c$ and $\alpha = \sqrt{k^2 - \mu_0^2}$. It is clear that a wave will not propagate in the elastic region unless $k \geq \mu_0$.

The reflection and transmission coefficients, R and T respectively, together with the constants A and B are determined by requiring continuity of displacement and slope at $x = \pm r$. It is a straightforward but rather lengthy matter to show that

$$R \exp(2ikr) = \frac{\mu_0}{D} \sin(2\alpha r) \tag{2.111}$$

$$T \exp(2ikr) = \frac{2i\alpha k}{D} \tag{2.112}$$

where

$$D = (k^2 + \alpha^2) \sin(2\alpha r) + 2ikr \cos(2\alpha r) \tag{2.113}$$

We see, from (2.111), that perfect transmission (i.e. $R = 0$) occurs when $\sin(2\alpha r) = 0$, that is, whenever an integral number of half-wavelengths of the wave on the elastic region fit into that region.

We also notice that R and T become unbounded at zeros of the denominator D. These will be identified as so-called **resonances** of the system.

2.8 Duhamel's Principle

So far, we have only been dealing with the homogeneous wave equation. We shall now show that this is really sufficient for many of our immediate purposes. That is, we shall show that the results we obtain when investigating the homogeneous equation can be used to generate solutions for the nonhomogeneous wave equation. As an illustration we consider in this section the IVP

$$\{\partial_t^2 - c^2\partial_x^2\}u(x, t) = f(x, t), \ (x, t) \in \mathbf{R} \times \mathbf{R} \tag{2.114}$$

$$u(x, 0) = \varphi(x), \ u_t(x, 0) = \psi(x) \tag{2.115}$$

For convenience of presentation we shall again write $A = -c^2\partial_x^2$.

We shall assume that the solution of (2.114), (2.115) has the form

$$u(x, t) = v(x, t) + w(x, t) \tag{2.116}$$

We now proceed as before by interpreting (2.114) as an ordinary differential equation. To this end we shall understand that u, the function defining the displacement $u(x, t)$, has the interpretation

$$u: t \to u(\cdot, t) =: u(t) \tag{2.117}$$

and similarly for v and w in (2.116).

With this in mind we assume that

$$\{d_t^2 + A\}v(t) = 0, \quad v(0) = \varphi, \quad v_t(0) = \psi \tag{2.118}$$

and that

$$\{d_t^2 + A\}w(t) = f(t), \quad w(0) = 0, \quad w_t(0) = 0 \tag{2.119}$$

We have already seen that the problem (2.118) has a solution of the form

$$v(t) = \left(\cos\left(tA^{1/2}\right)\right)\varphi + A^{-1/2}\left(\sin\left(tA^{1/2}\right)\right)\psi$$

which leads to

$$v(x, t) = \left(\cos\left(tA^{1/2}\right)\right)\varphi(x) + A^{-1/2}\left(\sin\left(tA^{1/2}\right)\right)\psi(x) \tag{2.120}$$

To obtain $w(x, t)$ in (2.119) we set

$$w(t) = \int_0^t \eta(t,\tau)d\tau \tag{2.121}$$

where $\eta(t, \tau)$ is assumed to satisfy

$$\{d_t^2 + A\}\eta(t, \tau) = 0, \ \eta(\tau, \tau) = 0, \ \eta_t(\tau, \tau) = f(\tau) \tag{2.122}$$

A straightforward calculation then shows that $w(t)$ defined by (2.121) is a solution of the IVP (2.119). Therefore, combining these results we obtain the required solution in the form

$$\begin{aligned} u(x,t) &= \left(\cos\left(tA^{1/2}\right)\right)\varphi(x) + A^{-1/2}\left(\sin\left(tA^{1/2}\right)\right)\psi(x) + \int_0^t \eta(t,\tau)d\tau \\ &= (B_t(t)\varphi)(x) + (B(t)\psi)(x) + \int_0^t \eta(t,\tau)d\tau \end{aligned} \tag{2.123}$$

where

$$B(t) := A^{-1/2}\left(\sin\left(tA^{1/2}\right)\right) \tag{2.124}$$

Bearing in mind how the solution form (2.120) was obtained and applying the same techniques to the IVP (2.122), where in this case the initial conditions are imposed at $t = \tau$ rather than $t = 0$, we obtain

$$\eta(t,\tau) = A^{-1/2}\left(\sin\left((t-\tau)A^{1/2}\right)\right)f(\tau) = B(t-\tau)f(\tau) \tag{2.125}$$

Hence

$$w(t) = \int_0^t \eta(t,\tau)d\tau = \int_0^t B(t-\tau)f(\tau)d\tau \tag{2.126}$$

and consequently

$$u(x,t) = (B_t(t)\varphi)(x) + (B(t)\psi)(x) + \int_0^t B(t-\tau)f(x,\tau)d\tau \tag{2.127}$$

Thus we see that if we can solve the homogeneous equation, that is, if we can interpret (2.124) in a practical, constructive manner, then we can solve the inhomogeneous equation in the form (2.127). This is known as **Duhamel's Principle** and (2.126) is said to define the **Duhamel Integral** for the inhomogeneous equation. With this section in mind we shall concentrate our attention, almost entirely, for the remainder of the book on the associated homogeneous equations.

2.9 On the Far Field Behaviour of Solutions

In this section we give a first indication of how solutions of FPs and PPs might be considered as being AE. This we shall do by considering the following IVP. Determine a quantity $w(x, t)$ which satisfies

$$\{\partial_t^2 + L(x)\}w(x, t) = 0 \tag{2.128}$$

$$w(x, s) = \varphi_s(x), \quad w_t(x, s) = \psi_s(x) \tag{2.129}$$

where $s \in \mathbf{R}$ is a fixed initial time and in (2.129) $\varphi_s(x)$ and $\psi_s(x)$ are given initial data. In (2.128)

$$L(x) = -c^2\partial_x^2 + V(x) \tag{2.130}$$

where c is the wave speed and V is a function characterising a perturbation of the (one-dimensional) Laplacian ∂_x^2.

Problems of the form (2.128)–(2.130) are typical of those which arise when investigating potential scattering, an example of which was given above when discussing wave motions on an elastically braced string. We would remark that (2.128) and (2.130) together provide a perturbation of the classical wave equation and such an equation is frequently referred to as the plasma wave equation.

When $V(x) \equiv 0$ we refer to (2.128)–(2.130) as a Free Problem (FP). We notice that this FP is governed by the familiar wave equation. When $V(x) \neq 0$ everywhere then (2.128)–(2.130) is referred to as a Perturbed Problem (PP).

We have seen in the previous sections that the wave equation, which governs the FP in this case, has solutions which can be written in the form

$$w(x, t) = f(x - ct) + g(x + ct) \tag{2.131}$$

where f and g are arbitrary functions which characterise waves, of constant profile, travelling with velocity c from left to right and right to left respectively.

In the particular case when both waves have the same time dependency, $\exp(-i\omega t)$, then the familiar separation of variables technique indicates that the wave equation has solutions, denoted $w(x, t, k)$, which we can write in the form

$$w(x, t, k) = \exp(-i\omega t)\{u_+(x, k) + u_-(x, k)\} \tag{2.132}$$

where the quantities $u_\pm(x, k)$ must satisfy

$$\left\{\frac{d^2}{dx^2} + k^2\right\}u_\pm(x, k) = 0, \quad k^2 = \frac{\omega^2}{c^2}, \quad x \in \mathbf{R} \tag{2.133}$$

However, although the $u_\pm(x, k)$ both satisfy (2.133) nevertheless they have different properties. This can be seen by first noticing that (2.133) yields

$$u_+(x,k) = \exp(ikx) \quad \text{and} \quad u_-(x,k) = \exp(-ikx) \tag{2.134}$$

Combining (2.132) and (2.134) we obtain

$$w(x,t,k) = \exp\left(-i\omega\left(t - \frac{x}{c}\right)\right) + \exp\left(-i\omega\left(t + \frac{x}{c}\right)\right) \tag{2.135}$$

Thus, recalling the form (2.131) we see that u_+ characterises a wave moving from left to right and u_- characterises a wave moving from right to left, both waves having the same time dependency $\exp(-i\omega t)$. Equivalently, we say that u_+ represents an **outgoing wave**, since in $\mathbf{R}$, for $x > 0$, it is moving away from the origin whilst u_- represents an **incoming wave** since, in $\mathbf{R}$, for $x < 0$, it is moving towards the origin. This feature of the wave motion leads quite naturally to the following definition which caters for more general situations [6].

Definition 2.3. Solutions $v(x,k)$ of the equation

$$\{\Delta + k^2\}v(x,k) = 0, \quad x \in \mathbf{R}, \quad n \geqslant 1 \tag{2.136}$$

are said to satisfy the **Sommerfeld radiation condition** if and only if for all k as $r = |x| \to \infty$

$$\left\{\frac{\partial}{\partial r} \mp k\right\} v(x,k) = o\left(\frac{1}{r}\right) \tag{2.137}$$

$$v(x,k) = O\left(\frac{1}{r^{1/2(n-1)}}\right) \tag{2.138}$$

The estimates (2.137), (2.138) are understood to hold uniformly with respect to the direction $x/|x|$. The estimate (2.137) taken with the minus (plus) sign is called the Sommerfeld outgoing (incoming) radiation condition.

With $u_\pm$ defined as in (2.134) it is clear that $u_+(x,k)$ is an outgoing solution and $u_-(x,k)$ is an incoming solution in the sense of Definition 2.3. A derivation of these radiation conditions can be found in the text cited in the Commentary.

In later sections we shall see that the $u_\pm(x,k)$ have the important property that *any* solution of (2.136) can be expressed as a linear combination of the $u_\pm(x,k)$ [1]. As a consequence we shall refer to the $u_\pm(x,k)$ as **fundamental solutions** of (2.133).

We now turn our attention to the case when the potential $V(x) \neq 0$, that is to the PP which is a perturbation of the FP we have just been discussing. A natural first step when dealing with this PP is to see if the time dependency in the problem can again be separated out. If we assume that this separation is possible then instead of (2.132) and (2.133) we now have

$$w(x,t,k)=\exp(-i\omega t)\{w_+(x,k)+w_-(x,k)\},\quad k^2=\frac{\omega^2}{c^2} \tag{2.139}$$

where

$$\left\{\frac{d^2}{dx^2}+k^2-Q(x)\right\}w_\pm(x,k)=0,\quad k^2=\frac{\omega^2}{c^2},\quad x\in\mathbf{R} \tag{2.140}$$

and

$$Q(x)=\frac{V(x)}{c^2} \tag{2.141}$$

Since (2.140) is clearly a perturbation of (2.133), which has certain fundamental solutions, $u_\pm(x, k)$, defined in (2.134), then it is natural to ask if (2.140) also has fundamental solutions and, if so, can they be regarded as perturbations of the $u_\pm(x, k)$? Furthermore, if such fundamental solutions of (2.140) exist then are they in some sense AE to the fundamental solutions, $u_\pm(x, k)$, of (2.133)? It turns out that for (2.140) there is a family of solutions, parameterised by k, for which the answer to both these questions is in the affirmative. These are the so-called Jost solutions.

2.9.1 Jost Solutions

Bearing in mind the remarks at the end of the last section the Jost solutions, whenever they exist, are those solutions of (2.140), denoted by $\psi^\pm$ and $\varphi^\pm$, which have the following uniform asymptotic behaviour:

as $x \to +\propto$

$$\psi^\pm(x,k) = u_\pm(x,k)\{1 + o(1)\} = \exp(\pm ikx)\{1 + o(1)\} \tag{2.142}$$

$$\psi_x^\pm(x,k)=\frac{\partial u_\pm(x,k)}{\partial x}\{1+o(1)\}=\exp(\pm ikx)\{\pm ik+o(1)\} \tag{2.143}$$

as $x \to -\infty$

$$\varphi^\pm(x,k) = u_\mp(x,k)\{1 + o(1)\} = \exp(\mp ikx)\{1 + o(1)\} \tag{2.144}$$

$$\varphi_x^\pm(x,k)=\frac{\partial u_\mp(x,k)}{\partial x}\{1+o(1)\}=\exp(\mp ikx)\{\pm ik+o(1)\} \tag{2.145}$$

Alternatively, these can be written in the more compact forms:

$$\lim_{x\to\infty}\{\exp(\mp ikx)\psi^\pm(x,k)\}=1 \tag{2.146}$$

$$\lim_{x\to\infty}\{\exp(\mp ikx)\psi_x^\pm(x,k)\}=\pm ik \tag{2.147}$$

$$\lim_{x\to-\infty}\{\exp(\pm ikx)\varphi^{\pm}(x,k)\}=1 \tag{2.148}$$

$$\lim_{x\to-\infty}\{\exp(\pm ikx)\varphi_x^{\pm}(x,k)\}=\mp ik \tag{2.149}$$

Thus we see that as $x \to \pm\infty$ the Jost solutions will have the behaviour of plane waves. Consequently, we can expect the perturbed equation (2.140) to have solutions which can be interpreted as **distorted plane waves** which are AE to solutions of the FP.

Perhaps one of the more convenient ways of settling the questions of existence and uniqueness of the Jost solutions is to represent them as solutions of certain integral equations. This will have the added bonus of yielding a constructive method. To this end we recall that solutions of an equation of the form

$$y''(x) + \{k^2 - q(x)\}y(x) = 0 \tag{2.150}$$

can be obtained by the variation of parameters method. Specifically, assume that (2.150) has a solution that can be written in the form

$$y(x) = A(x)\exp(ikx) + B(x)\exp(-ikx) \tag{2.151}$$

Substituting (2.151) into (2.150) it is natural to set

$$(2ik)A'(x)\exp(ikx) = q(x)y(x) \tag{2.152}$$

and so obtain

$$A(x)=\frac{1}{2ik}\int_0^x q(v)y(v)\exp(-ikv)dv+C_1 \tag{2.153}$$

Now substitute (2.152), (2.153) into (2.150) and obtain

$$B(x)=\frac{-1}{2ik}\int_0^x q(v)y(v)\exp(+ikv)dv+C_2 \tag{2.154}$$

The integration constants C_1 and C_2 have now to be chosen to ensure that asymptotic behaviours of the forms (2.142)–(2.145) are obtained as required. To this end, if we identify y in (2.150) with the Jost solutions ψ^+ and φ^+ and consider the form of $\exp(-ikx)\psi^+$and $\exp(+ikx)\varphi^+$ as $x \to \infty$ and $x \to -\infty$ respectively then the required asymptotic behaviour will be obtained if for ψ^+ we set

$$C_1=1-\frac{1}{2ik}=\int_0^\infty q(v)y(v)\exp(-ikv)dv \tag{2.155}$$

and

$$C_2 = \frac{1}{2ik} = \int_0^\infty q(v)y(v)\exp(+ikv)dv \tag{2.156}$$

For φ^+ we set

$$C_1 = \frac{1}{2ik} = \int_{-\infty}^0 q(v)y(v)\exp(-ikv)dv \tag{2.157}$$

$$C_2 = 1 - \frac{1}{2ik} = \int_{-\infty}^0 q(v)y(v)\exp(+ikv)dv \tag{2.158}$$

to obtain the required asymptotic behaviour.

Combining these several results we obtain

$$\psi^\pm(x,k) = \exp(\pm ikx) + \frac{1}{k}\int_x^\infty q(v)\sin(k(v-x))\psi^\pm(v,k)dv \tag{2.159}$$

$$\varphi^\pm(x,k) = \exp(\mp ikx) - \frac{1}{k}\int_{-\infty}^x q(v)\sin(k(v-x))\varphi^\pm(v,k)dv \tag{2.160}$$

which are Volterra integral equations of the second kind for the Jost solutions $\psi^\pm(x,k)$ and $\varphi^\pm(x,k)$.

Solutions of (2.159), (2.160) can be obtained by successive approximations in the form

$$\psi^\pm(x,k) = \exp(\pm ikx)\sum_{j=0}^\infty h_j^\pm(x,k) \tag{2.161}$$

$$\varphi^\pm(x,k) = \exp(\mp ikx)\sum_{j=0}^\infty g_j^\pm(x,k) \tag{2.162}$$

where, for example, with $\operatorname{Im} k \geqslant 0$

$$\begin{aligned} g_0^\pm(x,k) &= 1 \\ g_{j+1}^\pm(x,k) &= -\frac{1}{k}\int_x^\infty q(v)\exp(\mp ik(v-x))\sin(k(v-x))g_j(v,k)dv \end{aligned} \tag{2.163}$$

Similar results for the $h_j^\pm$ can be obtained.

These Jost solutions have been extensively studied and their principal features are conveniently summarised in the following theorem.

Theorem 2.4. *The Schrödinger equation*

$$\left\{\frac{\partial^2}{\partial x^2} + (k^2 - Q(x))\right\}y(x,k) = 0, \quad x \in \mathbf{R}$$

has Jost solutions $\psi^{\pm}$ and $\varphi^{\pm}$ which satisfy the Volterra integral equations

$$\psi^{\pm}(x,k) = \exp(\pm ikx) + \frac{1}{k}\int_{x}^{\infty} Q(v)\sin(k(v-x))\psi^{\pm}(v,k)dv \qquad (2.164)$$

$$\varphi^{\pm}(x,k) = \exp(\mp ikx) - \frac{1}{k}\int_{-\infty}^{x} Q(v)\sin(k(v-x))\varphi^{\pm}(v,k)dv \qquad (2.165)$$

provided

$$\int_{-\infty}^{\infty} \{1+v\}Q(v)dv < \infty \qquad (2.166)$$

Furthermore

(i) The Jost solutions $\psi^{\pm}(x, k)$ and $\varphi^{\pm}(x, k)$ are unique. Successive approximations to these solutions can be obtained in the forms (2.161) and (2.162).

(ii) For every $x \in \mathbf{R}$ the Jost solutions $\psi^{+}(x, k)$ and $\varphi^{+}(x, k)$ and their derivatives $\psi_x^{+}(x, k)$ and $\varphi_x^{+}(x, k)$ are

(a) continuous with respect to k for Im $k \geqslant 0$

(b) analytic with respect to k for Im $k > 0$.

Analogous properties for $\psi^{-}(x, k)$ and $\varphi^{-}(x, k)$ hold for Im $k \leq 0$.

(iii) The Jost solutions $\psi^{\pm}(x, k)$ and $\varphi^{\pm}(x, k)$ are inter-related as follows.

$$\psi^{-}(x, k) = \psi^{+}(x, k^*) = \psi^*(x, k^*)$$

$$\varphi^{-}(x, k) = \varphi^{+}(x, -k^*) = \varphi^*(x, k^*)$$

2.9.2 Some Scattering Aspects

The Wronskian of two solutions, $y_1(x, k)$ and $y_2(x, k)$, of the Schrödinger equation (2.150) is denoted $W(y_1, y_2)$ and defined by

$$W(y_1, y_2) = y_1(x, k)y_2'(x, k) - y_1'(x, k)y_2(x, k) \qquad (2.167)$$

where the prime denotes differentiation with respect to x. A fundamental property of the Wronskian is given by the following theorem.

Theorem 2.5. *Two solutions of the Schrödinger equation* (2.150) *are linearly independent if and only if their Wronskian is non-zero.*

For the Jost solutions $\psi^{\pm}$ and $\varphi^{\pm}$ we find that as $x \to \infty$

$$W(\psi^{+}, \psi^{-}) = -2ik + o(1) \qquad (2.168)$$

and as $x \to -\infty$

$$W(\varphi^{+}, \varphi^{-}) = +2ik + o(1) \qquad (2.169)$$

Consequently Theorem 2.5 indicates that (ψ^+, ψ^-) and (φ^+, φ^-) are two linearly independent pairs of solutions of (2.140).

Since any linear combination of the pair (φ^+, φ^-) and the pair (φ^+, φ^-) will also yield a solution of (2.140) we see that we can, in particular, write

$$\varphi^+(x, k) = c_{11}(k)\psi^+(x, k) + c_{12}(k)\psi^-(x, k) \tag{2.170}$$

$$\psi^+(x, k) = c_{21}(k)\varphi^-(x, k) + c_{22}(k)\varphi^+(x, k) \tag{2.171}$$

where the coefficients c_{ij} have yet to be determined.

We now recall that in the case when $Q(x) = 0$ then the Schrödinger equation (2.140) reduces to the familiar wave equation which has the associated solutions $\psi_0^\pm(x, k)$ and $\varphi_0^\pm(x, k)$ where

$$\psi_0^\pm(x, k) = \exp(\pm ikx), \quad \varphi_0^\pm(x, k) = \exp(\mp ikx) \tag{2.172}$$

In this particular case the coefficients c_{ij} in (2.170), (2.171) are such that

$$c_{11}(k) = c_{22}(k) = 0$$

Furthermore, we notice that ψ^+ and φ^- characterise plane waves moving from left to right whilst ψ^- and φ^+ characterise plane waves moving from right to left.

When $Q(x) \neq 0$ for all x then we obtain (2.170), (2.171). The limiting behaviour of the Jost solutions given in (2.142)–(2.145) indicates that (2.170) represents a solution of the Schrödinger equation which, by the properties of the left-hand side, reduces to $\exp(-ikx)$ as $x \to -\infty$ whilst the right-hand side reduces to $c_{11}(k)$ $\exp(ikx) + c_{12}(k)\exp(-ikx)$ as $x \to \infty$. Consequently (2.170) is a solution of the Schrödinger equation (2.140) which represents the scattering, by the potential $Q(x)$, of a plane wave of amplitude $c_{12}(k)$ incident from $x = +\infty$ and moving right to left. The scattering process gives rise to a reflected plane wave of amplitude c_{11} moving left to right towards $x = +\infty$ and to a transmitted wave with unit amplitude moving right to left towards $x = -\infty$. It is customary to normalise this process so that the incident wave has unit amplitude in which case (2.170) is rewritten in the form

$$T_R(k)\varphi^+(x, k) = -R_R(k)\psi^+(x, k) + \psi^-(x, k) \tag{2.173}$$

where

$$R_R(k) = -\frac{c_{11}(k)}{c_{12}(k)}, \quad T_R(k) = \frac{1}{c_{22}(k)} \tag{2.174}$$

and the subscript R refers to the fact that we are dealing with an incident wave from the right. The minus sign is included for later convenience.

Similarly, (2.171) can be interpreted as a solution of the Schrödinger equation which represents the scattering, by a potential $Q(x)$, of a plane wave incident from $x = -\infty$. The process gives rise as before to a scattered and transmitted wave and normalising as before we can write (2.171) in the form

$$T_L(k)\psi^+(x,k) = \varphi(x,k) + R_L(k)\varphi^+(x,k) \tag{2.175}$$

where

$$R_L(k) = -\frac{c_{22}(k)}{c_{21}(k)}, \quad T_L(k) = \frac{1}{c_{21}(k)} \tag{2.176}$$

and the subscript L denotes that we are dealing with a wave incident from the left. Again the minus signs are included for later convenience.

The equations (2.173) and (2.175) can be written conveniently in the matrix form

$$\Phi^-(k) = S(k)\Phi^+(k) \tag{2.177}$$

where

$$\Phi^-(k) = \begin{bmatrix} \varphi^- \\ \psi^- \end{bmatrix}(k), \quad \Phi^+(k) = \begin{bmatrix} \varphi^+ \\ \psi^+ \end{bmatrix}(k) \tag{2.178}$$

and

$$S(k) = \begin{bmatrix} R_L & T_L \\ T_R & R_R \end{bmatrix}(k) \tag{2.179}$$

The matrix $S(k)$ is called the **scattering matrix** for the problem. Its role in scattering theory will be discussed in later sections. For the time being we simply note that it provides a connection between the incident fields and the scattered fields.

2.10 Concluding Remarks

It turns out that the various techniques and strategies we have outlined so far can be extended to cater efficiently and constructively with more general problems than those dealing simply with wave motions on strings. These generalisations can be made in a relatively easy manner if we choose to work with the actual functions involved rather than with the numerical values of the functions. To be able to do this requires that we should work within a mathematical structure which generates an easily solvable, but abstract, version of the given physical problem and yet is one which always ensures that there is an easy path back to the required physical (numerical) results. We indicate in the following chapter a way of achieving this.

Essentially the main steps are the following.

1. Identify a collection of elements which contains those elements which can be used.

   Characterise, for example, wave motions which have some particular property such as finite energy. Call this collection X.

   The elements of X are abstract quantities which can be thought of, for example, as functions themselves rather than their numerical values.
2. Endow X with a set of rules which will allow the elements of X to be manipulated algebraically and geometrically. These rules would parallel and extend the familiar processes which are used in the Euclidean space $\mathbf{R}^n$, $n < \infty$ The collection X taken together with the structure defined in terms of these algebraic and geometric rules we shall call a **space** and denote it here, for the time being, by H.
3. Introduce the notion of an **operator** which, in its simplest form, maps (transforms) one element of H into some other element of H.
4. Use these several notions to represent (realise) a given physical problem, which is essentially a problem involving numerical values of functions, in the space H. This will yield an abstract problem involving the functions themselves rather than their numerical values.
5. Investigate the availability of associated inverse operators as a means of solving the abstract problem.
6. Settle questions of the existence and uniqueness of solutions to the abstract problem. That is, examine the well-posed nature of the abstract problem.
7. Solve the abstract problem.
8. Interpret the solution of the abstract problem in a manner which will allow the recovery of the required physical results.

The first four steps can be made by introducing the notion of a so-called Hilbert space structure and using the properties of (linear) operators on such spaces.

Step 5 can be made using results from the spectral theory of (linear) operators on a Hilbert space.

Step 6 can be made efficiently and constructively using results from the elegant theory of semigroups of operators.

Step 7 can be achieved using results from the theory of ordinary differential equations but it must be remembered that the work is in the abstract space, H, rather than $\mathbf{R}^n$.

The final step here will be made by introducing the notion of a generalised Fourier transform and proving generalised Fourier inversion theorems modelled on the Fourier–Plancherel theory which can be established quite independently of any scattering phenomena.

With this preparation we will then be well placed to develop a scattering theory which would highlight constructive methods for analysing echo field phenomena. This we shall demonstrate by investigating in the following chapters a number of specific, physically relevant, problems. However, before embarking on this we shall first introduce, in the next chapter, a number of basic ideas and results from general mathematical analysis which will be sufficient to create a suitable mathematical structure in which to work and settle items 1–8 above.

References and Further Reading

[1] W.O. Amrein, J.M. Jauch and K.B. Sinha: *Scattering Theory in Quantum Mechanics*, Lecture Notes and Supplements in Physics, Benjamin, Reading, 1977.

[2] G.R. Baldock and T. Bridgeman: *The Mathematical Theory of Wave Motion*, Mathematics and its Applications, Ellis Horwood, London, 1981.

[3] A. Kufner and J. Kadlec: *Fourier Series*, Illiffe, London, 1971.

[4] M. Reed and B. Simon: *Methods of Modern Mathematical Physics*, vols 1–4, Academic Press, New York, 1975.

[5] G.F. Roach: *Greens Functions* (2nd Edn), Cambridge Univ. Press, London, 1970/1982.

[6] G.F. Roach: *An Introduction to Linear and Nonlinear Scattering Theory*, Pitman Monographs and Surveys in Pure and Applied Mathematics, Vol. 78, Longman, Essex, 1995.

[7] W.A. Strauss: *Partial Differential Equations, an introduction*, Wiley and Sons, New York, 1992.

[8] E.C. Titchmarsh: *Introduction to the Theory of Fourier Integrals*, Oxford University Press, 1937.

[9] H.F. Weinberger, *A First Course in Partial Differential Equations*, Blaisdell, Waltham, MA, 1965.

[10] C.H. Wilcox: *Scattering Theory for the d'Alembert Equation in Exterior Domains*, Lecture Notes in Mathematics, No. 442, Springer, Berlin, 1975.

3

Preliminary Mathematical Material

3.1 Introduction

This rather long chapter is provided mainly for the benefit of those who are interested in studying wave phenomena but whose mathematical background does not necessarily include modern functional analysis and operator theory. For such scientists this chapter is intended to be a guide through the material available whilst for those with more mathematical background it can act as a source. Virtually no proofs are given. Details of these can be found in the references cited. Despite the fact that much of the material might, at first sight, give the impression of being unnecessarily abstract, nevertheless, it will be seen to be of considerable use in justifying various approaches to the development of constructive methods of solving physical problems.

3.2 Notations

Some standard mathematical notations which are frequently used throughout the book are the following.

Let X and Y be any two sets. The **inclusion** $x \in X$ denotes that the quantity x is an element in the set X. The **(set) inclusion** $Y \subset X$, denotes that the set Y is contained in the set X in which case the set Y is said to be a **(proper) subset** of the set X. The possibility that Y might actually be the same as X exists and in this case we write $Y \subseteq X$ and simply refer to Y as a **subset** of X. The set which consists of elements that belong to either X or Y or both is called the **union** of X and Y and is denoted by $Y \cup X$. The set consisting of all elements belonging to X and Y simultaneously is called the **intersection** of X and Y and is denoted by $X \cap Y$. The set consisting of all the elements of X which do not belong to Y is called the **difference** of X and Y is denoted by $X \setminus Y$. In particular, if $Y \subset X$ then $X \setminus Y$ is called the **complement** of Y in X.

We write $X \times Y$ to denote the set of elements of the form (x, y) where $x \in X$ and $y \in Y$. We recall that a simple example of this notation is the Euclidean space $\mathbf{R}^2$ where we identify $X = Y = \mathbf{R}$.

It is clear that $X \subset X \cup Y$, $Y \subset X \cup Y$, $X \cap Y \subset X$, $X \cap Y \subset Y$ and that if $Y \subset X$ then $X = Y \cup (X \setminus Y)$.

3.3 Vector Spaces

Vector spaces will provide a framework within which abstract quantities such as functions themselves rather than their numerical values can be manipulated *algebraically* in a meaningful manner.

Definition 3.1. A **vector space (linear space)** over a set of **scalars K** is a non-empty set, X, of elements $x, y, \ldots$ called **vectors**, together with two algebraic operations called **vector addition** and **multiplication by a scalar** which satisfy

L1: $x + y + z = (x + y) + z = x + (y + z), \quad x, y, z \in X$

L2: There exists a **zero element** $\theta \in X$ such that

$$x + \theta = x, \quad x \in X$$

L3: If $x \in X$ then there exists an element $(-x) \in X$ such that

$$x + (-x) = \theta, \quad x \in X$$

L4: $x + y = y + x, \quad x, y \in X$

L5: $(\alpha + \beta)x = \alpha x + \beta y, \quad \alpha, \beta \in \mathbf{K}, \quad x \in X$

L6: $\alpha(x + y) = \alpha x + \alpha y, \quad \alpha \in \mathbf{K}, \quad x, y \in X$

L7: $\alpha(\beta x) = \alpha\beta x, \quad \alpha, \beta \in \mathbf{K}, \quad x \in X$

L8: There exists a unit element $I \in \mathbf{K}$ such that

$$Ix = x, \quad x \in X$$

In following chapters, **K** will usually be either **R** or **C**.

The notion of the distance between abstract quantities can be introduced by mimicking familiar processes in Euclidean geometry

Definition 3.2. A **metric space** M is a set X and a real-valued function, d, called a **metric or distance function** defined on $X \times X$ such that for all $x, y, z \in X$

M1: $d(x, y) \geqslant 0$

M2: $d(x, y) = 0$ if and only if $x = y$

M3: $d(x, y) = d(y, x)$

M4: $d(x, z) \leq d(x, y) + d(y, z)$

The axiom M4 is known as the **triangle inequality**.

We emphasise that a set X can be made into a metric space in many different ways simply by employing different metric functions. Consequently, for the sake of clarity we shall sometimes denote a metric space in the form (X, d) in order to make explicit the metric employed.

Example 3.3. Let $X = \mathbf{R}^2$ and d be the usual Euclidean distance function (metric). Then if $x = (x_1, x_2, x_3, \ldots, x_n)$ and $y = (y_1, y_2, y_3, \ldots, y_n)$ denote any two points in $\mathbf{R}^n$ the distance between these two points is given by

$$d(x, y) = \left\{ \sum_{i=1}^{n} (x_i - y_i)^2 \right\}^{1/2}$$

It is this particular distance function and its properties that we mimicked when formulating Definition 3.2. Clearly, (X, d) in this example is a metric space.

Example 3.4. Let $X = C[0, 1]$, the set of all real-valued continuous functions defined on the subset $[0, 1] \subset \mathbf{R}$. This set can be made into two metric spaces, M_1 and M_2, where

$$\begin{aligned} M_1 &:= (X, d_1) \\ d_1(f, g) &:= \max_{x \in [0,1]} |f(x) - g(x)|, \quad f, g \in X \\ M_2 &:= (X, d_2) \\ d_2(f, g) &:= \int_0^1 |f(x) - g(x)| dx, \quad f, g \in X \end{aligned}$$

In constructing M_1 and M_2 we must, of course, **prove** that d_1, and d_2 each satisfy the metric space axioms M1 to M4 in Definition 3.2. In these two examples M1 to M3 are obviously satisfied. The axiom M4, usually the hardest property to establish, is seen to hold in these two cases by virtue of well-known properties of the modulus and of classical Riemann integrals.

We remark that the metric space $C[0, 1]$ is an example of an infinite dimensional space and is one which we shall frequently have occasion to use.

Once we have introduced the notion of distance in abstract spaces then we are well placed to give a precise meaning to what is meant by convergence in such spaces. This will be a necessary ingredient when we come to develop constructive, approximation methods.

Definition 3.5. A sequence of elements $\{x_n\}_{n=1}^{\infty}$ in a metric space $M := (X, d)$ is said to **converge** to an element $x \in X$ if

$$d(x, x_n) \to 0 \quad \text{as} \quad n \to \infty$$

In this case we write either $x_n \to x$ as $n \to \infty$ or $\lim_{n \to \infty} x_n = x$ where it is understood that the limit is taken with respect to the distance function d.

We would emphasise that the use of different metrics can induce different convergence results. For instance in Example 3.4 we have

$$d_2(f, g) \leq d_1(f, g), \quad f, g \in X$$

Therefore, given a sequence $\{f_n\}_{n=1}^{\infty} \subset X$ such that $f_n \to f$ with respect to d_1 then it follows that we also have $f_n \to f$ with respect to d_2. However, if we are given that the sequence converges with respect to d_2, then it does not follow that the sequence also converges with respect to d_1.

Definition 3.6. A sequence $\{f_n\}_{n=1}^{\infty} \subset (X, d)$ is called a **Cauchy sequence** if, for all $\varepsilon > 0$, there exists a number $N(\varepsilon)$ such that $n, m \geqslant N(\varepsilon)$ implies $d(f_n, f_m) < \varepsilon$.

The following is a standard result.

Theorem 3.7. *In any metric space $M := (X, d)$ every convergent sequence*
(i) *has a unique limit*
(ii) *is a Cauchy sequence.*

It must be emphasised that the converse of (ii) is false as it is possible that there are Cauchy sequences in (X, d) which might not converge to a limit element $x \in X$. This difficulty can be avoided by restricting attention to certain preferred classes of metric spaces.

Definition 3.8. A metric space in which all Cauchy sequences converge to an element of that space is called a **complete metric space**.

It can be shown that for the metric spaces M_1, M_2 in Example 3.4 the metric space M_1 is complete but M_2 is incomplete.

Many of the strategies adopted when analysing problems will be seen to rely on the following notion.

Definition 3.9. Given a metric space $M := (X, d)$ a set $Z \subset X$ is said to be **dense** in X if every element $y \in X$ is the limit, with respect to d, of a sequence of elements in Z.

Consequently, in most practical situations we try to work with “nice” elements belonging to a set, such as Z, which is dense in some larger set, X, whose elements can provide more general results.

To give an indication that this enlargement, or **completion** as it is more properly called, can be made available we first need to introduce the following notions.

Definition 3.10. (i) Let X_1, X_2 be sets and let $M \subseteq X_1$ be a subset. A **mapping** f from M into X_2 is a rule that assigns to an element $x \in M \subseteq X_1$ an element $f(x) =: y \in X_2$ and we write

$$f : X_1 \supseteq M \to X_2$$

The element $y = f(x) \in X_2$ is the **image** of x with respect to f. The set M is called the **domain** of Definition of f which is denoted $D(f)$. Consequently, we write

$$f : X_1 \supseteq D(f) \to X_2$$

and

$$x \to f(x) = y \in X_2 \quad \text{for all} \quad x \in D(f) \subseteq X_1$$

The set of all images with respect to f is the **range** of f, denoted $R(f)$, where

$$R(f) := \{y \in X_2 : \mathrm{y} = f(x) \quad \text{for} \quad x \in D(f) \subseteq X_1\}$$

(ii) A mapping f is called either **injective** or an **injection** or **one-to-one** (denoted 1-1) if for every $x_1, x_2 \in D(f) \subseteq X_1$ we have that

$$x_1 \neq x_2 \quad \text{implies} \quad f(x_1) \neq f(x_2)$$

This means that different elements in $D(f)$ have different images in $R(f)$.

(iii) A mapping f is called either **surjective** or a **surjection** or a mapping of $D(f)$ **onto** X_2 if $R(f) = X_2$.

(iv) For an injective mapping $T : X_1 \supseteq D(T) \to X_2$ the **inverse mapping**, T^{-1}, is defined to be the mapping $R(T) \to D(T)$ such that $y \in R(T)$ is mapped onto that $x \in D(T)$ for which $Tx = y$. Less generally, but more conveniently here, we define T^{-1} only if T is 1-1 and onto X_2.

We notice that in (iv) we have written Tx rather than T (x). This anticipates the notation used in later sections.

In the following chapters we will deal with a variety of different mappings. For example, a **function** is usually understood to be mapping of one (real or complex) number into some other number. It is a rule that assigns to a number in one set a number in some other set. The term **operator** will be reserved for mappings between sets of abstract elements such as functions themselves rather than their numerical values.

Definition 3.11. A mapping, f, from a metric space (X_1, d_1) to a metric space (X_2, d_2) is said to be **continuous**, if for $\{x_n\}_{n=1}^{\infty} \subset X_1$ we have $f(x_n) \to f(x)$ with respect to the structure of (X_2, d_2) whenever, $x_n \to x$ with respect to the structure of (X_1, d_1).

Definition 3.12. Let $M_j = (X_j, d_j)$, $j = 1, 2$ be metric spaces. A mapping, f, which

(i) satisfies $f : X_1 \to X_2$ is one-to-one and onto (**bijection**)

(ii) preserves metrics in the sense

$$d_2(f(x), f(y)) = d_1(x, y), \quad x, y \in X_1$$

is called an **isometry** and M_1, M_2 are said to be **isomorphic**.

It is clear that an isometry is a continuous mapping. Furthermore, isometric spaces are essentially identical as metric spaces in the sense that any result which holds for a metric space $M = (X, d)$ will also hold for all metric spaces which are isometric to M. It is for this reason that we always try to work in a framework of isometric spaces. This will be particularly important when we come to develop some of the finer points of scattering theory.

We now state a fundamental result which indicates in what sense an incomplete metric space can be made complete.

Theorem 3.13. *If* $\mathrm{M} = (X, d)$ *is an incomplete metric space then it is possible to find a complete metric space* $\tilde{M} = (\tilde{X}, \tilde{d})$ *so that* M *is isometric to a dense subset of* $\tilde{M}$.

The familiar concepts of **open and closed sets** on the real line extend to arbitrary metric spaces according to the following Definition.

Definition 3.14. If $M = (X, d)$ is a metric space then

(i) the set

$$B(y, r) := \{x \in X : d(x, y) < r\}$$

is called the **open ball** in M, of radius $r > 0$ and centre y.

(ii) a set $G \subset X$ is said to be **open with respect to** d if, for all $y \in G$, there exists $r > 0$ such that $B(y, r) \subset G$.

(iii) a set $N \subset G$ is called a **neighbourhood** of $y \in N$ if $B(y, r) \subset N$ for some $r > 0$.

(iv) a point x is called a **limit point** of a subset $Y \subset X$ if

$$B(x, r) \cap \{Y \setminus \{x\}\} \neq \phi \quad \text{for all} \quad r > 0$$

where ϕ denotes the empty set.

(v) a set $F \subset X$ is said to be **closed** if it contains all its limit points.

(vi) the union of F and all its limit points is called the **closure** of F and will be denoted $\bar{F}$.

(vii) a point $x \in Y \subset X$ is an **interior point** of Y if Y is a neighbourhood of x.

These various abstract notions can be quite simply illustrated by considering subsets of the real line.

A particularly important class of metric spaces is the following.

Definition 3.15. **A normed linear space** is a vector space X, defined over $\mathbf{K} = \mathbf{R}$ or $\mathbf{C}$, together with a mapping $\|\cdot\| : X \to \mathbf{R}$, known as a **norm** on X, satisfying

N1: $\|x\| \geqslant 0$ for all $x \in X$
N2: $\|x\| = 0$ if and only if $x = \theta$, the zero element in X
N3: $\|\lambda x\| = |\lambda|\ \|x\|$, for all $\lambda \in \mathbf{K}$ and $x \in X$
N4: $\|x + y\| \leq \|x\| + \|y\|$ (triangle inequality).
The pair $(X, \|\cdot\|)$ is referred to as a real or complex normed linear (vector) space depending on whether the underlying field $\mathbf{K}$ is $\mathbf{R}$ or $\mathbf{C}$.

Example 3.16. (i) $X = \mathbf{R}^n$ is a real normed linear space with a norm defined by

$$\|x\| = \|(x_1, x_2, x_3, \ldots, x_n)\| := \left\{ \sum_{k=1}^{n} |x_k|^2 \right\}^{1/2}$$

(ii) $X = C[0, 1]$, the set of all continuous functions defined on $[0, 1]$, is a real normed linear space with a norm defined by either

$$\|f\| := \sup_{x \in [0,1]} |f(x)|, \quad f \in X$$

or

$$\|f\| := \int_0^1 |f(y)| dy, \quad f \in X$$

We notice that any normed linear space $(X, \|\cdot\|)$ is also a metric space with the metric (distance function) d defined by

$$d(x, y) = \|x - y\|$$

This is the so-called **induced metric** on X. With this understanding we see that such notions as convergence, continuity, completeness, open and closed sets which we introduced earlier for metric spaces carry over to normed linear spaces. Typically we have.

Definition 3.17. Let $(X, \|\cdot\|)$ be a normed linear space. A sequence $\{x_n\}_{n=1}^{\infty} \subset X$ is said to **converge** to $x \in X$ if, given $\varepsilon > 0$ there exists $N(\varepsilon)$ such that

$$\|x_n - x\| < \varepsilon \quad \text{whenever} \quad n \geqslant N(\varepsilon)$$

in which case we write either $\|x_n - x\| \to 0$ as $n \to \infty$ or $x_n \to x$ as $n \to \infty$.

Definition 3.18. (i) The normed linear space $(X, \|\cdot\|)$ is **complete** if it is complete as a metric space in the induced metric.
(ii) A complete normed linear space is called a **Banach space**.

We would emphasise that there are metric spaces which are not normed linear spaces. A comparison of Definitions 3.2 and 3.15 clearly indicates this.

3.4 Distributions

A distribution is a generalisation of the concept of a classical function. It is a powerful mathematical tool for at least three reasons. First, in terms of the theory of distributions, it is possible to give a precise mathematical description of such idealised physical quantities as, for example, point charges and instantaneous impulses. Second, distribution theory provides a means for interchanging limiting operations when such interchanges might not be valid for classical functions. For instance, in contrast to classical analysis, in distribution theory there are no problems arising from the existence of non-differentiable functions. Indeed, all distributions, or generalised functions as they are sometimes called, can be treated as being infinitely differentiable. Third, distribution theory enables us to use series which in classical analysis would be considered as being divergent.

Distribution theory arises as a result of the following observation. A continuous, real or complex valued function f of the real variable $x = (x_1, x_2, x_3, \ldots, x_n) \in \mathbf{R}^n$ can be defined on $\mathbf{R}^n$ in one or other of two distinct ways. First, we can prescribe its value, $f(x)$, at each point $x \in \mathbf{R}^n$. Alternatively, we could prescribe the value of the integral

$$I_f(\varphi) := \int_{\mathbf{R}^n} f(x)\varphi(x)dx$$

for *each* continuous, complex-valued function φ whose value, $\varphi(x)$, is zero for sufficiently large $|x|$ (this latter to ensure that the integral exists).

These two definitions have distinct and quite different characterisations. In the first a function is considered as a rule which assigns *numbers to numbers*. In the second, which leads to something we will eventually call a distribution, we have a rule which assigns *numbers to functions*.

When working with the second approach, instead of dealing with the pointwise values, $f(x)$, of the function f we consider the *functional* I_f and its "values", $I_f(\varphi)$, at each of the so-called *test functions* φ.

These two descriptions of f are equivalent. To see this assume that for two continuous functions f and g the functionals I_f and I_g, defined as indicated above, are equal. That is, for any test function φ we have $I_f(\varphi) = I_g(\varphi)$. It then follows from elementary properties of the integral that $f(x) = g(x)$ for all $x \in \mathbf{R}^n$. Hence the required equivalence is established.

A partial rationale for introducing this second way of defining a function can be given as follows. A distributed physical quantity cannot be characterised by its value at a point but rather by its averaged value in a sufficiently close neighbourhood of that point. For instance, it is impossible to measure the density of a material at a point. In practice we can only measure the average density of the material in a small neighbourhood of the point and then call this the density at the point. Thus we can think of a generalised function as being defined by its "average values" in a neighbourhood of the point of interest. Consequently, from a physical standpoint it is more convenient to consider continuous functions as functionals of the form indicated above.

To clarify matters we give some examples and introduce a little more notation.

Example 3.19. Let $C(\Omega)$ denote the set of all complex-valued functions which are continuous on the region Ω. For $f \in C(\Omega)$ let

$$\|f\|_{\infty} := \sup\{|f(x)| : x \in \Omega\}$$

It is readily verified that $C(\Omega)$ is a complex vector space with respect to the usual pointwise operations on functions. Furthermore, $C(\Omega)$ is a complete, normed linear space with respect to $\|\cdot\|_{\infty}$.

Definition 3.20. Let a real or complex valued function f, defined on a domain $D \subset \mathbf{R}^n$, be nonzero only for points belonging to a subset $\Omega \subset D$. Then the closure of Ω, denoted $\overline{\Omega}$, is called the **support** of f and is denoted $\operatorname{supp} f$.

If Ω is a **compact set**, that is, a closed and bounded set, then the function f is said to have **compact support** in D.

In order to simplify the notation when working in more than one dimension we shall frequently use **multi-index notation**. This can be easily introduced if we consider a partial differential expression of the form

$$L = \sum_{q_1+\ldots+q_N \leq p} a_{q_1 \ldots q_N}(x) \partial_1^{q_1} \partial_2^{q_2} \ldots \partial_N^{q_N}, \quad p \geq 0$$

where $\partial_j := \partial/\partial x_j$, $j = 1, 2, \ldots, N$ and the q_j, $j = 1, 2, \ldots, N$ are non-negative integers with $a_{q_1, \ldots, q_N}$ denoting differentiable (to sufficient order) functions.

Any set $q := \{q_1, q_2, \ldots, q_N\}$ of non-negative integers is called a **multi-index**. The sum $|q| = q_1 + \ldots + q_N$ is called the **order** of the multi-index. We denote $a_{q_1, \ldots, q_N}(x)$ by $a_q(x)$ and $\partial_1^{q_1} \partial_2^{q_2} \ldots \partial_N^{q_N}$ by D^q. Consequently the above partial differential expression can be written as

$$L = \sum_{|q| \leq p} a_q(x) D^q.$$

Definition 3.21. Let Ω be an open set in $\mathbf{R}^n$.

(i) $C_0^{\infty} := \{\varphi \in C^{\infty}(\Omega) : \operatorname{supp} f \subset \Omega\}$.

Here $\varphi \in C^{\infty}(\Omega)$ indicates that φ and *all* its partial derivatives of all orders exist and are continuous. An element $\varphi \in C^{\infty}(\Omega)$ is referred to as a **smooth element** or as an **infinitely differentiable element**.

(ii) The set $C_0^{\infty}(\Omega)$ is called the set of **test functions** defined on Ω.

(iii) A sequence of test functions, $\{\varphi_n\}_{n=1}^{\infty}$, is said to be **convergent** to a test function φ if

(a) φ_n and φ are defined on the same compact set

(b)

$$\|\varphi_n - \varphi\|_\infty \to 0 \qquad as \quad k \to \infty$$

$$\|D^\alpha \varphi_n - D^\alpha \varphi\|_\infty \to 0 \quad as \quad k \to \infty$$

for all multi-indices α. Here, $\|\varphi\|_\infty := \sup_{x \in \Omega} |\varphi(x)|$.

(iv) The set $C_0^\infty(\mathbf{R}^n)$ together with the topology (a concept of convergence) induced by the convergence defined in (iii) is called the **space of test functions** and is denoted by $\mathcal{D}(\mathbf{R}^n)$.

Unless certain subsets of $\mathbf{R}^n$ have to be emphasised then we shall write $\mathcal{D} \equiv \mathcal{D}(\mathbf{R}^n)$ unless otherwise stated.

(v) **A linear functional** f on $\mathcal{D}$ is a *mapping* $f: \mathcal{D} \to \mathbf{K} = \mathbf{R}$ or $\mathbf{C}$ such that

$$f(a\varphi + b\psi) = af(\varphi) + bf(\psi)$$

for all $a, b \in C$ and $\varphi, \psi, \in \mathcal{D}$.

A mapping $f: \mathcal{D} \to \mathbf{K}$ is a *rule* which given any $\varphi \in \mathcal{D}$ produces a *number* $z \in \mathbf{K}$ and we write

$$z = f(\varphi) \equiv \langle f, \varphi \rangle$$

This indicates the **action** of the functional f on φ.

We remark that if $\varphi \in C_0^\infty(\Omega)$, $\Omega \subset \mathbf{R}^n$, then φ vanishes on $\partial\Omega$, the "boundary" of the set Ω.

Example 3.22. Let φ be the function defined on $\Omega := \mathbf{R}$. by

$$\varphi(x) = \begin{cases} 0, & |x| \geq a \\ \exp\left\{\dfrac{1}{x^2 - a^2}\right\}, & |x| < a \end{cases}$$

It is readily shown that φ is infinitely, continuously differentiable on $\mathbf{R}$ and that supp $\varphi = [-a, a]$. Hence $\varphi \in C_0^\infty(\mathbf{R})$.

Definition 3.23. (i) A functional f on $\mathcal{D}$ is continuous if it maps every convergent sequence in $\mathcal{D}$ into a convergent sequence in $\mathbf{K} = \mathbf{R}$ or $\mathbf{C}$; that is,

$$f(\varphi_n) \to f(\varphi) \quad \text{whenever} \quad \varphi_n \to \varphi \text{ in } \mathcal{D}$$

(ii) A **continuous, linear functional** on $\mathcal{D}$ is called a **distribution** or **generalised function**.

Notation 3.24. *For convenience at this stage we shall use a bold type face to indicate a distribution and will write* $\langle \mathbf{f}, \varphi \rangle$ *to denote the action of the distribution* $\mathbf{f}$ *on the test function* φ. *We will be able to relax this notation later.*

Definition 3.25. (Convergence of distributions)

A sequence $\{\mathbf{f}_n\}$ of distributions is said to be **convergent** if the sequence of numbers $\{\langle \mathbf{f}_n, \varphi \rangle\}$ is convergent for all $\varphi \in \mathcal{D}$.

This definition implies that if $\{\mathbf{f}_n\}$ is a convergent sequence of distributions then there is a distribution $\mathbf{f}$ such that for $n \to \infty$

$$\langle \mathbf{f}_n, \varphi \rangle \to \langle \mathbf{f}, \varphi \rangle \quad \text{for all} \quad \varphi \in \mathcal{D}$$

In this case $\mathbf{f}_n$ is said to **converge weakly** to $\mathbf{f}$ as $n \to \infty$.

Definition 3.26. (i) A function f which is integrable on every open, bounded subset $\Omega \subset \mathbf{R}^n$ is said to be **locally integrable** on $\mathbf{R}^n$.

(ii) For every locally integrable function f there is a distribution $\mathbf{f}$ defined by

$$\langle \mathbf{f}, \varphi \rangle = \int_{\mathbf{R}^n} f(s)\varphi(s)ds =: \mathbf{f}(\varphi)$$

The distribution $\mathbf{f}$ is said to be **generated** by the function f.

(iii) A distribution generated by a locally integrable function is called a regular distribution. Distributions which are not **regular distributions** are called **singular distributions**.

We remark that Definition 3.26 (ii) is meaningful since $B := \text{supp } \varphi \subset \mathbf{R}^n$. Consequently we then have

$$|\mathbf{f}(\varphi)| = \left| \int_B f(s)\varphi(s)ds \right| \leq \sup_{x \in B} |\varphi(x)| \int_B |f(s)| ds$$

The right-hand side is bounded since f is locally integrable. Hence we can conclude that $\mathbf{f}(\varphi)$ is well defined.

Thus we see that the class (set) of all distributions will contain elements which correspond to ordinary (classical) functions as well as singular distributions which do not.

Definition 3.27. The set of all distributions on $\mathcal{D}$ together with the topology indicated in Definition 3.25 is called the **dual** of $\mathcal{D}$ and denoted $\mathcal{D}'$.

For the sake of our convenience in later sections we summarise here the main properties of a distribution which we have introduced so far.

(i) Linearity:
For any test functions φ, ψ and complex numbers α, β

$$\mathbf{f}(\alpha\varphi + \beta\psi) = \alpha\mathbf{f}(\varphi) + \beta\mathbf{f}(\varphi)$$

(ii) Continuity:

$$\mathbf{f}(\varphi_n) \to \mathbf{f}(\varphi) \quad \text{whenever} \quad \varphi_n \to \varphi$$

(iii) Equality:
Two distributions **f** and **g** are equal provided $\mathbf{f}(\varphi) = \mathbf{g}(\varphi)$ for any test function φ.
They are said to be different if there exists a test function φ such that $\mathbf{f}(\varphi) \neq \mathbf{g}(\varphi)$.

Strictly speaking, equality here means equality almost everywhere (ae), that is, if the set $\{x: \mathbf{f}(\varphi) \neq \mathbf{g}(\varphi)\}$ has measure zero then $\mathbf{f}(\varphi) = \mathbf{g}(\varphi)$. The notion of measure is introduced in the next chapter. For the time being it is sufficient to think of the situation in $\mathbf{R}^1$ when measure can be identified with the length of an interval and a set of measure zero is a point.

(iv) Linear combinations:
The linear combination $(\alpha\mathbf{f} + \beta\mathbf{g})$ of two distributions **f** and **g** is defined as

$$\langle \alpha\mathbf{f} + \beta\mathbf{g}, \varphi\rangle = \alpha\langle \mathbf{f}, \varphi\rangle + \beta\langle \mathbf{g}, \varphi\rangle$$

(v) The product of a distribution **f** and a smooth function h is defined in a natural manner as

$$\langle h\mathbf{f}, \varphi\rangle = \langle \mathbf{f}, h\varphi\rangle \ \forall \ \varphi \in \mathcal{D}$$

If $\varphi \in \mathcal{D}$ and h is a smooth function then $h\varphi$ is also a test function. However, if h is not smooth then $h\varphi \notin \mathcal{D}$. Therefore we cannot define the product of a distribution with a function which is discontinuous or has discontinuous derivatives.

We notice that our definition is a generalisation of the familiar identity

$$\int_\Omega \{h(x)f(x)\}\varphi(x)dx = \int_\Omega h(x)\{f(x)\varphi(x)\}dx$$

which always holds when f is locally integrable.

We have already pointed out that a generalised function does not have values *at a point*. However, it is possible to give meaning to the statement that a distribution becomes zero in a *region*. The distribution **f** becomes zero in the region Ω if $\langle \mathbf{f}, \varphi\rangle = 0$ for all $\varphi \in \mathcal{D}(\Omega)$, and we write $\mathbf{f} = \boldsymbol{\theta}$ in Ω.

(vi) The set of all points such that in no neighbourhood of each point does $\mathbf{f} \neq \boldsymbol{\theta}$ is known as the support of the distribution **f**. We denote the **support** of **f** by supp **f**. If supp **f** is bounded then the distribution **f** is said to have **compact support**.

We now turn to the differential calculus of distributions. Our plan here, as indeed it will be when developing most properties of distributions, is to start with regular distributions and then generalise the results, whenever possible, to all distributions. For ease of presentation at this stage we restrict attention to processes in $\mathbf{R}^1 = \mathbf{R}$.

If f is a differentiable function which generates a regular distribution $\mathbf{f}$ and if $d\mathbf{f}/dx$ denotes the distribution generated by f' then we obtain, by integration by parts,

$$\left\langle \frac{d\mathbf{f}}{dx}, \varphi \right\rangle = \int_{\mathbf{R}} f'(s)\varphi(s)ds = -\int_{\mathbf{R}} f(s)\varphi'(s)ds$$

The integrated terms in the above vanish since φ is a test function and as such vanishes at infinity. Consequently, for regular distributions, corresponding to differentiable functions, we have

$$\left\langle \frac{d\mathbf{f}}{dx}, \varphi \right\rangle = -\langle \mathbf{f}, \varphi' \rangle$$

Example 3.28. [8]. For any distribution $\mathbf{f}$ the functional $\varphi \to -\langle \mathbf{f}, \varphi' \rangle$ with $\varphi \in \mathcal{D}$ is a distribution.

The following definition now follows quite naturally.

Definition 3.29. The derivative of a generalised function $\mathbf{f}$ is the generalised function $\mathbf{f}'$ defined by

$$\langle \mathbf{f}', \varphi \rangle = -\langle \mathbf{f}, \varphi' \rangle \ \forall \ \varphi \in \mathcal{D}$$

We see from the above that the distribution generated by the derivative, f', of a differentiable function f is the same as the derivative of the distribution $\mathbf{f}$. These two ways of interpreting the symbol $\mathbf{f}'$ for a differentiable function f are consistent with classical calculus [8].

The advantage of distribution theory over classical calculus is that *every* generalised function is differentiable; this follows from Example 3.28.

If a function f is locally integrable but not differentiable, in the classical sense, the associated distribution $\mathbf{f}'$ is called the **generalised derivative** of f.

It is a straightforward matter to obtain corresponding results when the underlying space is $\mathbf{R}^n$. For instance, let $\Omega \subset \mathbf{R}^n$ be an open set and let $\mathcal{D}(\Omega)$ denote the space of test functions defined on Ω. Let α denote a multi-index. The αth distributional derivative of a distribution $\mathbf{f}$ on $\mathcal{D}(\Omega)$ is the distribution $D^\alpha \mathbf{f}$ defined by

$$D^\alpha \mathbf{f}(\varphi) = \langle D^\alpha \mathbf{f}, \varphi \rangle = (-1)^{|\alpha|} \langle \mathbf{f}, D^\alpha \varphi \rangle = (-1)^{|\alpha|} \mathbf{f}\,(D^\alpha \varphi)$$

This follows using integration by parts (i.e. Green's theorem).

We now give a number of examples to illustrate these various ideas.

Example 3.30. The Dirac delta, δ, defined in $\mathbf{R}^1$ by

(i) $\delta(x) = 0,\ x \neq 0$

(ii) $\int_{-\infty}^{\infty} \delta(x)dx = 1$

(iii) $\int_{-\infty}^{\infty} \delta(x)\varphi(x)dx = \langle \boldsymbol{\delta}, \varphi \rangle = \varphi(0), \quad \varphi \in \mathcal{D}$

generates a continuous, linear functional, $\boldsymbol{\delta}$. That it is a linear functional follows immediately from (iii). The continuity of $\boldsymbol{\delta}$ follows from

$$|\boldsymbol{\delta}(\varphi)| = |\langle \boldsymbol{\delta}, \varphi \rangle| = |\varphi(0)| \leq \sup|\varphi(x)| = \|\varphi\|_\infty$$

However, $\boldsymbol{\delta}$ is a singular distribution. To see this let φ be the test function defined by

$$\varphi(x) = \begin{cases} 0, & b > |x| \geq a \\ \exp\left\{\dfrac{1}{x^2 - a^2}\right\}, & |x| < a \end{cases}$$

where $b > a > 0$.

If we assume that $\boldsymbol{\delta}$ is a regular distribution then we can easily obtain

$$\exp\left\{\frac{-1}{a^2}\right\} = |\varphi(0)| = \left|\int_{-b}^{b} \delta(s)\varphi(s)ds\right| \leq \frac{1}{e}\int_{-a}^{a} \delta(s)ds$$

and by taking the limit a $\to 0$ we obtain a contradiction.

The derivative of $\boldsymbol{\delta}$ is, following Definition 3.29, defined by

$$\langle \boldsymbol{\delta}', \varphi \rangle = -\ \langle \boldsymbol{\delta}, \varphi' \rangle = -\varphi'(0)$$

Similarly the nth derivative of $\boldsymbol{\delta}$ is given by

$$\langle \boldsymbol{\delta}^{(n)}, \varphi \rangle = (-1)^{(n)}\ \langle \boldsymbol{\delta}, \varphi^{(n)} \rangle = (-1)^{(n)}\varphi^{(n)}(0)$$

A particularly useful example is provided by the functional $\mathbf{f}\boldsymbol{\delta}$ with $f(x) = x \in (-1, 1) =: \Omega$. In this case we have, for all $\varphi \in C_0^\infty(\Omega)$

$$(\mathbf{f}\boldsymbol{\delta})(\varphi) = (\mathbf{x}\boldsymbol{\delta})(\varphi) = \langle \mathbf{x}\boldsymbol{\delta},\ \varphi \rangle = \langle \boldsymbol{\delta},\ x\varphi \rangle = [x\varphi]_{x=0} = 0$$

Thus $\mathbf{x}\boldsymbol{\delta} = \boldsymbol{\theta} \in (C_0^\infty(\Omega))'$ is the zero distribution on $\mathcal{D}(\Omega)$.

Every continuous function is locally integrable and hence generates a distribution (see Definition 3.26). However, there are many discontinuous functions which are also locally integrable.

Example 3.31. Consider the function f defined by

$$f(x) = |x|^{-1/2}, \quad x \in [-1, 1]$$

This function has a singularity at the origin. However, it is locally integrable since

$$\int_a^b |f(s)|\, ds = \int_a^b |s|^{-1/2}\, ds$$

is bounded in each interval $(a, b) \subset [-1, 1]$. Thus f generates a distribution $\mathbf{f}$ defined by

$$\mathbf{f}(\varphi) = \langle \mathbf{f}, \varphi \rangle = \int_{-1}^{1} |s|^{-1/2} \varphi(s) ds, \quad \varphi \in C_{\mathbf{0}}^{\infty}[-1, 1]$$

Hence $\mathbf{f}$ is a regular distribution.

Example 3.32. The function defined by $|x|$ is a locally integrable function which is differentiable for all $x \neq 0$. However, it is not differentiable at $x = 0$. Nevertheless, it has a generalised derivative which is calculated as follows. For any test function, φ, we have

$$\begin{aligned}
\left\langle |\mathrm{X}|', \varphi \right\rangle &= -\langle |\mathrm{X}|, \varphi' \rangle \\
&= -\int_{-\infty}^{\infty} |s| \varphi'(s) ds \\
&= -\int_{-\infty}^{0} |s| \varphi'(s) ds - \int_{0}^{\infty} |s| \varphi'(s) ds \\
&= -\int_{-\infty}^{0} \varphi(s) ds + \int_{0}^{\infty} \varphi(s) ds
\end{aligned}$$

where we have integrated by parts and used the fact that the test function φ vanishes at infinity. We now introduce a function sgn defined by

$$\operatorname{sgn}(x) = \begin{cases} -1 & \text{for} \quad x < 0 \\ 1 & \text{for} \quad x < 0 \end{cases}$$

It is not necessary here for us to specify sgn(0) since it can be shown, as an easy exercise that the above function will generate the same distribution **sgn** for any choice of sgn(0). Consequently, we have from the above that for all $\varphi \in \mathcal{D}$

$$\begin{aligned}
\left\langle |\mathrm{X}|', \varphi \right\rangle &= \int_{-\infty}^{\infty} \operatorname{sgn}(s) \varphi(s) ds \\
&= \langle \mathbf{sgn}, \varphi \rangle
\end{aligned}$$

Therefore, we say that $|X|' = \mathbf{sgn}$, *in the sense of distributions.*

We remark that since the generalised derivative is a distribution it is meaningless to talk about its value at a point; it can only be given any meaning over an interval.

Example 3.33. The Heaviside function, H, defined on $[-a, a]$ by

$$H(x) = \begin{cases} 0, & -a \le x < 0 \\ 1, & 0 \le x < a \end{cases}$$

is locally integrable and generates the distribution $\mathbf{H}$ according to

$$\mathbf{H}(\varphi) = \langle \mathbf{H}, \varphi \rangle = \int_{-a}^{a} H(s)\varphi(s)ds = \int_{0}^{a} \varphi(s)ds, \quad \varphi \in C_0^{\infty}(-a, a)$$

Hence, $\mathbf{H}$ is a regular distribution.

The distributional derivative of $\mathbf{H}$ is calculated, in the now familiar manner, as follows, bearing in mind that the test function, φ, vanishes at infinity.

$$\langle \mathbf{H}', \varphi \rangle = -\int_{0}^{\infty} \varphi'(s)ds = \varphi(0) = \int_{-\infty}^{\infty} \delta(s)\varphi(s)ds = \langle \boldsymbol{\delta}, \varphi \rangle = \boldsymbol{\delta}(\varphi)$$

Therefore, we see that $\mathbf{H}' = \boldsymbol{\delta}$, in the sense of distributions.

Although the previous discussion has been conducted with respect to ordinary derivatives nevertheless similar results can be obtained for partial derivatives using the multi-index notation introduced earlier.

When we come to deal with differential equations having the typical form

$$Lu = f$$

where L is a given differential expression then various types of solution must be considered. If f is a regular distribution generated by a function which is locally integrable but not continuous then the above equation cannot be expected to have any meaning in the classical sense. A similar observation holds if f is a singular distribution. In these cases we say that the equation holds in the *sense of distributions.* The solutions to this equation which might be obtained in these cases will be distributions and are known as **weak** or **generalised solutions** of the equation.

Example 3.34. Consider the differential equation

$$xu'(x) = 0, \quad x \in (-1, 1)$$

This equation has a classical solution

$$u(x) = \text{constant}$$

However if we regard the equation as a distributional differential equation then it has a weak or generalised solution of the form

$$\mathbf{u} = c_1\mathbf{H} + c_2$$

where $\mathbf{H}$ is the distribution generated by the Heaviside function, H, and c_1, c_2 are constants. To see that this is indeed the case we first notice that (see Example 3.33)

$$\mathbf{u}' = c_1\boldsymbol{\delta}$$

This, in turn, indicates that for any test function φ

$$\mathbf{xu}'(\varphi) = \langle \mathbf{xu}', \varphi \rangle = \langle \mathbf{u}', x\varphi \rangle = \langle c_1\boldsymbol{\delta}, x\varphi \rangle = c_1\{(x\varphi)(0)\} = 0$$

and we conclude that $\mathbf{xu}' = \mathbf{0}$, or equivalently, $xu'(x) = 0$ in the sense of distributions.

Again, there are various kinds of solution of an equation of the form $Lu = f$ when it is considered as an equation involving a generalised function f. These are classified as follows.

Definition 3.35. (i) A distribution, $\mathbf{u}$, satisfying the equation $L\mathbf{u} = \mathbf{f}$ is called a **distributional** or **generalised solution** of the equation.

(ii) A function u which is sufficiently continuously differentiable and thus generates a regular distribution $\mathbf{u}$ which satisfies $L\mathbf{u} = \mathbf{f}$ in the generalised sense is called a **classical solution** of the equation $Lu = f$.

(iii) A function u which is not n-times continuously differentiable, and therefore cannot be a classical solution of $Lu = f$, but which generates a regular distribution, $\mathbf{u}$, which is a generalised solution of $L\mathbf{u} = \mathbf{f}$ is called a **weak solution** of $Lu = f$.

(iv) A **distributional solution** of $L\mathbf{u} = \mathbf{f}$ is a solution $\mathbf{u}$ which is a **singular distribution.**

Finally, in this subsection, we briefly sketch, for later use, the notion of the **convolution** of two distributions **f** and **g**.

The classical formula for the convolution of two continuous functions f and g defined on $\mathbf{R}$ is given by

$$(f*g)(x) := \int_{\mathbf{R}} f(x-y)g(y)dy = \int_{\mathbf{R}} f(y)g(x-y)dy$$

In the distributional case we adopt the following.

Definition 3.36. The convolution, $\mathbf{f} * \mathbf{g}$ of two distributions $\mathbf{f}$ and $\mathbf{g}$ is defined to be

$$\langle (\mathbf{f} * \mathbf{g})(x), \varphi(x) \rangle = \langle \mathbf{f}(x), \langle \mathbf{g}(y), \varphi(x + y) \rangle \rangle$$

where $\varphi \in \mathcal{D}$. We have abused notation slightly in order to emphasise the "integration" variable. Furthermore, we have assumed in this definition that the distribution $\mathbf{g}$ has compact support. Consequently, the right-hand side of this expression is well defined since

$$\varphi(x) := \langle \mathbf{g}(y), \varphi(x + y) \rangle$$

is a test function.

We notice that when $\mathbf{f}$ and $\mathbf{g}$ are regular distributions then $\mathbf{f} * \mathbf{g}$ is also a regular distribution and we recover the classical formula for the convolution of two functions.

Example 3.37. The Dirac delta concentrated at $x = a$ is denoted $\boldsymbol{\delta}_a(x)$ and defined to be

$$\boldsymbol{\delta}_a(x) = \delta(x - a)$$

This has the property (see Definition 3.25)

$$\langle \boldsymbol{\delta}_a(x), \varphi(x) \rangle = \int_R \boldsymbol{\delta}_a(x) \varphi(x) dx = \varphi(a)$$

Furthermore, if $\mathbf{f}$ is a regular distribution then

$$\begin{aligned} \langle (\mathbf{f} * \boldsymbol{\delta}_a)(x), \varphi(x) \rangle &= \langle \mathbf{f}(x), \langle \boldsymbol{\delta}_a(y), \varphi(x + y) \rangle \rangle \\ &= \langle \mathbf{f}(x), \varphi(x + a) \rangle \\ &= \langle \mathbf{f}(x - a), \varphi(x) \rangle \end{aligned}$$

Thus, remembering that the integration variables are only written in for convenience, we have

$$\mathbf{f}(x) * \boldsymbol{\delta}_a(x) = \mathbf{f}(x - a)$$

in a distributional sense.

Since it can readily be shown that different locally integrable functions define different distributions it follows that the set of locally integrable functions can be embedded in the set of all distributions. An even more powerful result can be obtained, namely, that $\mathcal{D}$ is dense in $\mathcal{D}'$. Hence every distribution is a weak limit of test functions. For more detailed discussion of these remarks see the Commentary and the references cited there.

With these various observations in mind we see that when a function f is considered as a distribution then it is identical to all functions which can be obtained by changing the values of $f(x)$ on isolated points, more precisely, on sets of measure zero. Hence, a distribution is **not** associated with **a** function but with an **equivalence class** of functions which are equal almost everywhere (ae), that is, everywhere except on sets of measure zero.

The use of bold type to indicate a distribution will be suppressed from now onwards. Whether a quantity f is to be regarded as either a function or as a distribution will usually be clear from the text. However, the bold type will be restored if clarification is needed.

3.5 Fourier Transforms and Distributions

Integral transforms play an important role in mathematical analysis and its applications. Perhaps one of their most impressive properties is that they can transform differentiation into the algebraic operation of multiplication by a scalar. Consequently, if we are faced with having to solve an ordinary differential equation for an unknown function u then we can use a suitable integral transform to obtain an equivalent algebraic equation for $\hat{u}$, the transform of u. We then solve the algebraic equation for $\hat{u}$ and recover the required solution function u by means of an associated inverse integral transform. Similarly, a partial differential equation involving an unknown function v can be transformed into an equivalent ordinary differential equation for the transformed function $\hat{v}$. The required function v is recovered by solving for $\hat{v}$ and then using an associated inverse integral transform.

The prototype of such transforms is the Fourier transform. Here we give a *very* brief sketch of some of the more important aspects of the classical Fourier transform. Full details of the theory of Fourier transforms and their applications can be found in the references cited in the Commentary. In this connection we would particularly recommend [3], [10] and, for the mathematical theory, [11]. Once we have introduced the classical Fourier transform we shall turn attention to the notion of generalised Fourier transforms which cater for distributions.

Definition 3.38. A function $f\colon \mathbf{R} \to \mathbf{C}$ is said to be **absolutely integrable** if $\int_{-\infty}^{\infty} |f(x)|\,dx$ exists.

Examples 3.39. (i) Every test function is absolutely integrable.

(ii) Every continuous function which tends to zero faster than $|x|^{-(1+\alpha)}$, where $\alpha > 0$, as $0 < \alpha \leq |x| \to \infty$ is absolutely integrable.

(iii) No polynomials, other than the trivial polynomial which is everywhere zero, are absolutely integrable.

The **classical Fourier transform** is defined as follows.

Definition 3.40. For any absolutely integrable function, f, the (classical) Fourier transform of f, denoted $\hat{f}$, is defined by

$$\hat{f}(p) = \frac{1}{\sqrt{2\pi}} \int_{-\infty}^{\infty} f(x) e^{-ipx} dx = (\hat{F} f)(p)$$

The integral is convergent since f is absolutely integrable. Indeed, it can be shown to be uniformly convergent with respect to p.

Definition 3.41. A function $f: \mathbf{R} \to \mathbf{C}$ is said to be **piecewise smooth** if

(i) all its derivatives exist and are continuous except possibly at a set of points $x_1, x_2, x_3, \ldots$ such that any finite interval contains only a finite number of the x_i.

(ii) the function and all its derivatives have at most a finite number of jump discontinuities.

(iii) The function is said to be n-times continuously differentiable if (i), (ii) are satisfied with "all" replaced by "its first n".

Associated with Definition 3.40 is the **inverse transform**.

Definition 3.42. If f is absolutely integrable, continuous and piecewise smooth then

$$f(x) = \frac{1}{\sqrt{2\pi}} \int_{\mathbf{R}} \hat{f}(p) e^{+ipx} dp = (\hat{f} f)(x) \equiv \hat{F}^*(x) f$$

where $\hat{F}^*$ is the transform inverse to $\hat{F}$.

Definitions 3.40 and 3.42 can be combined to provide what we will come to call an **inversion theorem** of the form

$$\hat{f}(p) = \frac{1}{\sqrt{2\pi}} \int_{\mathbf{R}} f(x) e^{-ipx} dx = (\hat{f} f)(p) \tag{3.1}$$

$$f(x) = \frac{1}{\sqrt{2\pi}} \int_{\mathbf{R}} \hat{F}(p) e^{+ipx} dp = (\hat{F}^* f)(x) \tag{3.2}$$

In practice we will always have to demonstrate that Definitions 3.40 and 3.42 are available; that is, we have to **prove** the inversion theorem for the problem being considered. The proof of such a theorem is not easy. Nevertheless, Fourier transforms play a central rôle in modern applied mathematical analysis. Their important property is the reciprocal nature of the transform as indicated by the proved inversion theorem. A glance at Definition 3.40 and Definition 3.42 indicates that f is related to $\hat{f}$ in the same way that $\hat{f}$ is related to f apart from a minus sign.

We shall return to this aspect in subsequent chapters. In these later chapters we will see that it is possible to construct an integral transform which is particularly appropriate for the specific problem under consideration in that it has an associated inverse theorem.

The Fourier transform we have just introduced has the following properties.

Example 3.43. Let f be an absolutely integrable and piecewise smooth function.

(i) $f(-x) = (\hat{\hat{f}})(x)$

(ii) $(\hat{f'})(p) = ip\hat{f}(p)$

(iii) $\frac{d}{dp}[(\hat{F}\,f)(p)] = -i[\hat{F}\,(xf)](p)$

(iv) If $f_a(x) := f(\mathrm{x} - a)$ then

$$(\hat{F}f_a) = \exp(-ipa)(\hat{F}f)(p)$$

The details are left as an exercise.

An alternative Definition of the Fourier transform which is frequently used is given by

$$\tilde{f}(p) = \int_{\mathbf{R}} f(x)e^{ipx}dx = (\tilde{F}f)(p) \equiv \tilde{F}(p)f$$

$$f(x) = \frac{1}{2\pi}\int_{\mathbf{R}} \tilde{f}(p)e^{-ipx}dp = (\tilde{F}*f) \equiv \tilde{F}^*(x)f$$

The two transforms are related in the following way

$$\hat{F}(p) = \frac{1}{\sqrt{2\pi}}\tilde{F}(-p)$$

$$\hat{F}^*(x) = \sqrt{2\pi}\tilde{F}^*(-x)$$

This follows directly from the definitions. Corresponding to the properties outlined in Example 3.43 we have

Example 3.44. For f as in Example 3.43 we have

(i) $2\pi f(-x) = (\tilde{\tilde{f}})(x)$

(ii) $(\tilde{f'})(p) = -ip\tilde{f}(p)$

(iii) $\frac{d}{dp}[(\tilde{F}f)(p)] = i[\tilde{F}(xf)](p)$

(iv) If $f_a(x) := f(x - a)$ then

$$(\tilde{F}f_a) = \exp(-ipa)(\tilde{F}f)(p)$$

In the following chapters we shall use the Fourier transform in the form $\hat{F}$. This transform can be interpreted by saying that any (sufficiently "nice") function

f can be regarded as a superposition, either as a sum or as an integral, of an infinite number of sinusoidal waves (characterised by $\exp(-ipx)$) with different frequencies p where the wave of frequency p has amplitude $\overline{\hat{f}(p)\sqrt{2\pi}}$. For this reason the integral relation

$$f(x) = \frac{1}{\sqrt{2\pi}} \int_{\mathbf{R}} \hat{f}(p) e^{+ipx} dp = (\hat{F} * f)(x)$$

is called the **spectral resolution** of the function f and $\hat{f}$ is referred to as the **spectral density** of f.

The development of the theory of Fourier transforms based on Definition 3.40 and Definition 3.41 is inadequate for our purposes. This is because the theory is restricted to absolutely integrable functions and these are not suitable for the analysis we have in mind. To see this simply recall that we are mainly interested here in differential equations and these have solutions that have the form of either polynomials or trigonometric functions or exponential functions and none of these are absolutely integrable. This means that the forms we are most interested in do *not* have Fourier transforms in the classical sense of Definition 3.40 and Definition 3.42. Consequently, we might expect to get a better theory of the Fourier transforms if we work in terms of generalised functions and this indeed proves to be the case.

The intuitively natural way to define the Fourier transform of a generalised function $\mathbf{f}$ is to define $\hat{\mathbf{f}}$, the Fourier transform of $\mathbf{f}$ by

$$\langle \hat{\mathbf{f}}, \varphi \rangle = \langle \mathbf{f}, \hat{\varphi} \rangle \tag{3.3}$$

for all $\varphi \in \mathcal{D}$. However, the right-hand side of (3.3) does not make any sense because, in general, $\hat{\varphi}$ is not a test function since the Fourier transform of a function of bounded support (see Definition 3.21) is not usually also a function of bounded support. To overcome this difficulty and at the same time retain as much as possible of the methodology we have so far outlined we could try to introduce a new space of test functions, denoted by $\mathbf{S(R)}$, with the property that Fourier transforms of functions in $\mathbf{S(R)}$ would also be in $\mathbf{S(R)}$. This means that we would have to develop another version of distribution theory, using exactly the same methodology as before, but with certain technical modifications that would ensure that Fourier transforms do indeed fit into the theory. With this in mind we introduce

Definition 3.45. A smooth function $f: \mathbf{R} \to \mathbf{C}$ such that for all $n, r \geq 0$

$$x^n \varphi^{(r)}(x) \to 0 \quad \text{as} \quad |x| \to \infty$$

is called a **function of rapid decay**.

The set of all functions of rapid decay is denoted by $\mathbf{S(R)}$.

The collection $\mathbf{S(R)}$ can readily be shown to have the following properties.

Properties 3.46. (i) *Every test function is a function of rapid decay, that is* $\mathcal{D} \subseteq \mathbf{S(R)}$.

(ii) *If* φ, $\varphi \in \mathbf{S(R)}$ *then* $(a\varphi + b\varphi) \in \mathbf{S(R)}$ *for all constants, a, b.*

(iii) *If* $\varphi \in \mathbf{S(R)}$ *then* $x^n\varphi^{(r)}(x) \in \mathbf{S(R)}$ *for all* $n, r \geq 0$.

(iv) *If* $|x^n\varphi^{(r)}(x)|$ *is bounded for each* $n, r \geq 0$ *then* φ *is a function of rapid decay.*

(v) *Every function of rapid decay is absolutely integrable.*

Because of its particular importance in applications we prove the following result.

Theorem 3.47. *If* $\varphi \in \mathbf{S(R)}$ *then* $\hat{\varphi} \in \mathbf{S(R)}$

Proof.

- Property 3.46(v) $\Rightarrow$ φ has a Fourier transform.
- Properties 3.46(iii) and 3.46 (v) $\Rightarrow x^n\varphi(x)$ is absolutely integrable.
- Example 3.43(iii) applied n times shows that $\hat{\varphi}$ is differentiable n times for any n.
- Applying Example 3.43(iii) r times we obtain

$$\begin{aligned}\left|p^n\widehat{\varphi^{(r)}}(p)\right| &= \left|p^n\int_{\mathbf{R}}(-ix)^r\varphi(x)e^{-ipx}dx\right| \\ &= \left|\int_{\mathbf{R}}x^r\varphi(x)\left(\frac{d}{dx}\right)^n e^{-ipx}dx\right| \\ &= \left|\int_{\mathbf{R}}e^{-ipx}\left(\frac{d}{dx}\right)^n(x^r\varphi(x))dx\right|\end{aligned}$$

where we have integrated by parts n times. Hence

$$\left|p^n\widehat{\varphi^{(r)}}(p)(p)\right| \leq \int_{\mathbf{R}}\left|\left(\frac{d}{dx}\right)^n(x^r\varphi(x))\right|dx$$

Properties 3.46(iii) and 3.46(v) guarantee the convergence of the integral on the right-hand side and thus we obtain a bound on the left-hand side. It then follows by Property 3.46(iv) that $\hat{\varphi}$ is a function of rapid decay. ■

A notion of convergence in $\mathbf{S(R)}$ is introduced as follows.

Definition 3.48. If φ, φ_1, φ_2 are functions of rapid decay then we say that $\varphi_m \to \varphi$ in $\mathbf{S(R)}$ as $m \to \infty$ provided that for all integers r and n we have, uniformly in x

$$x^n\varphi_m^{(r)}(x) \to x^n\varphi^{(r)}(x) \quad \text{as} \quad m \to \infty$$

With this preparation we can now introduce the following.

Definition 3.49. A **distribution of slow growth** is a continuous linear functional on the space **S**(**R**). Alternatively, we say that it is a linear functional which maps every convergent sequence in **S**(**R**) into a convergent sequence in **C**.

We remark that every distribution of slow growth is a distribution in the sense introduced earlier. The converse, however, is not true.

It is generally true that functions of slow growth generate distributions of slow growth and that functions which do not grow slowly, in the sense described above, generate distributions which are *not* of slow growth.

Ordinary functions which grow too rapidly at infinity do not belong to the set of distributions of slow growth. This we can express more precisely in the following manner.

Definition 3.50. (i) $f(x) = O(x^n)$ as $|x| \to \infty$ means that there exist numbers A and R such that $|f(x)| \leq A\ |x|^n$ whenever $|x| > R$.

(ii) A function $f{:}\,\mathbf{R} \to \mathbf{C}$ which is locally integrable and such that $f(x) = O(x^n)$ for some n as $|x| \to \infty$ is called a **function of slow growth**.

Example 3.51. (i) Every nth degree polynomial is $O(x^n)$.

(ii) e^{-x} is not a function of slow growth since $x^{-n}e^{-x} \to \infty$ as $|x| \to \infty$ for any n.

(iii) e^{iax} is a function of slow growth if x and a are real.

Definition 3.52. To each locally integrable function of slow growth, f, there corresponds a distribution of slow growth, **f**, a **regular distribution**, defined by

$$\langle \mathrm{f}, \varphi \rangle = \int_{\mathbf{R}} f(x)\varphi(x)dx, \quad \varphi \in \mathbf{S}(\mathbf{R})$$

We have introduced the above statement as a definition. However, it is frequently presented in the form of a theorem in which it is proved that the above functional has all the properties required to ensure that **f** is a distribution of slow growth.

With these several results and remarks in mind, a theory of distributions of slow growth can now be constructed in a similar manner to that used when dealing with ordinary distributions. We leave the details as an exercise.

The reason for introducing the space **S**(**R**) was to be able to define Fourier transforms of distributions in the same way as already discussed for ordinary functions.

We define the Fourier transform $\hat{\mathbf{f}}$ of a generalised function **f** by

$$\langle \hat{\mathbf{f}}, \varphi \rangle = \langle \mathrm{f}, \hat{\varphi} \rangle, \quad \text{for} \quad \varphi \in \mathbf{S}(\mathbf{R})$$

In order that this definition should make sense we need the following technical result (see [2], [11]).

Theorem 3.53. *If* $\mathbf{f}$ *is a distribution of slow growth then the functional* $\hat{\mathbf{f}}: \varphi \to \langle \mathbf{f}, \hat{\varphi} \rangle$ *is a distribution of slow growth.*

With this result available we can introduce the following.

Definition 3.54. (i) If $\mathbf{f}$ is a distribution of slow growth then its Fourier transform is the distribution of slow growth, $\hat{\mathbf{f}}$, defined by

$$\langle \hat{\mathbf{f}}, \varphi \rangle = \langle \mathbf{f}, \hat{\varphi} \rangle, \quad \text{for} \quad \varphi \in \mathbf{S}(\mathbf{R})$$

(ii) If f is a locally integrable function of slow growth then the distribution, $\hat{\mathbf{f}}$, is called the **generalised Fourier transform** of f.

We now have available, symbolically, the same structure as that used for ordinary functions. Indeed, Examples 3.43 and 3.44 will hold with f replaced by $\mathbf{f}$ appropriately.

We conclude this section by giving examples of the Fourier transforms of some frequently occurring functions.

Example 3.55. The simplest function is the constant function. Write $\mathbf{1}$ for the distribution generated by the constant function whose value everywhere is 1. Then

$$\langle \hat{1}, \varphi \rangle = \langle \mathbf{I}, \hat{\varphi} \rangle = \int_{\mathbf{R}} 1 \cdot \hat{\varphi}(x) dx = \int_{\mathbf{R}} \hat{\varphi}(x) dx$$

Now

$$(\varphi)(q) = \frac{1}{\sqrt{2\pi}} \int_{\mathbf{R}} e^{-ipx} \cdot \varphi(x) dx$$

which implies

$$(\hat{\hat{\varphi}})(0) = \frac{1}{\sqrt{2\pi}} \int_{\mathbf{R}} \hat{\varphi}(x) dx$$

Using Example 3.43(i)

$$(\hat{\hat{\varphi}})(0) = \varphi(-0) = \varphi(0)$$

Combining these several results we obtain

$$\langle \hat{\mathbf{I}}, \varphi \rangle = \frac{1}{\sqrt{2\pi}} \int_{\mathbf{R}} \hat{\varphi}(x) dx = \frac{1}{\sqrt{2\pi}} (\hat{\hat{\varphi}})(0) = \frac{1}{\sqrt{2\pi}} \varphi(0) = \frac{1}{\sqrt{2\pi}} \int_{\mathbf{R}} \hat{\varphi}(x) \delta(x) dx$$

and conclude that

$$\hat{\mathbf{I}} = \sqrt{2\pi} \delta$$

Example 3.56. (i) Write **x** for the distribution generated by the function f defined by $f(x) = x$.

Then

$$\hat{x} = -\sqrt{2\pi}\delta' i$$

(ii)

$$\hat{\boldsymbol{\delta}} = \frac{1}{\sqrt{2\pi}}\mathbf{1}$$

(iii)

$$\delta\left(\widehat{x-a}\right) = \frac{1}{\sqrt{2\pi}}e^{-iax}$$

$$\left(\widehat{e^{-iax}}\right) = \sqrt{2\pi}\delta(x+a)$$

The details are left as an exercise.

References and Further Reading

[1] N.I. Akheiser and L.M. Glazman: *Theory of Linear Operators in Hilbert Space*, Pitman-Longman, London, 1981.

[2] J. Arsac: *Fourier Transforms and the Theory of Distributions*, Prentice Hall, New York, 1966.

[3] J.W. Dettman: *Mathematical Methods in Physics and Engineering*, McGraw-Hill, New York, 1962.

[4] G. Helmberg: *Introduction to Spectral Theory in Hilbert Space*, Elsevier, New York, 1969.

[5] E. Kreyszig: *Introductory Functional Analysis with Applications*, Wiley, New York, 1978.

[6] F. Riesz and B. Sz-Nagy: *Functional Analysis*, Ungar, New York, 1955.

[7] G.F. Roach: *Greens Functions* (2nd Ed.), Cambridge Univ. Press, London, 1970/1982.

[8] G.F. Roach: *An Introduction to Linear and Nonlinear Scattering Theory*, Pitman Monographs and Surveys in Pure and Applied Mathematics, Vol 78, Longman, Essex 1995.

[9] W. Rudin: *Principles of Mathematical Analysis* (3rd Ed.), McGraw-Hill, New York, 1976.

[10] I.N. Sneddon: *Fourier Transforms*, McGraw-Hill, New York, 1951.

[11] E.C. Titchmarsh: *Introduction to the Theory of Fourier Integrals*, Oxford Univ. Press, 1937.

4

Hilbert Spaces

4.1 Introduction

In this chapter we generalise the familiar concepts of algebra and geometry that we used in a Euclidean space setting when we have always been dealing with numbers and the numerical values of functions. We now want to extend these ideas in such a way that we can discuss the functions themselves rather than their numerical values. It turns out that a particular type of normed linear space called a **Hilbert space** provides an ideal setting for this purpose.

We begin by defining an inner product which is an abstract version of the familiar scalar product of elementary vector algebra. This will allow us to define, just as in finite dimensional Euclidean spaces, the notion of angles, particularly right angles, in abstract vector spaces. We then go on to introduce the idea of two elements of an (abstract) vector space being perpendicular. This is followed by a discussion of how sets of perpendicular elements of a vector space can form a basis for an infinite dimensional space in a similar way that the x, y and z-axes form a basis for $\mathbf{R}^3$.

We end this chapter with a brief outline of the salient features of operators on Hilbert spaces.

Definition 4.1. An **inner product** on a vector space X is a rule which assigns to elements $x, y \in X$ a real or complex number, denoted by (x, y), called the inner product of $x, y \in X$ which has the properties

(i) (x, x) is real and positive for all $x \neq \theta$ the zero element in X and $(\theta, \theta) = 0$

(ii) $(x, y) = \overline{(y, x)}$ for all $x, y \in X$ where the bar denotes complex conjugate

(iii) $(ax, y) = a(x, y)$ for all $x, y \in X$ and any scalar a

(iv) $(x + y, z) = (x, z) + (y, z)$ for all $x, y, z \in X$.

A vector space X together with an inner product $(\cdot, \cdot)$ is called an **inner product space.**

We notice that for a real inner product space the bar is redundant since all the inner products are real. Furthermore, we see that (ii) and (iv) imply that we also have $(x + y, z) = (x, z) + (y, z)$. Thus, the algebraic properties of an inner product space would appear to be the same as for the scalar product in ordinary vector algebra. However, there is one important difference. In a complex space the inner product is *not* linear in both arguments. It is, in fact, *conjugate linear* in the sense that $(x, ay) = \bar{a}(x, y)$ for any scalar a and $x, y \in X$. This follows from Definition 4.1(ii) and (iii)

$$(x, ay) = \overline{a(x, y)} = \bar{a}(x, y)$$

Examples 4.2. (i) On $\mathbf{C}^n$, the set of n-tuples of complex numbers $x = (x_1, x_2, \ldots, x_n)$, $y = (y_1, y_2, \ldots, y_n)$ where $x_k, y_k \in \mathbf{C}$ for $k = 1, 2, \ldots, n$, we can define an inner product

$$(x, ay) := \sum_{k=1}^{n} x_k \overline{y_k}$$

For the corresponding real space, $\mathbf{R}^n$, we can define an inner product with the same symbolic form but now all quantities are real.

(ii) On $C[a, b]$, the set of complex-valued, continuous functions defined on the interval $[a, b] \in \mathrm{R}$ we can define an inner product of the form

$$(f, g) = \int_a^b f(x)\overline{g(x)}\, dx, \quad f, g \in C[a, b]$$

Definition 4.3. On any inner product space, X, we can define a norm by

$$\|x\|^2 = (x, x)$$

Exercise 4.4. (i) *Verify that the inner products introduced in Example* 4.2 *satisfy the inner product axioms in Definition* 4.1.

(ii) *Verify that the quantity* $\|\cdot\|$ *introduced in Definition* 4.3 *is indeed a norm.*

The norm induced by the inner product introduced in Example 4.2(ii) is referred to as the **square integrable norm** since

$$\|f\|^2 = (f, f) = \int_a^b f(x)\overline{f(x)}\, dx = \int_a^b |f(x)|^2\, dx$$

The set of all continuous functions defined on the interval $[a, b]$ endowed with the square integrable norm is a normed space (verify). Convergence in this normed space is understood as follows.

Definition 4.5. Let $F, f_1, f_2, f_3, \ldots$ be functions with action either $\mathbf{R} \to \mathbf{R}$ or $\mathbf{R} \to \mathbf{C}$. We say that $f_n \to F$ **in the mean** on $[a, b]$ as $n \to \infty$ if

$$\int_a^b |f_n(x) - F(x)|^2 \, dx \to 0$$

Clearly, this definition only holds for functions that are sufficiently well behaved for the integral to exist.

Convergence in the mean is less demanding than uniform convergence. It is called convergence in the mean because it is the mean value of $(f_n - F)$ that tends to zero and not the value at particular points.

The idea of convergence in the mean is particularly important in mathematical physics. The equations in mathematical physics are often solved by series expansion methods, for example Fourier series. These expansions do not always converge uniformly but they often can be shown to converge in the mean.

Some of the differences between the types of convergence mentioned above are given by the following example.

Example 4.6. Define functions f_n, $n = 1, 2, \ldots$ by $f_n(x) = \exp(-nx)$. It is clear that $f_n \in C[0, 1]$. Furthermore, for each $x \in [0, 1]$ we have

$$f_n(x) \to F(x) \quad \text{as} \quad n \to \infty \tag{4.1}$$

where F is defined by

$$F(x) = 0 \quad \text{for} \quad x \neq 0$$
$$F(0) = 1$$

We emphasise that the result (4.1) is a statement about the convergence of a sequence of **numbers**, $f_n(x)$, for any $x \in [0, 1]$. It is *not* a statement about the sequence of functions f_n, $n = 1, 2, \ldots$ The statement (4.1) indicates that $f_n \to F$ **pointwise**. It is also the case that $f_n \to F$ in the sense of convergence in the mean. The integral involved is $\int_0^1 \exp(-2nx)dx$ which is easily evaluated and can be shown to tend to zero as $n \to \infty$.

In this example $F \notin C[0, 1]$ since it is discontinuous. To ease matters we may be tempted to consider the function F_0 defined by

$$F_0(x) = 0 \quad \text{for all} \quad x \in [0, 1]$$

Clearly $F_0 \in C[0, 1]$. We can now ask if $f_n \to F_0$ in the sense of convergence in $C[0, 1]$. Different answers can be obtained depending on the norm used on $C[0, 1]$.

We have seen that the collection $C[0, 1]$ can be turned into a normed space in a number of ways. We consider two cases.

Case 1:

$C[0, 1]$ is endowed with the **uniform norm** defined by

$$\|f\| := \sup\{|f(x)| : 0 \leq x \leq 1\}$$

This is also referred to as the sup. norm.

Case 2:
$C[0, 1]$ is endowed with the **square integrable norm** defined by

$$\|f\|^2 = \int_0^1 |f(x)|^2 \, dx$$

Using the uniform norm we conclude that $f \nrightarrow F_0$. This follows because $|f_n(0) - F_0(0)| = 1$ for all n and hence sup $|f_n(x) - F_0(x)|$ cannot possibly tend to zero. However, if we use the norm in Case 2 then a simple integration indicates that $f_n \to F_0$.

Limits in the mean are not unique. This is because the value of an integral is unaffected by changing the value of the integrand at a number of isolated points.

We see from the above that there are two important types of convergence in *function spaces*. The first is uniform convergence, which is convergence with respect to the **sup. norm** introduced in Case 1 above. The second is **convergence in the mean** which is convergence with respect to the integral norm introduced in Case 2.

We have seen above that a sequence of continuous functions can converge in the mean to a discontinuous function. However, such a sequence cannot converge uniformly to a discontinuous function. Yet we notice that any uniformly convergent sequence in $C[a, b]$ can be integrated term by term and as such converges in the mean. Therefore, we see that uniform convergence implies convergence in the mean but the converse does not hold.

One of the powerful features of applied functional analysis is that it provides a means of introducing and working with a norm which is best suited to the problem in hand. In much of the analysis in this monograph we shall be working in a Hilbert space rather than a Banach space structure. A particularly important and frequently occurring Hilbert space is $L_2[a, b]$.

Definition 4.7. $L_2[a, b]$ is the completion of $C[a, b]$ with respect to the square integrable norm.

Thus $L_2[a, b]$ contains all functions which are the limits of continuous functions in the sense of mean convergence.

Example 4.8. We have seen that $L_2[a, b]$ contains discontinuous functions. Indeed, it can contain functions that have infinite discontinuities provided that the discontinuity is "nice" enough for it to be square integrable. For instance the function f defined by $f(x) = x^{-1/3}$ is an element of $L_2[0, 1]$. However, the function g defined by $g(x) = x^{-2/3}$ is not. We can conclude that every function f which is such that the integral $\int_a^b |f(x)|^2 \, dx$ exists is an element of $L_2[a, b]$.

There is an important aspect of $L_2[a, b]$ which much always be borne in mind, namely the space $L_2[a, b]$ contains functions which are *not* integrable according to the theory of Riemann integration.

For example the function F defined by $F(x) = 1$ for rational values of x and zero for all other values of x is a discontinuous function and is such that $\int_a^b |F(x)|^2 \, dx$ does not exist in the Riemann sense. Now, Riemann integration is the "usual" method of integration which we use in practical problems but this little example shows that Riemann integration is not always adequate. Fortunately, a more powerful integration theory is available, namely Lebesgue integration theory [5], [9], [11], by means of which we can show that the above integral does exist and hence that $F \in L_2[a, b]$.

In this monograph there will be no need to have a detailed knowledge of Lebesgue theory. It will be sufficient simply to know it is available. Consequently, $L_2[a, b]$ can be regarded as a Banach space which contains all ordinary, that is Riemann, square integrable functions together with other functions which are highly discontinuous, just like the function F introduced above, which must be included to make the space complete.

This situation is eased by recalling that $L_2[a, b]$ is the completion of $C[a, b]$ with respect to the square integrable norm. Thus, every element of $L_2[a, b]$ can be approximated, arbitrarily well, by elements in $C[a, b]$. That is to say, $C[a, b]$ is dense in $L_2[a, b]$ (see Definition 3.9). For this reason we really need only work with elements of $C[a, b]$ initially and then, if necessary, use denseness arguments to obtain more general results.

We see that the notion of an inner product as introduced here is an abstract version of the familiar scalar or dot-product of finite dimensional vector algebra. In this connection simply think of the algebra used in n-dimensional Euclidean space $\mathbf{R}^n$.

We have already mentioned that when analysing problems it is often more profitable and easier to work with the actual functions involved rather than with the numerical values of such functions. Working in the structure of inner product spaces provides a means of doing this. In order to complete this introduction of an analytical structure that will enable us to work meaningfully with functions rather than with their numerical values, whilst still retaining the familiar methods used in finite dimensional Euclidean space, we need the additional ingredient of **orthogonality**. Once this concept has been introduced it will lead naturally to the meaning of the angle between two functions and of a basis of an infinite dimensional space.

We notice from the above that, should we so wish, we could build up essentially all Euclidean geometry in the context of an inner product space structure. With this in mind we recall the following familiar result [5].

Theorem 4.9. *(Cauchy–Schwarz inequality)*
For any complex numbers x_k, y_k, $k = 1, 2, \ldots, n$

$$\left| \sum_{k=1}^{n} x_k \overline{y_k} \right|^2 \le \left(\sum_{k=1}^{n} |x_k|^2 \right) \left(\sum_{k=1}^{n} |y_k|^2 \right)$$

where the bar denotes complex conjugate.

Corollary 4.10. *(Cauchy–Schwarz inequality for integrals)*

$$\left|\int_a^b f(x)g(x)dx\right|^2 \leq \left(\int_a^b |f(x)|^2\right)\left(\int_a^b |g(x)|^2\right)$$

for any functions f and g that ensure the integrals exist.

An abstract version of these two results is available for general inner product spaces. To indicate this we need a little notational preparation.

Definition 4.11. Let V be a vector space and $S \subset V$ a subset of elements $x_1, x_2, \ldots, x_p$. Consider the relation

$$\sum_{k=1}^{p} a_k x_k = 0 \tag{4.2}$$

where a_k, $k = 1, \ldots, p$ are scalars. If (4.2) only holds for $a_k = 0, k = 1, \ldots, p$ then the elements x_k, $k = 1, \ldots, p$ are said to be **linearly independent** and S is a **linearly independent subset** of V. The elements x_k, $k = 1, \ldots, p$ are said to be **linearly dependent** if they are not linearly independent. That is, (4.2) will hold for some p-tuple of scalars not all of which are zero. Similarly S is a **linearly dependent subset** of V if it is not linearly independent.

Definition 4.12. A vector space V is said to be **finite dimensional** if there is a positive integer n such that V contains a linearly independent set of n elements but any set of $(n + 1)$ or more elements of V is linearly dependent. The integer n is the **dimension** of V and is denoted dim $V = n$. By definition the vector space $V = \{\theta\}$, where θ is the zero element, is finite dimensional and dim $V = 0$.

If V is not finite dimensional then it is **infinite dimensional**.

Infinite dimensional vector spaces are of greater interest than the finite dimensional ones. This is particularly true for practical problems defined in terms of partial differential equations. For instance $C[a, b]$ is infinite dimensional whereas $\mathbf{R}^n$ and $\mathbf{C}^n$ are finite dimensional.

With this preparation the abstract version of Theorem 4.9, suitable for use in inner product spaces, can be stated in the following form [5].

Theorem 4.13. *(Schwarz inequality)*
Let V be an inner product space.

(i) $|(x, y)|^2 \leq (x, x)(y, y)$ *for any* $x, y \in V$.

(ii) *Equality in* (i) *only holds if* x *and* y *are linearly dependent.*

We have already seen that in any inner product space we can define a norm by

$$\|x\|^2 := (x, x)$$

Consequently, the Schwarz inequality can be written in the form

$$|(x,y)| \leq \|x\| \, \|y\| \tag{4.3}$$

This important result indicates that in a real inner product space we have

$$-1 \leq \frac{(x, y)}{\|x\|\|y\|} \leq 1$$

If we compare this result with the expression

$$\cos\varphi = \frac{x \cdot y}{|x||y|}$$

where φ denotes the angle between two vectors x and y in three-dimensional geometry then we can define an angle between two elements of an inner product space by

$$\varphi := \cos^{-1}\left\{\frac{(x, y)}{\|x\|\|y\|}\right\} \tag{4.4}$$

4.2 Orthogonality, Bases and Expansions

In the previous section we defined, in (4.4), the angle between two elements of an inner product space. A particularly useful notion which comes out of this is that of **orthogonality**.

Definition 4.14. (i) Two elements x, y of an inner product space, X, are said to be orthogonal if (x, y) = 0.

(ii) A set of elements $\{x_k\}_{k=1}^{\infty} \subset X$ is called **orthonormal** if

$$(x_k, x_m) = \begin{cases} 1, & k - m \\ 0, & k \neq m \end{cases}$$

This definition applies to both real and complex inner product spaces.

Examples 4.15. (i) For $\mathbf{C}^2$ endowed with the inner product

$$(x, y) = \sum_{k=1}^{2} x_k \overline{y_k}, \quad x, y \in \mathrm{C}^2$$

the elements $x = [1,i]$ and $y = [1, -i]$ are orthogonal because

$$(x, y) = ([1, i], [1, -i]) = 1.1 + i\overline{(-i)} = 1 + i^2 = 0$$

(ii) For $C[0, \pi]$ with inner product

$$(f, g) := \int_0^\pi f(x)\overline{g(x)}\, dx$$

the functions defined by $f(x) = \sin(mx)$ and $g(x) = \sin(nx)$ are orthogonal for any positive integers m, n with $m \neq n$.

An abstract version of Pythagoras' theorem is as follows.

Theorem 4.16. *If x and y are orthogonal elements of an inner product space then*

$$\|x + y\|^2 = \|x\|^2 + \|y\|^2$$

Proof.

$$\|x + y\|^2 = (x + y, x + y) = (x, x) + (y, y) + (x, y) + (y, x) = \|x\|^2 + \|y\|^2 \quad \blacksquare$$

It is perhaps interesting to remember the amount of labour involved in proving this result in elementary geometry classes!

We recall that all vectors in $\mathbf{R}^3$ can be expressed in terms of three unit vectors one each in the direction of the x-, y- and z-axis respectively. The three unit vectors are said to form a basis for $\mathbf{R}^3$ and the coefficients attached to each, when expressing an arbitrary vector, v, in terms of these basis elements, are referred as the *components* of v with respect to the basis. We want to parallel this situation when working in infinite dimensional spaces. To this end we introduce the following.

Definition 4.17. A set of elements $\{x_k\}$ of an inner product space is called an **orthogonal set** if

(i) $(x_j, x_k) = 0$ whenever $j \neq k$

(ii) for each k we have $x_k \neq \theta =$ zero element.

Part (ii) of this definition excludes the zero element. The effect of this is indicated by the following.

Lemma 4.18. *A finite orthogonal set is linearly independent.*

Proof. Let $\{x_k\}_{k=1}^n$ be an orthogonal set. We want to show that $\Sigma_{k=1}^n c_k x_k = 0$ implies that the scalars c_k, $k = 1, 2, 3, \ldots, n$ are all zero.

If $\Sigma_{k=1}^n c_k x_k = 0$ then for any j

$$0 = \left(\sum_{k=1}^n c_k x_k, x_j \right) = \sum_{k=1}^n c_k (x_k, x_j) = c_k \|x_k\|^2$$

the last equality following by orthogonality. Therefore we conclude that $c_k = 0$, $k = 1, 2, 3, \ldots, n$ provided $\|x_k\| \neq 0$. This is assured by part (ii) and the required result follows. ■

This result indicates that any set containing the zero element is a linearly dependent set. We shall rely on this result later when we work with various types of expansions which are particularly useful in applications.

Definition 4.19. An **orthogonal basis** for an inner product space V is an orthogonal set $\{e_n\}$ which is such that for any $x \in V$ there are scalars c_n such that

$$x = \sum_{k=1}^{\infty} c_k e_k$$

It should be noted that not every (inner product) space has a basis. It might be that a space is so large that infinite linear combinations such as those in Definition 4.19 do not account for all elements of the space. However, those spaces, and especially inner product spaces, which occur in applications usually do have a basis. Furthermore, in applications it turns out the spaces involved are usually complete. Consequently, for the remainder of this monograph we shall adopt the following definition.

Definition 4.20. A complete inner product space with a basis is a **Hilbert space**.

When we are working in finite dimensional spaces a basis is a very useful commodity because instead of manipulating the vectors or elements of the finite dimensional space we can manipulate their components. These components are numbers, either real or complex, and as such are often more amenable to computation than the elements themselves. It turns out that much the same thing is true in infinite dimensional spaces. Consequently, it is of prime importance to know how bases can be constructed for infinite dimensional spaces. The following offers a systematic and constructive method for doing this.

Theorem 4.21. *(Gram–Schmidt orthogonalisation process)*
Given a sequence $\{f_n\}$ in an inner product space there is an orthogonal sequence $\{g_n\}$ such that every finite linear combination of the f_n is a finite linear combination of the g_n.

The proof of this theorem is straightforward and constructive. However, it is rather lengthy and the details can be found in standard texts such as [5]. It is sufficient for us at the moment to know that this process exists.

Elements of a Hilbert space can be represented as an infinite series of basis vectors. The ability to do this will enable us to develop a number of powerful, constructive methods for solving problems we encounter in applications. What we are about to do is very similar to developing a classical Fourier series and it turns out that the required coefficients can be obtained in much the way as the classical Fourier coefficients are determined. To see this we require the following property of inner products.

Theorem 4.22. *Let V denote an inner product space with inner product* $(\cdot, \cdot)$.

(i) *If* $\{x_n\}$ *is a sequence in V such that* $x_n \to x \in V$ *as* $n \to \infty$ *then* $(x_n, y) \to (x, y)$ *as* $n \to \infty$ *for all* $y \in V$.

(ii) *If* $\{u_n\}$ *is a sequence in V and if* $S := \Sigma_{k=1}^{\infty} u_k$ *then*

$$\sum_{k=1}^{\infty} (u_k, y) = (S, y) \quad \text{for any} \quad y \in V.$$

Proof. Using the Schwarz inequality we obtain

$$|(x_n, y) - (x, y)| = |(x_n - x, y)| \leq \|x_n - x\| \, \|y\|$$

and the right-hand side tends to zero by hypothesis.

The second part follows immediately by setting $x = S$ and $x_n = \Sigma_{k=1}^{n} u_k$ in the above. ∎

This result indicates that for fixed $y \in V$ the inner product is a continuous function of x and vice versa.

Theorem 4.23. *Let* $\{e_n\}$ *be an orthogonal basis for an inner product space V. Any* $x \in V$ *can be expanded in the form* $x = \Sigma_{k=1}^{\infty} (c_k, e_k)$ *where the coefficients* c_k *are determined by*

$$c_n = \frac{(x, e_n)}{\|e_n\|^2} \tag{4.5}$$

Proof. Using the definition of basis we see that there are scalars (numbers) c_n such that for any $x \in V$ we have a result of the form $x = \Sigma_{k=1}^{\infty} (c_k, e_k)$. Consequently, using Theorem 4.22 we have that for all n

$$(x, e_n) = \sum_{k=1}^{\infty} c_k (e_k, e_n)$$

All terms on the right-hand side of this expression vanish except when $k = n$ and hence (4.5) follows immediately. ∎

The importance of this last result is that it offers a generalisation of the usual Fourier series expansion to inner product spaces.

Definition 4.24. Let $\{e_n\}$ be an orthogonal set in an inner product space V. For any $x \in V$ the series $\Sigma_{k=1}^{\infty} c_k e_k$ is called a **generalised Fourier series** and the numbers c_n defined in (4.5) are called the **generalised Fourier coefficients** or **expansion coefficients** of x with respect to the set $\{e_n\}$.

Once determined these generalised Fourier expansions will enable us to decompose a given, hard problem into a number of simpler problems. This aspect we shall deal with in detail in the chapter dealing with spectral theory. Another benefit of having available a generalised Fourier expansion is that it provides a means of developing constructive approximation procedures for use in practical applications. To see this notice first that the coefficients (4.5) have the property that $\Sigma_{k=1}^{N} c_k e_k$ can be made as close as we wish to x simply by taking N large enough. However, in practical, numerical computations we do not have the luxury of having the ability to work with an infinite number of elements and consider what happens as $N \to \infty$. Instead we are constrained to working with a finite number of elements. Consequently, we have to answer questions of the following form. Given a finite set of elements $e_1, e_2, \ldots, e_N$ of a Hilbert space then what linear combination of these elements is the best approximation to a given element x in the Hilbert space? The answer will be an expression of the form

$$\sum_{k=1}^{N} c_{kN} e_k \tag{4.6}$$

In (4.6) the coefficient c_{kN} depends on N, hence the double subscripts in (4.6). This means that if we use the coefficients c_{kN} when using the finite set e_1, $e_2, \ldots, e_N$ then we can expect to have to change *all* the coefficients to get the best approximation to x if we work with a finite set of the form $e_1, e_2, \ldots, e_N$, e_{N+1}. However, it turns out that when the elements $e_1, e_2, \ldots, e_N$ are pairwise orthogonal the coefficients in (4.6) are independent of N and are just the coefficients used (4.5). For this reason we will always make determined efforts in applications to work with orthogonal sets and the associated orthogonal expansions.

This remarkable feature is encapsulated in the following theorem.

Theorem 4.25. *Let $\{e_k\}_{k=1}^{N}$ be an orthogonal set in an inner product space V. For any $x \in V$ the coefficients c_k which minimize $\left\| x - \Sigma_{k=1}^{N} c_k e_k \right\|$ are given by* (4.5).

Proof. Set $c_k = (x, e_k) \| e_k \|^2 + d_k$ and expand $\left\| x - \Sigma_{k=1}^{N} c_k e_k \right\|$ as an inner product. Direct calculation of the inner product terms then clearly indicates that the required minimum is obtained when $d_k \equiv 0$ for all k. ∎

In this overview of the notion of a basis we have been using the intuitively obvious concept of the countability of a set. For the sake of completeness we make this more precise as follows.

Definition 4.26. A set is said to be **countably infinite** if it can be put in a one-to-one correspondence with the set of all positive integers.

A set is said to be **countable** if it is either finite or countably infinite. That is, the elements of the set can be put into one-to-one correspondence with either a finite set of integers or the set of all integers.

This definition simply means that we can label uniquely each element in the set with an integer. For more details see the texts cited in the reference, in particular see [5], [9].

In many books dealing with Hilbert space theory the following result is proved.

Theorem 4.27. *An inner product space has a basis if and only if it contains a countable dense set.*

We remark that in applications this result is often taken as axiomatic.

A natural question to ask about the coefficients in the generalised Fourier expansion of an element of an inner product space concerns their behaviour as $k \to \infty$. It turns out that they tend to zero as indicated by the following theorem.

Theorem 4.28. *(Bessel's inequality)*
Let $\{e_k\}$ be an orthonormal set in an inner product space V. For any $x \in V$

$$\sum_{k=1}^{\infty} |c_k|^2 \leq \|x\|$$

where $c_k = (x, ek)$, $k = 1, 2, \ldots$

This result leads to the following criterion for deciding whether or not a given orthonormal sequence is a basis.

Theorem 4.29. *(Parseval's relation)*
Let $\{e_k\}$ be an orthonormal sequence in an inner product space V. This set is a basis for V if and only if for each $x \in V$

$$\sum_{k=1}^{\infty} |c_k|^2 = \|x\|$$

where $c_k = (x, e_k)$, $k = 1, 2, \ldots$ are the expansion coefficients for $x \in V$ with respect to the set $\{e_k\}$.

The proofs of these last two results can be found in any of the texts on functional analysis cited in the References.

An important result which will lead the way to the spectral theorem and associated decomposition results in a Hilbert space setting is the following.

Theorem 4.30. *(Riesz–Fischer theorem)*
Let $\{e_k\}$ be an orthonormal basis for an infinite dimensional Hilbert space H.

If $\{c_k\}$ is a sequence of numbers with the property that the series $\Sigma_{k=1}^{\infty}|c_k|^2$ is convergent then there exists an $x \in H$ such that $x = \Sigma_{k=1}^{\infty} c_k e_k$ with $c_k = (x, e_k)$.

Here H could be a real or complex Hilbert space. Correspondingly, the numbers c_k could be either real or complex.

Again the proof of this theorem is to be found in standard texts on functional analysis [5].

We now provide some additional geometric aspects of abstract Hilbert spaces. We begin with two frequently used notational features.

Definition 4.31. (i) An inner product space is called a **pre-Hilbert space.**

(ii) A complete inner product space is called a **Hilbert space.**

Definition 4.32. (i) A subset G of a Hilbert space H is called a **linear manifold** (or simply a manifold) if it is invariant with respect to linear operations, that is, if λ_1, λ_2 are scalars then $\lambda_1 g_1 + \lambda_2 g_2 \in G$ whenever $g_1, g_2 \in G$.

(ii) A closed linear manifold is called a **subspace** of H.

Warning: Always check the definition of "subspace" being used by an author; they do vary!

If G_1, G_2 are two manifolds in a Hilbert space H then the subset of H, denoted $G_1 + G_2$, consisting of all elements $g \in H$ of the form $g = g_1 + g_2$ with $g_1 \in G_1$, $g_2 \in G_2$ is called the **sum** of G_1 and G_2.

The subset $G_1 \cap G_2 \subset H$ consisting of elements of H that are simultan-eously elements of both G_1 and G_2 is a manifold called the **intersection** of G_1 and G_2.

Definition 4.33. Let G_1, G_2 be two manifolds in a Hilbert space H. The **direct sum** of G_1 and G_2, denoted $G_1 \oplus G_2$ is defined by

$$G_1 \oplus G_2 = (G_1 + G_2 : G_1 \cap G_2 = \{\theta\})$$

where θ denotes the (unique) zero element in H.

Definition 4.34. (i) Let H_k, $k = 1, 2$ be Hilbert spaces with inner products $(\cdot, \cdot)_k$, $k = 1, 2$ respectively. The set of ordered pairs

$$\langle x, y\rangle, \quad x \in H_1, \quad y \in H_2$$

is a Hilbert space, H, with linear operations and inner product defined by

$$\begin{aligned}\lambda_1\langle x_1, y_1\rangle + \lambda_2\langle x_2, y_2\rangle &= \langle \lambda_1 x_1 + \lambda_2 x_2, \lambda_1 y_1 + \lambda_2 y_2\rangle(\langle x_1, y_1\rangle, \langle x_2, y_2\rangle) \\ &= (x_1, x_2)_1 + (y_1, y_2)_2\end{aligned}$$

where $x_1, x_2 \in H_1$, $y_1, y_2 \in H_2$ and λ_1, λ_2 are scalars, either real or complex.

The Hilbert space H is called the **direct product** of H_1 and H_2 and is denoted $H := H_1 \times H_2$.

(ii) Let H_k, $k = 1, 2, \ldots$ be a sequence of Hilbert spaces with structure $(\cdot, \cdot)_k$, $\|\cdot\|_k$, $k = 1, 2, \ldots$ respectively. Let H denote the set of sequences $\{x_k\}_{k=1}^{\infty}$ $x_k \in H_k$, $k = 1, 2, \ldots$ which satisfy $\Sigma_{k=1}^{\infty} \|x_k\|_k^2 < \infty$. The set H is a Hilbert space with inner product defined by

$$(x, y) = \sum_{k=1}^{\infty} (x_k, y_k)_k$$

with $x, y \in H$ and $x_k, y_k \in H_k$, $k = 1, 2, \ldots$. We write

$$H = \mathop{\times}_{k=1}^{\infty} H_k$$

Hilbert spaces provide a very clear and easily workable extension of the familiar algebra and geometry of finite dimensional spaces to infinite dimensional spaces. In applications we want to know that such an extension is always available. This will mean that for much of the time in applications we can work as though we were making the analysis in finite dimensional spaces. A limiting process could then provide the more general results.

Let G be a subspace (or just a manifold) of a Hilbert space H. The elements of H that are orthogonal to G, that is orthogonal to all elements of G, constitute a subspace of H called the **orthogonal complement** of G in H which is denoted $G^{\perp}$. The dimension of $G^{\perp}$ is called the **co-dimension** of G and we write co-dimension $G = \dim G^{\perp}$.

One of the more powerful results of Hilbert space theory for use in applications is that a Hilbert space H has an orthogonal decomposition into the orthogonal direct sum of subspaces G and $G^{\perp}$. This feature is encapsulated in the following celebrated theorem.

Theorem 4.35. (*Projection Theorem*)
Let

(i) *H be a Hilbert space with structure $(\cdot, \cdot)$, $\|\cdot\|$*
(ii) *M be a subspace of H*
(iii) *$M^{\perp}$ be the orthogonal complement of M in H.*

Then every element $x \in H$ can be written uniquely in the form

$$x = y + z, \quad y \in M, \quad y \in M^{\perp}$$

This theorem can be written compactly in the form

$$H = M \oplus M^{\perp} = \{y + z : y \in M, z \in M^{\perp}\}$$

Furthermore, direct calculation shows that

$$\|x\|^2 = \|y\|^2 + \|z\|^2$$

Theorem 4.35 is the bedrock for much of the analysis in this monograph and indeed for the majority of practical applications. It provides a means for decomposing not only Hilbert spaces but also various operations that are performed on them. This will often enable quite difficult problems to be broken down into simpler and more easily manageable components.

4.3 Linear Functionals and Operators on Hilbert Spaces

A **mapping** is a generalisation to vector spaces of the notion of a function (see Definition 3.10).

Definition 4.36. Let V_1, V_2 be vector spaces and X_1, X_2 subsets of V_1, V_2 respectively. A **mapping**, T, with action denoted by $T: X_1 \to X_2$ is a *rule* which, given any $x \in X_1$ associates with it an element of X_2 denoted Tx.

We shall write $x \to Tx$ to denote that $x \in X_1$ is mapped into the element $Tx \in X_2$.

We now restrict attention to Hilbert spaces. Let H be a Hilbert space with structure $(\cdot, \cdot)$, $\|\cdot\|$.

A mapping f of a manifold $D(f) \subset H$ into a manifold $R(f) \subset \mathbf{K} = \mathbf{R}$ or $\mathbf{C}$ is called a **functional** on H and we write

$$f: D(f) \subset H \to R(f) \subset K$$

This notation indicates that f can be either a real or complex valued function on H.

This mapping will be a **linear functional** if

$$f(\lambda_1 h_1 + \lambda_2 h_2) = \lambda_1 f(h_1) + \lambda_2 f(h_2)$$

for any $h_1, h_2 \in D(f)$ and scalars λ_1, λ_2. The manifold $D(f)$ is called the **domain (of definition)** of the mapping f whilst $R(f)$ is the **range** of the mapping f.

A more general type of mapping is an **operator**.

A mapping L of a manifold $D(L) \subset H$ onto a manifold $R(L)$ with action denoted by

$$L: H \supset D(L) \to R(L) \subset H$$

is called a **linear operator** in H if

$$L(\lambda_1 h_1 + \lambda_2 h_2) = \lambda_1 L(h_1) + \lambda_2 L(h_2)$$

for any $h_1, h_2 \in D(L)$ and scalars λ_1, λ_2. As above $D(L)$ and $R(L)$ are called respectively the **domain** and **range** of the **operator** L.

A linear operator $L: D(L) \subset H \to R(L) \subset H$ is said to be **densely defined** on H if $D(L)$ is dense in H, that is, if $\overline{D(L)} = H$.

Two of the simplest but quite important linear operators on H are

(i) the **trivial operator** Θ which is such that

$$\Theta h = \theta = \text{zero element in } H, \quad \text{for all} \quad h \in H$$

(ii) the identity operator I which is such that

$$Ih = h \in H \quad \text{for all} \quad h \in H$$

Given a linear operator L defined in a Hilbert space H then all solutions of the equation $Lh = \theta \in H$ form a manifold $N(L) \subset H$ known as the **null space** or **kernel** of the operator L.

When $N(L) = \{\theta\}$ then the equation $Lh = \theta \in H$ has only the trivial solution $h = \theta$. In this case it is possible to define an **inverse operator**, L^{-1}, which is such that [5], [9]

(i) $D(L^{-1}) = R(L)$ and $R(L^{-1}) = D(L)$

(ii) $L^{-1}g = h$, where $g \in D(L^{-1}) = R(L)$ and $h \in H$ is the unique solution of $Lh = g$

(iii) the property (ii) implies

$$L^{-1}(Lh) = h \in D(L)$$

$$L(L^{-1}g) = g \in D(L^{-1})$$

An important feature of operators on a Hilbert space is provided by the following definition which offers a natural extension to infinite dimensional spaces of familiar aspects of geometry in $\mathbf{R}^3$.

Definition 4.37. Let H be a Hilbert space and $L: H \to H$ a linear operator. The set of all elements $\langle h, Lh \rangle \in H \times H$ is denoted $\Gamma(L)$ and called the **graph** of L.

A linear manifold Γ in $H \times H$ is the graph of some linear operator L if and only if it does not contain elements of the form $\langle \theta, g \rangle$, $g \neq \theta$.

Two linear operators L_1, L_2 on a Hilbert space H are said to be **equal** if $D(L_1) = D(L_2) =: D \subset H$ and $L_1 h = L_2 h$ for any $h \in D$. Equivalently the operators are equal if $\Gamma(L_1) = \Gamma(L_2)$.

Remark 4.38. We would strongly emphasise that it should always be borne in mind that the operators L_1, L_2 with $D(L_1) \neq D(L_2)$ are **different** operators even if there holds $L_1h = L_2h$ for any $h \in D(L_1) \cap D(L_2)$.

An operator L' is said to be an **extension** of an operator L (alternatively, L is a **restriction** of L') if $D(L) \subset D(L')$ and $Lh = L'h$ for all $h \in D(L)$. Hence, an extension L' of L is *any* operator which agrees with L when applied to elements of $D(L)$ but is arbitrary elsewhere. This arbitrariness is removed in practice by requiring that the extension should have certain properties of the original operator, for instance, continuity. In the particular case when $D(L)$ is dense in H it can be shown that the extension is unique.

Let L be a linear operator in a Hilbert space H and let $G_1 \subset H$ be a subspace. Let $G_2 = G_1^{\perp}$ be the orthogonal complement of G_1 in H. Further, set

$$D_1 = G_1 \cap D(L)$$

$$D_2 = G_2 \cap D(L) = G_1^{\perp} \cap D(L)$$

The subspace G_1 is called a **reducing subspace** of L if

$$D(L) = D_1 \oplus D_2, \quad LD_1 \subset G_1, \quad LD_2 \subset G_2$$

It is clear that $G_2 = G_1^{\perp}$ is a reducing subspace (see Theorem 4.35).

We shall denote the restriction of L to G_1 and G_2 by L_1 and L_2 respectively and refer to L_1 and L_2 as the **parts (components)** of L. It is clear that $D(L) = D_1 \oplus D_2$ and that

$$Lh = L_1g_1 + L_2g_2, \quad h \in D(L)$$

where g_k, $k = 1, 2$ is the projection of h onto the subspace G_k, $k = 1, 2$. Consequently, $R(L) = R(L_1) \oplus R(L_2)$. We shall refer to the operator L as the **direct sum** of its parts L_1, L_2 and we write $L = L_1 \oplus L_2$.

It is clear from the above remarks that a study of the operator L is equivalent to a study of its components L_1 and L_2.

A decomposition of an operator into more than two parts can be similarly defined.

The above remarks provide the first indications of how a given operator might be decomposed into several more manageable parts. We shall return to this aspect in detail when we deal with the topic of spectral theory.

Let

$$L_k : H \supset D(L_k) \to R(L_k) \subset H, \quad k = 1, 2$$

be linear operators. Bearing in mind that usually these operators are not defined on all of H we define their sum, $L_1 + L_2$, and their product, L_1L_2, as follows

$$D(L_1 + L_2) = D(L_1) \cap D(L_2)$$
$$(L_1 + L_2)h = L_1 h + L_2 h, \quad h \in D(L_1 + L_2) \tag{4.7}$$

and

$$D(L_1 L_2) = \{h \in D(L_2) : L_2 h \in D(L_1)\}$$
$$(L_1 L_2)h = L_1(L_2 h), \quad h \in D(L_1 L_2) \tag{4.8}$$

We remark that $D(L_1 + L_2) = D(L_1 L_2) = D(L_2)$ when $D(L_1) = H$. Also, in general, $L_1 L_2 \neq L_2 L_1$, that is L_1 and L_2 do not commute.

A linear **functional**, f, on a Hilbert space H is said to be **bounded** on H if

$$\|f\| := \sup\left\{\frac{\|f(h)\|}{\|h\|} : h \in D(f) \subset H, h \neq \theta\right\} < \infty \tag{4.9}$$

where $\|f\|$ is the norm of the functional f.

Similarly an **operator** L on H is said to be **bounded** on H if

$$\|L\| := \sup\left\{\frac{\|Lh\|}{\|h\|} : h \in D(L) \subset H, h \neq \theta\right\} < \infty \tag{4.10}$$

where $\|L\|$ is the norm, or more fully, the **operator norm** of the operator L.

We remark that the operator L has a bounded inverse, L^{-1}, if and only if

$$\inf\left\{\frac{\|Lh\|}{\|h\|} : h \in D(L) \subset H, h \neq \theta\right\} > 0 \tag{4.11}$$

and

$$\|L^{-1}\| = \left\{\inf\left\{\frac{\|Lh\|}{\|h\|} : h \in D(L) \subset H, h \neq \theta\right\}\right\}^{-1} \tag{4.12}$$

It is possible to describe the general form of linear functionals on a Hilbert space. Before doing this we need to introduce the following two results.

Theorem 4.39. *If a linear functional is continuous at any one point then it is uniformly continuous.*

Proof. Let H be a Hilbert space. If $f : H \to \mathbf{K} = \mathbf{R}$ or $\mathbf{C}$ is continuous at $g \in H$ then for any $\varepsilon > 0$ there is a $\delta > 0$ such that $|f(g + h) - f(g)| < \varepsilon$, $h \in H$ whenever $\|h\| < \delta$. It then follows from the linearity of f that for any $x \in H$

$$|f(x+h) - f(x)| = |f(h)| = |f(g+h) - f(g)| < \varepsilon$$

whenever $\|h\| < \delta$. Since δ is independent of x the required uniformity follows. ■

This result indicates that a linear functional is either discontinuous everywhere or uniformly continuous everywhere. Hence we shall simply call a uniformly continuous functional a continuous functional.

Continuity and boundedness are essentially the same property for linear functionals but not for functionals in general. Indeed the following result can be obtained [5], [9].

Theorem 4.40. *A linear functional is continuous if and only if it is bounded.*

We now turn to a description of the general form of bounded linear functionals defined on either a whole Hilbert space H or on a dense manifold D of H.

Consider the function F_h defined by

$$Fh(g) = (g, h), \quad g \in H \tag{4.13}$$

where $h \in H$ is given. This functional is linear by virtue of the properties of an inner product and also continuous for the same reason. Alternatively, we can use the Schwarz inequality to show that it is bounded with

$$\|F_h\| = \sup\left\{\frac{\|F_h(g)\|}{\|g\|} : g \neq \theta\right\} \leq \|h\|$$

Substituting $g = h$ in this inequality we can conclude that $\|F_h\| = \|h\|$.

A general result in this connection is the following [5], [9].

Theorem 4.41. *(Riesz representation theorem)*

If F is a bounded linear functional defined on either all of a Hilbert space, H, or a dense manifold $D(F) \subset H$ then there exists a unique vector $h \in H$ such that

$$F = F_h = (\cdot, h) \quad \text{with} \quad \|F\| \doteq \|F_h\| = \|h\|$$

[*An alternative, slightly more transparent, statement of this is as follows: For every continuous linear functional F on a Hilbert space H there is a unique $h \in H$ such that $F(g) = (g, h)$ for all $g \in H$.*]

A proof of this important result can be found in the standard texts cited in the References and Commentary.

The set of linear bounded functionals on H is denoted by H^* and is referred to as the **dual** of H (see Definition 3.27). We define linear operations and an inner product on this set in the following manner.

Definition 4.42. (Structure of H^*)

(i) $\lambda_1 F_{h_1} + \lambda_1 F_{h_2} = F_{h_3}, h_3 := \overline{\lambda}_1 h_1 + \overline{\lambda}_2 h_2$

(ii) $(F_{h_1}, F_{h_2}) - (h_1, h_2)$ where $\lambda_k \in \mathbf{K}$ and $h_k \in$ H, $k = 1, 2$.

As a consequence of Definition 4.42 we consider H^* as a Hilbert space. To see that this makes sense we keep in mind (4.7). Then recognising that an inner product is conjugate linear we have

$$\begin{aligned}(\lambda_1 F_{h_1} + \lambda_2 F_{h_2})(g) &= \lambda_1 F_{h_1}(g) + \lambda_2 F_{h_2}(g) = \lambda_1 (g, h_1) + \lambda_2 (g, h_2) \\ &= (g, \overline{\lambda}_1 h_1) + (g, \overline{\lambda}_2 h_2) = F_{h_3}(g),\ h_3 = \overline{\lambda}_1 h_1 + \overline{\lambda}_2 h_2.\end{aligned}$$

Furthermore, the Riesz representation theorem indicates that there is a one-to-one mapping of H onto H^* such that $h \to F_h$. This mapping is isometric because $\|F_h\| = \|h\|$. Finally, it is conjugate linear because from the above

$$\lambda_1 h_1 + \lambda_2 h_2 \to \overline{\lambda}_1 F_{h_1} + \overline{\lambda}_2 F_{h_2}$$

These several remarks indicate that to regard H^* as a Hilbert space does indeed make sense.

We now turn our attention to the corresponding features of bounded linear operators defined on the whole Hilbert space H.

We denote by $B(H)$ the set of all linear, bounded operators on H that map H into itself. This set is invariant with respect to linear operations, that is, for scalars $\lambda_k \in \mathbf{K}, k = 1, 2$ we have $(\lambda_1 L_1 + \lambda_2 L_2) \in B(H)$ whenever $L_1, L_2 \in B(H)$. Consequently, $B(H)$ can be considered as a linear space. We also notice that $B(H)$ is invariant with respect to products of operators, that is, $L_1 L_2 \in B(H)$ whenever $L_1, L_2 \in B(H)$.

Much of the analysis that follows will involve approximation processes. Consequently, we need an understanding of the convergence processes which are available.

In Hilbert spaces there are two principal notions of convergence.

Definition 4.43. Let H be a Hilbert space with structure $(\, , \,), \|\cdot\|$

(i) A sequence $\{x_k\}_{k=1}^{\infty} \subset H$ is said to be **strongly convergent** to an element $x \in H$ if

$$\lim_{k \to \infty} \|x_k - x\| = 0$$

in which case we write

$$s - \lim_{k \to \infty} x_k = x$$

(ii) A sequence $\{x_k\}_{k=1}^{\infty} \subset H$ is said to be **weakly convergent** to an element $x \in H$ if for all $y \in H$

$$\lim_{k \to \infty} (x_k, y) = (x, y)$$

in which case we write

$$w-\lim_{k\to\infty} x_k = x$$

Bounded linear operators acting in Hilbert spaces will be of particular interest to us in later chapters. Consequently, it will be convenient to introduce the following convergence concepts in $B(H)$.

Definition 4.44. Let H be a Hilbert space and $\{T_k\}_{k=1}^{\infty} \subset B(H)$.

(i) The sequence $\{T_k\}_{k=1}^{\infty}$ is **strongly convergent** to T if, for all $h \in H$

$$\lim_{k\to\infty} \|T_k h - Th\| = 0$$

and we write

$$s-\lim_{k\to\infty} T_k = T$$

(ii) The sequence $\{T_k\}_{k=1}^{\infty}$ is **weakly convergent** to T if $\{T_k h\}_{k=1}^{\infty}$ is weakly convergent to Th for all $h \in H$ and we write

$$w-\lim_{k\to\infty} T_k = T$$

(iii) The sequence $\{T_k\}_{k=1}^{\infty}$ is **uniformly convergent** to T if

$$\lim_{k\to\infty} \|T_k - T\| = 0$$

and we write

$$u-\lim_{k\to\infty} T_k = T$$

We remark that uniform convergence is sometimes called **convergence in norm**.

The following results will be useful later.

Theorem 4.45. *Let H be a Hilbert space and $T_k, T, S_k, S \in B(H)$, k = 1, 2, . . .*

(i) *If $s-\lim_{k\to\infty} T_k = T$ and $s-\lim_{k\to\infty} S_k = S$ then $s-\lim_{k\to\infty} T_k S_k = TS$.*

(ii) *If $u-\lim_{k\to\infty} T_k = T$ and $u-\lim_{k\to\infty} S_k = S$ then $u-\lim_{k\to\infty} T_k S_k = TS$.*

For a proof of this see the references cited in the Commentary, in particular [5], [6], [9]. We remark that a similar result does *not* hold for weak convergence.

4.4 Some Frequently Occurring Operators

In this section we gather together the salient features of some linear operators which we will often meet in the following chapters.

Definition 4.46. A bounded linear operator T on a Hilbert space H is **invertible** if there exists a bounded linear operator T^{-1} on H such that

$$TT^{-1} = T^{-1}T = I$$

where I is the identity operator on H.

We remark that in the above definition T is on H, that is $D(T) = H$. When $D(T) \subset H$, that is when T is **in** H, then the definition will have to be modified. We will return to this aspect when it is needed later in some applications.

The following result can be established [5], [6], [9]. Indeed, it is often used as the definition of an inverse operator.

Theorem 4.47. *An invertible operator $T \in B(H)$ is* 1-1 *and maps H onto H. The inverse of T is unique.*

The next result, which provides a generalisation of the geometric series $(1 - a)^{-1}$, is fundamental to much of the material in later chapters.

Theorem 4.48. *(Neumann series) Let H be a Hilbert space and* let $T \in B(H)$ *have the property* $\|T\| < 1$. *Then*

(i) $(I - T)$ is invertible
(ii) $(I - T)^{-1} \in B(H)$
(iii) $(I - T)^{-1} = \Sigma_{k=0}^{\infty} T_k$ (Neumann series) is uniformly convergent
(iv) $\|(I - T)^{-1}\| \leq (1 - \|T\|)^{-1}$

The proof of this result is entirely straightforward but rather lengthy [5], [6].

The existence of an inverse operator is of fundamental importance when solving operator equations. Such operators are not always easy either to determine or to work with. However, associated with a given operator on a Hilbert space is an operator that has some of the flavour of an inverse operator, namely an **adjoint operator**.

Let H be a Hilbert space and $T \in B(H)$. Then, forgiven $h \in H$ the functional (Tg, h), $g \in H$ is linear and bounded. Consequently, by the Riesz representation Theorem, there is a unique element h^* such that $(Tg, h) = (g, h^*)$
The mapping $h \to h^*$ is defined on all of H and is readily seen to be linear and bounded. Consequently, we write

$$T^*h = h^* \quad \text{for any} \quad h \in H$$

so that

$$(Tg, h) = (g, T^*h), \quad g, h \in H$$

These various observations can be written compactly in the following form (see Commentary).

Theorem 4.49. *Let H be a Hilbert space with structure $(\ ,\ )$ and let $T \in B(H)$. There exists a unique, linear, bounded operator T^* on H, called the* **adjoint** *of T, defined by*

$$(Tg, h) = (g, T^*h), \quad \text{for all} \quad g, h \in H$$

which is such that

$$\|T^*\| = \|T\|$$

Some elementary properties of T^* are contained in the following.

Theorem 4.50. *Let H be a Hilbert space and let $T, S \in B(H)$.*

(i) $T^{**} := (T^*)^* = T$

(ii) $(\lambda T)^* = \bar{\lambda} T^*, \qquad \lambda \in \mathrm{C}$

(iii) $(T + S)^* = T + S^*$

(iv) $(TS)^* = S^*T^*$

(v) If T is invertible then so also is T^* and

$$(T^*)^{-1} = (T^{-1})^*$$

(vi) $\|TT^*\| = \|T^*T\| = \|T\|^2$.

We emphasise, as before, that these several results have to be modified when $D(T) \neq H$.

Definition 4.51. Let H be a Hilbert space and $T \in B(H)$.

(i) If $T = T^*$ then T is **self-adjoint** or **Hermitian**.

(ii) If $T^* = T^{-1}$ then T is unitary.

(iii) If $TT^* = T^*T$ then T is **normal**.

(iv) If T is unitary and $R, S \in B(H)$ are such that

$$R = TST^{-1}$$

then R and S are **unitarily equivalent** with respect to T.

(v) T is self-adjoint if and only if

$$(Tf, g) = (f, Tg) \quad \text{for all} \quad f, g \in H$$

(vi) T is unitary if and only if T is invertible and

$$(Tf, g) = (f, T^{-1}g) \quad \text{for all} \quad f, g \in H.$$

Remarks: (i) It is instructive to examine the compatibility of (v) and (vi) above with the various results mentioned above in this section.

(ii) If T is self-adjoint then (Tf, f) is real.

Theorem 4.52. Let H be a Hilbert space with structure (,). If $T \in B(H)$ is self-adjoint then $\|T\|$, the norm of T, can be determined by

$$\|T\| = \sup_{\|f\|=1} |Tf, f| = \max\{|m|, |M|\}$$

where

$$M := \sup\{(Tf, f) : \|f\| = 1\}$$

$$m := \inf\{(Tf, f) : \|f\| = 1\}$$

and $f \in H$.

The numbers M and m are, respectively, the **upper** and **lower bounds** of the operator T.

A means of comparing the "size" of various operators on H is afforded by the following.

Definition 4.53. Let H be a Hilbert space with structure (,) and $T \in B(H)$. If the operator T is self-adjoint then it is said to be **non-negative**, and we write $T \geqslant 0$, if and only if for all $f \in H$ we have $(Tf, f) \geqslant 0$. When the inequality is strict then T is said to be **positive**.

It now follows that

(a) T is non-negative if $m \geqslant 0$.
(b) Operators T_1, T_2 are such that $T_1 \leq T_2$ if and only if $(T_2 - T_1) \geqslant 0$.

An important class of bounded operators are the so-called **finite dimensional** operators T which map H onto $R(T)$ where $\dim R(T) < \infty$. A finite dimensional operator, or an **operator of finite rank** as it is sometimes called, can be expressed in the form

$$Th = \sum_{k=1}^{N} a_k(h, f_k)g_k \tag{4.14}$$

where $\dim R(T) = N$. The real positive numbers a_k are ordered in the form $a_1 \geqslant a_2 \geqslant \ldots \geqslant a_N > 0$ and the sets of vectors $\{f_k\}$, $\{g_k\} \subset H$ are orthonormal.

In the general theory of operators a significant role is played by the so-called **compact** operators whose properties are similar to those of finite dimensional operators.

Definition 4.54. An operator $T \in B(H)$ is **compact** if it is the limit of a uniformly convergent sequence $\{T_k\}$ of operators of finite rank, that is, $\|T - T_k\| \to 0$ as $k \to \infty$.

An equivalent definition is:

An operator $T \in B(H)$ is **compact** if for every bounded sequence $\{f_k\}_{k=1}^{\infty} \subset H$ the sequence $\{Tf_k\}_{k=1}^{\infty}$ has a strongly convergent subsequence.

Example 4.55. (i) If T_1 is a compact operator and T_2 is a bounded operator on the same Hilbert space H then T_1T_2 and T_2T_1 are also compact operators.

(ii) If T_k, λ_k, $k = 1, 2, \ldots, N$ are, respectively, compact linear operators and complex-valued coefficients then

$$T = \sum_{k=1}^{N} \lambda_k T_k$$

is a compact operator.

(iii) Operators T, T^*, TT^*, T^*T are simultaneously compact or non-compact.

Compact operators have a relatively simple structure similar to that of operators of finite rank. Specifically, a compact operator T can be written in the form

$$Th = \sum_{k=1}^{\infty} a_k (h, f_k) g_k \tag{4.15}$$

where $\{a_k\}$ is a non-increasing sequence of real positive numbers which tends to zero, that is $a_k \to 0$ as $k \to \infty$. The numbers a_k are called the **singular numbers** of the operator T in (4.15). The sets $\{f_k\}$, $\{g_k\}$ are orthonormal.

If in (4.15) we have $a_k = 0$ for sufficiently large $k \geqslant N$ then T is a finite dimensional operator as in (4.14); that is, T has a **decomposition** of the form (4.14).

If T is self-adjoint then the sets $\{f_k\}$ and $\{g_k\}$ can be chosen to satisfy, for all k, either $f_k = g_k$ or $f_k = -g_k$.

An operator $T \in B(H)$ can also be usefully decomposed in the form

$$T = T_R + iT_I \tag{4.16}$$

where

$$T_R := \frac{1}{2}(T + T^*) = T_R^*$$

$$T_I := \frac{1}{2i}(T - T^*) = T_I^*$$

This decomposition is similar to the decomposition of complex numbers.

One of the simplest self-adjoint operators is the **projection operator** which is defined in the following way.

Let H be a Hilbert space with structure (,) and let M be a subspace. The projection theorem (Theorem 4.35) states that every element $f \in H = M \oplus M^{\perp}$ can be expressed *uniquely* in the form

$$f = g + h, \quad g \in M, \quad h \in M^{\perp} \tag{4.17}$$

The uniqueness of this representation allows us to introduce a linear operator $P: H \to H$ defined by

$$Pf = g \in M, \quad f \in H \tag{4.18}$$

Such an operator is called a **projection operator** or **projector** as it provides a projection of H onto M.

We notice that (4.17) can also be written in the form

$$f = g + h = Pf + (I - Pf) \tag{4.19}$$

which indicates that

$$(\mathrm{I} - \mathrm{P}): H \to M^{-}$$

is a projection of H onto M^{-}.

When it is important to emphasise the subspace involved in a decomposition of the form (4.19) we will write $P = P(M)$.

Projection operators play a very important part in much of the analysis that is to follow. For convenience we collect together here their main properties. The proof of these various properties is a standard part of courses on the general theory of linear operators on Hilbert spaces; the Commentary indicates a number of sources for the interested reader, in particular [1], [4].

Theorem 4.56. Let H be a Hilbert space with structure (,).

(a) Abounded linear operator $P: H \to H$ is a projection on H if and only if

$$P = P^2 = P^*$$

An operator P such that $P = P^2$ is called **idempotent**.

(b) Let $M \subset H$ be a subspace (closed linear manifold). The projection $P: H \to M$ has the properties

(i) $(Pf, f) = \|Pf\|^2$
(ii) $P \geqslant 0$
(iii) $\|P\| \leq 1$, $\|P\| = 1$ if $R(P) \neq \theta$
(iv) $(I - P) =: \mathrm{P}^{\perp}: H \to M^{\perp}$ is a projection.

Theorem 4.57. *Let H be a Hilbert space with structure $(\ ,\ )$ and let $M_k \subset H$, $k = 1, 2$ be subspaces in H. The projections*

$$P\,(M_k)\colon H \to M_k, \quad k = 1, 2$$

have the following properties.

(i) *The operator $P := P(M_1)P(M_2)$ is a projection on H if and only if*

$$P(M_1)P(M_2) = P(M_2)P(M_1)$$

When this is the case then

$$P\colon H \to R(P) = M_1 \cap M_2$$

(ii) *M_1, M_2 are orthogonal if and only if*

$$P(M_1)P(M_2) = 0$$

in which case $P(M_1)$ and $P(M_2)$ are said to be mutually ***orthogonal projections***.

(iii) *The operator $P := P(M_1) + P(M_2)$ is a projection on H if and only if M_1 and M_2 are orthogonal. In this case*

$$P\colon H \to R(P) = M_1 \oplus M_2$$

that is $P(M_1) + P(M_2) = P(M_1 \oplus M_2)$.

(iv) *The projections $P(M_1)$, $P(M_2)$ are* ***partially ordered*** *in the sense that the following statements are equivalent.*

(a) $P(M_1)P(M_2) = P\,(M_2)P\,(M_1) = P\,(M_1)$

(b) $M_1 \subseteq M_2$

(c) *$N(P(M_2)) \subseteq N(P(M_1))$ where $N(P(M_k))$ denotes the null space of $P(M_k)$, $k = 1, 2$*

(d) *$\|P(M_1)f\| \leq \|P\,(M_2)f\|$ for all $f \in H$*

(e) *$P(M_1) \leq P(M_2)$.*

(v) *A projection $P(M_2)$ is a* ***part*** *of a projection $P(M_1)$ if and only if $M_2 \subset M_1$.*

(vi) *The difference $P(M_1) - P(M_2)$ of two projections $P(M_1)$, $P(M_2)$ is a projection if and only if $P(M_2)$ is a part of $P(M_1)$. When this is the case then*

$$P := (P(M_1) - P(M_2))\colon H - R(P) = M_1 \cap M_2^{\perp}$$

that is $(P(M_1) - P(M_2)) = P(M_1 \cap M_2^{\perp})$.

(vii) *A series of mutually orthogonal projections $P(M_k)$, $k = 1, 2, \ldots$ on H, denoted by $\sum_k P\left(M_k\right)$ is strongly convergent to the projection $P(M)$ where $M = \bigoplus_k M_k$.*

(viii) *A linear combination of projections* $P(M_k)$, $k = 1, 2, \ldots, N$ *on* H, *denoted by* $P := \Sigma_{k=1}^{N} \lambda_k P(M_k)$, *where* λ_k *are real-valued coefficients, is self-adjoint on* H.

(ix) *Let* $\{P(M_k)\}_{k=1}^{\infty}$ *denote a monotonically increasing sequence of projections* $P(M_k)$, $k = 1, 2, \ldots$ *on* H. *Then*

(a) $\{P(M_k)\}_{k=1}^{\infty}$ *is strongly convergenttoaprojection* P *on* H; *that is* $P(M_k)f \to Pf$ *as* $k \to \infty$ *for all* $f \in H$

$$\text{(b)}\quad P: H \to R(P) = \bigcup_{k=1}^{\infty} R(P(M_k))$$

$$\text{(c)}\quad N(P) = \bigcap_{k=1}^{\infty} N(P(M_k)).$$

We remark that there exists an intimate connection between general self-adjoint operators and projections. This we will discuss later under the heading of spectral theory. As a consequence we will be able to establish the various decomposition results and expansion theorems we have already mentioned.

In subsequent chapters we will often encounter the following particular type of compact operator.

Definition 4.58. Let H be a Hilbert space with structure $(\ ,\ )$. An operator $T \in B(H)$ is called a **Hilbert Schmidt operator** if

$$\|T\|^2 = \sum_{k=1}^{\infty} \|Te_k\|^2 < \infty$$

where $\{e_k\}_{k=1}^{\infty}$ is an orthonormal basis for H.

If T is a compact operator then it is a **compact Hilbert–Schmidt operator** if its singular numbers $a_k(T)$ tend to zero sufficiently rapidly so that the series $\Sigma_{k=1}^{\infty} a_k^2(T)$ is convergent; that is $\Sigma_{k=1}^{\infty} a_k^2(T) < \infty$.

It is clear from this definition that any finite-dimensional operator is a compact Hilbert–Schmidt operator.

Some of the best known classes of Hilbert–Schmidt operators are integral operators. It will be convenient, for later use, to introduce these operators as operators on weighted L_2-spaces.

Definition 4.59. The Hilbert space of functions which are square integrable on the whole space R^N with respect to some weight function ρ is denoted $L_{2,\rho}(\mathrm{R}^N)$. Here ρ is a real-valued positive function in each open, bounded region $\Omega \subset \mathrm{R}^N$. The structure $(\ ,\ )_\rho$, $\|\cdot\|_\rho$ is defined by

$$(f_1, f_2)_\rho := \int_{\mathbf{R}^N} f_1(x)\overline{f_2(x)}\rho(x)dx, \quad f_1, f_2 \in C_0^{\infty}(\mathbf{R}^N)$$

$$\|f\|_\rho^2 := \int_{\mathbf{R}^N} |f(x)|^2 \rho(x)dx, \qquad f \in C_0^{\infty}(\mathbf{R}^N)$$

The space $L_{2,\rho}(\mathrm{R}^N)$ is defined as the completion of $C_0^\infty(\mathrm{R}^N)$ with respect to this structure.

The space $L_{2,\rho}(\Omega)$, $\Omega \subset \mathrm{R}^N$ is defined similarly.

We define an **integral operator**, T, on $L_{2,\rho}(\Omega)$ according to

$$(Tf)(y) := \int_\Omega K(x,y) f(x) \rho(x) dx \tag{4.20}$$

where K is called the **kernel** of T. The kernel is square integrable on $\Omega \times \Omega$ in the sense

$$\int_\Omega \int_\Omega |K(x,y)|^2 \rho(x)\rho(y) dx dy = k_0^2 < \infty \tag{4.21}$$

The operator T defined in (4.20) is clearly linear and bounded. Further, its norm can be estimated according to $\|T\| \le k_0$. Moreover, it can be shown that T in (4.20) is a compact Hilbert–Schmidt operator.

The operator T^* which is adjoint to T in (4.20) is defined by

$$(T^*f)(x) := \int_\Omega \overline{K(x,y)} f(y) \rho(y) dy \tag{4.22}$$

The integral operator T in (4.20) is self-adjoint if and only if the kernel is symmetric in the sense that $K(x, y) = \overline{K(y, x)}$.

An integral operator is finite dimensional if and only if its kernel $K(x, y)$ can be expressed in the form

$$K(x,y) = \sum_{k=1}^{N} g_k(x) f_k(y)$$

where f_k, g_k, $k = 1, 2, \ldots, N$ are square integrable functions. Kernels of this type are called **degenerate**. For an integral operator T with a degenerate kernel K it can be shown that $\dim(R(T)) \le N$.

4.5 Unbounded Linear Operators on Hilbert Spaces

Not all linear operators are bounded. However, it turns out that practically all the operators we encounter in applications are so-called closed operators and they retain much of the flavour of the continuity property displayed by bounded linear operators.

Definition 4.60. Let X_1, X_2 be normed linear spaces and $T: X_1 \to X_2$ a linear operator with domain $D(T) \subset X_1$. If

(i) $x_k \in D(T)$ for all k

(ii) $x_k \to x$ in X_1

(iii) $Tx_k \to y$ in X_2

when *taken together*, imply $x \in D(T)$ and $Tx = y$, then T is said to be a **closed operator**.

We would emphasise that in Definition 4.60 we require the *simultaneous* convergence of the sequences $\{x_k\}_{k=1}^{\infty}$ and $\{Tx_k\}_{k=1}^{\infty}$.

An alternative definition of a closed operator can be profitably given in terms of the graph of an operator.

Definition 4.61. Let X_1, X_2 be normed linear spaces with norms $\|\cdot\|_1$, $\|\cdot\|_2$ respectively. A linear operator

$$T: D(T) \subseteq X_1 \rightarrow X_2$$

is a closed linear operator if and only if its graph,

$$G(T) := \{(x_1, x_2) \in X_1 \times X_2 : x_1 \in D(T), x_2 = Tx_1\}$$

is a closed subset of $X_1 \times X_2$. We recall that $X_1 \times X_2$ is a normed linear space where the structure is defined in the usual component-wise manner. For example the norm is

$$\|(x, x_2)\|_{X_1 \times X_2} := \|x_1\|_1 + \|x_2\|_2$$

In applications it frequently turns out that a closed operator is in fact a bounded operator. To establish whether or not this is the case we need some if not all of the following standard results (see [5], [6], [9] and the Commentary).

Theorem 4.62. *(Uniform boundedness theorem)*
Let I be an index set and let

(i) *X, Y be Banach spaces*

(ii) *$\{T_\alpha\}_{\alpha \in I}$ be such that*

$$\sup_{\alpha \in I} \{\|T_\alpha x\|_Y\} < \infty \quad \text{for all} \quad x \in X$$

Then $\sup_{\alpha \in I} \{\|T_\alpha\| < \infty$. That is if $\{T_{\alpha x}\}_{\alpha \in I}$ is a bounded set in Y for all $x \in X$ then $\{T_\alpha\}$ is bounded in $B(X, Y)$. We recall that $B(X, Y)$ denotes the class of all bounded linear operators with action $X \rightarrow Y$.

Theorem 4.63. *(Bounded inverse theorem) Let X, Y be Banach spaces and let $T \in B(X, Y)$ be 1-1 and onto, then T^{-1} exists as a bounded operator.*

Theorem 4.64. *(Closed graph theorem) Let X, Y be Banach spaces and $T: X \supset D(T) \rightarrow Y$ be a closed linear operator. If $D(T)$ is a closed set then T is bounded.*

Example 4.65. We shall show that the process of differentiation can be realised as a closed linear operator on the space of continuous functions. This will be in a form which will be rather more general than we will require in this monograph as we will be principally concerned with working in Hilbert spaces. Nevertheless, this example will serve as a prototype.

Let $X = Y = C[a, b]$, $-\infty < a < b < \infty$ be endowed with the usual supremum norm which for convenience and brevity we sometimes denote by $\|\cdot\|_\infty$. Define, for example,

$$T: X \supset D(T) \to Y$$

$$(Tg)(x) = \frac{dg(x)}{dx} \equiv g'(x), \quad g \in D(T), \quad x \in [a, b]$$

$$D(T) := \{g \in X : g' \in X, \ g(a) = 0\}$$

Let $\{g_k\} \subset D(T)$ be a sequence with the properties that as $k \to \infty$ we have $g_k \to g$ and $Tg_k \to h$ with respect to $\|\cdot\|_\infty$. We can establish that T is closed if we can show that $g \in D(T)$ and that $Tg = h$. To this end consider

$$f(x) = \int_a^x h(t)dt, \quad x \in [a, b] \tag{4.23}$$

The convergence $Tg_k \to h$ with respect to $\|\cdot\|_\infty$ implies that the convergence is uniform with respect to x. Since $Tg_k \in C[a, b]$ for all k it follows that $h \in C[a, b]$. Consequently, by the Fundamental Theorem of Calculus we deduce that f in (4.23) is continuous and differentiable with $f'(x) = h(x)$ for all $x \in [a, b]$. Furthermore, the properties of the Riemann integral in [9] indicate that $f(a) = 0$. Collecting these results we can conclude that $f \in D(T)$.

Since $g_k \in D(T)$, the Fundamental Theorem of Calculus implies

$$g_k(x) = \int_a^x g_k'(t)dt$$

and we have

$$\begin{aligned} |g_k(x) - f(x)| &= \left| \int_a^x (g_k'(t) - h(t))dt \right| \\ &\leq \|g_k' - h\|_\infty (b - a) \end{aligned}$$

Now the right-hand side of this expression tends to zero by virtue of the convergence $Tg_k \to h$ and it follows that $g_k \to f \in D(T)$. Furthermore, we will also have that $g' = f' = h$. Hence T, as defined, is a closed operator.

If an operator, T, is not closed then it is sometimes possible to associate with T an operator that is closed. This parallels the process of associating with a set M in a metric space a closed set $\overline{M}$ called the closure of M.

Definition 4.66. A linear operator T is **closable** if whenever

(i) $f_n \in D(T)$

(ii) $f_n \to \theta$ as $n \to \infty$

(iii) Tf_n tends to a limit as n $\to \infty$ then $Tf_n \to \theta$ as $n \to \infty$.

If T is a closable operator defined on a normed linear space X then we may define an extension of T (see the comments following Definition 4.37), denoted $\overline{T}$, and called the **closure** of $\overline{T}$, in the following manner.

(i) Define

$$D(\overline{T}) := \{f \in X : \exists \{f_k\} \subset D(T) \text{ with } f_k \to f \text{ and } \{Tf_k\} \text{ a Cauchy sequence}\}$$

Equivalently we define $D(\overline{T})$ to be the closure of $D(T)$ with respect to the graph norm

$$\|f\|_G^2 = \|f\|_X^2 + \|Tf\|_X^2$$

(ii) For $f \in D(\overline{T})$ set

$$\overline{T}f = \lim_{k \to \infty} Tf_k$$

where $\{f_k\}$ is defined as in (i).

There are many connections between closed, bounded and inverse operators. The following theorem draws together a number of results which will be used frequently in later sections. Proofs of these results can be found in the texts cited in the Commentary; we would particularly mention [5], [8], [9], [6].

Theorem 4.67. *Let X be a Banach space. An operator $T \subset B(X)$ is closed if and only if $D(T)$ is closed.*

Theorem 4.68. *Let*

(i) *X_1, X_2 be normed linear spaces*

(ii) *$T := X_1 \supset D(T) \to X_2$ be a linear, one-to-one operator. Then T is closed if and only if the operator*

$$T^{-1} := X_2 \supset R(T) \to X_1$$

is closed.

Theorem 4.69. *Let*

(i) *X_1, X_2 be normed linear spaces*

(ii) *$T := X_1 \supset D(T) \to X_2$ be a linear operator.*

(a) *If there exists a constant $m \geqslant 0$ such that*

$$\|Tf\|_2 \geqslant m\|f\|_1, \quad f \in D(T)$$

then T is a one-to-one operator. Furthermore, T is closed if and only if $R(T)$ is closed in X_2.

(b) *Let T be one-to-one and closed. Then $T \in B(X_1, X_2)$ if and only if $R(T)$ is dense in X_2 and there exists a constant $m > 0$ such that*

$$\|Tf\|_2 \geqslant m\|f\|_1, \quad f \in D(T)$$

(c) *If T is closed then*

$$N(T) := \{f \in D(T): Tf = \theta\}$$

is a closed subset of X_1.

Finally in this section we highlight a number of important results that we will have to recognise when dealing with unbounded operators.

We have seen that two operators, T_1, T_2 are said to be equal, denoted $T_1 = T_2$, if $D(T_1) = D(T_2)$ and $T_1 f = T_2 f$ for all $f \in D(T_1) = D(T_2)$.

The **restriction** of an operator $T: X \supset D(T) \to Y$ to a subset $M \subset D(T)$ is denoted by $T|_M$ and is the operator defined by

$$\mathrm{T}|_M: M \to Y, T|_M f = Tf \quad \text{for all} \quad f \in M \tag{4.24}$$

An **extension** of T to a set $S \supset D(T)$ is an operator $\tilde{T}: S \to Y$ such that

$$\tilde{T}|_{D(T)} = T \tag{4.25}$$

This last definition indicates that $\tilde{T}f = Tf$ for all $f \in D(T)$. Hence T is a restriction of $\tilde{T}$ to $D(T)$.

In this monograph we say that T is an operator **on** a Hilbert space H if its domain, $D(T)$, is all of H. We say that T is an operator in H if $D(T)$ is not necessarily all of H.

Furthermore, if T_1 and T_2 are two linear operators then we shall use the notation

$$T_1 \subset T_2 \tag{4.26}$$

to denote that $T_1(T_2)$ is a restriction (extension) of $T_2(T_1)$.

A particularly useful feature of a bounded linear operator T on a Hilbert space H is that it is associated with a bounded linear operator T^*, the adjoint of T, defined according to the equation (see Theorem 4.49)

$$(Tf, g) = (f, T^*g) \quad \text{for all} \quad f, g \in H$$

where (,) denotes the inner product on H. The proof of the existence of such an adjoint operator makes use of the Riesz representation theorem (Theorem 4.41) and the proof breaks down when T is either unbounded or not defined on all of H. However,even when the proof fails it may happen that for some element $g \in H$ there is an element $g^* \in H$ such that

$$(Tf, g) = (f, g^*) \quad \text{for all} \quad f \in D(T) \tag{4.27}$$

If for some fixed $g \in H$ there is only one $g^* \in H$ such that (3.30) holds then we can write

$$g^* = T^*g \tag{4.28}$$

and consider the operator T^* as being well defined for at least this $g \in H$. However, it remains to determine the conditions under which (4.27) yields a unique $g^* \in H$. Results in this direction are as follows.

Theorem 4.70. *Let*
 (i) *T be a linear operator on a Hilbert space H*
 (ii) *there exist elements $g, g^* \in H$ such that $(Tf, g) = (f, g^*)$ for all $f \in D(T)$*
Then g^* is uniquely determined by g and (4.27) *if and only if $D(T)$ is dense in H.*

Theorem 4.71. *Let*
 (i) *T be a linear operator in a Hilbert space H with $\overline{D(T)} = H$*
 (ii) *$D(T^*) = \{g \in H$: there exists $g^* \in H$ satisfying* (4.27)*\}. Then $D(T^*)$ is a subspace of H and the operator T^* with domain $D(T^*)$ and defined by*

$$T^*g = g^* \quad \text{for all} \quad g \in D(T)$$

is a linear operator.

These two results lead naturally to the following.

Definition 4.72. Let T be a linear operator in a Hilbert space with $D(T) = H$. The operator T^* with domain $D(T^*)$ defined as in Theorem 4.71 is called the **adjoint** of T.

When T is a bounded operator then this definition of an adjoint operator coincides with that given in Theorem 4.49. However, we emphasise that in the general case when the operator may be unbounded then more attention must be given to the rôle and importance of domains of operators. The following results illustrate this aspect.

Theorem 4.73. *Let T,S be linear operators in a Hilbert space H.*

(i) *If $T \subset S$ and $\overline{D(T)} = H$ (which, incidentally, implies that $\overline{D(S)} = H$) then $T^* \supset S^*$.*

(ii) *If $\overline{D(T)} = D(T^*) = H$ then $T \subset T^{**}$.*

(iii) *If T is one-to-one and such that $\overline{D(T)} = D(\overline{T^{-1}}) = H$ then T^* is one-to-one and $(T^*)^{-1} = (T^{-1})^*$*

The following types of operators are important in applications.

Definition 4.74. Let T be a linear operator in a Hilbert space H and assume $D(\overline{T}) = H$.

(i) If $T = T^*$ then T is **self-adjoint** (in H).

(ii) If $T \subset T^*$ then T is **symmetric** (in H).

Some properties of these operators are indicated in the following.

Theorem 4.75. *Let T be a linear operator in a Hilbert space H.*

(i) *If T is one-to-one and self-adjoint then $\overline{D(T^{-1})} = H$ and T^{-1} is self-adjoint.*

(ii) *If T is defined everywhere on H then T^* is bounded.*

(iii) *If T is self-adjoint and defined everywhere on H then T is bounded.*

(iv) *If $\overline{D(T)} = H$ then T^* is closed.*

(v) *Every self-adjoint operator is closed.*

(vi) *If T is closable then $(\overline{T})^* = T^*$ and $\overline{T} = T^{**}$.*

(vii) *If T is symmetric then it is closable.*

For proofs of these various results see the references cited in the Commentary and in particular [8], [9].

Self-adjoint operators play a particularly important rôle in the following chapters. Here we give an indication how to decide whether or not a given operator has this property. First we need the following notion.

Definition 4.76. Let T be a linear operator in a Hilbert space H with $\overline{D(T)} = H$. The operator T is said to be **essentially self-adjoint** if $\overline{T} - T^*$.

We notice that if T is essentially self-adjoint then necessarily it must be closed since T^* is a closed operator (Theorem 4.75(iv)). Furthermore, from Theorem 4.75(vi) we conclude that $T^* = (\overline{T})^*$. Therefore T is essentially self-adjoint if and only if $\overline{T} = (\overline{T})^*$, that is, if and only if $\overline{T}$ is self-adjoint.

From Definition 4.74 we see that if T is symmetric then it is a restriction of its adjoint T^*. However, we would emphasise that a symmetric operator need not be self-adjoint. It may well be that a symmetric operator *might* have a self-adjoint extension. Nevertheless we must always remember that the extension is *not* unique; the extension process is usually followed in order to preserve, or even to provide, such properties as for example linearity, boundedness, self-adjointness etc. We shall be particularly interested in the case when a symmetric operator has exactly one self-adjoint extension. In this connection we have the following results [8].

Theorem 4.77. *Let T be a linear, symmetric operator in a Hilbert space H. Then*

(i) *T is closable*
(ii) *$R(\overline{T} \pm I) = \overline{R(T \pm I)}$ where I is the identity on H.*

Theorem 4.78. *A symmetric operator $T{:}H \to H$ is self-adjoint if and only if*

$$R(T + iI) = H = R(T - iI)$$

Finally, in this section we give two examples of operators that are unbounded. These are the operators of multiplication by the independent variable and the differentiation operators. These operators frequently occur in mathematical physics and perhaps the quickest way to see this is to recall that when integral transform methods are used differentiation can be replaced by multiplication by the independent variable (see Example 3.43).

Example 4.79. Let $H := L_2(-\infty, \infty) = L_2(R)$ and let M denote the "rule" of multiplying by the independent variable, that is, $(Mf)(x) = xf(x)$. Consider the operator T defined by

$$T{:}H \supset D(T) \to H$$

$$Tf = Mf, \quad f \in D(T)$$

$$D(T) := \{f \in H{:}Mf \in H\}$$

Let

$$f_n(x) = \begin{cases} 1, & n \leq x < n+1 \\ 0, & \text{elsewhere} \end{cases}$$

Then clearly $\|f_n\| = 1$, where $\|\cdot\|$ denotes the usual norm in $L_2(\mathbf{R})$. Furthermore we have

$$\|Tf_n\|^2 = \int_{\mathrm{R}} |xf(x)|^2\, dx = \int_n^{n+1} x^2 dx > n^2$$

The two results imply that

$$\frac{\|Tf_n\|}{\|f_n\|} > n$$

Consequently, since we can choose n as large as we please, it follows that T is unbounded on H.

Example 4.80. Let $H = L_2(\mathbf{R})$ and consider the operator T defined by

$$T:H \supset D(T) \to H$$

$$Tf = i\frac{df}{dx} = if', \quad f \in D(T)$$

$$D(T) := \{f \in H : if' \in H\}$$

Further, let T be an extension of the operator T_0 defined to be

$$T_0 = T|_S$$

where $S = D(T) \cap L_2[0, 1]$ and $L_2[0, 1]$ is a subspace of $L_2(\mathbf{R})$. Consequently, if T_0 is unbounded then so is T. To show that T_0 is indeed unbounded consider the sequence, $\{fn\}$, defined by

$$f_n(x) = \begin{cases} 1 - nx, & 0 \leq x \leq \dfrac{1}{n} \\ 0, & \dfrac{1}{n} < x \leq 1 \end{cases}$$

The derivative of this function is

$$f_n'(x) = \begin{cases} -n, & 0 < x < \dfrac{1}{n} \\ 0, & \dfrac{1}{n} < x < 1 \end{cases}$$

Straightforward calculation yields

$$\|f_n\|^2 = \int_0^1 |f_n(x)|^2 \, dx = \frac{1}{3n}$$

and

$$\|T_0 f_n\|^2 = \int_0^1 |f_n'(x)|^2 \, dx = n$$

Hence

$$\frac{\|T_0 f_n\|}{\|f_n\|} = n\sqrt{3} > n$$

It follows that T_0, and hence T, is unbounded on H.

We shall return to these examples in later chapters.

References and Further Reading

[1] N.I. Akheiser and L.M. Glazman: *Theory of Linear Operators in Hilbert Space.* Pitman-Longman, London, 1981.

[2] J. Arsac: *Fourier Transforms and the Theory of Distributions*, Prentice Hall, New York, 1966.

[3] J.W. Dettman: *Mathematical Methods in Physics and Engineering*, McGraw-Hill, New York, 1962.

[4] G. Helmberg: *Introduction to Spectral Theory in Hilbert Space.* Elsevier, New York, 1969.

[5] E. Kreyszig: *Introductory Functional Analysis with Applications*, Wiley, NewYork, 1978.

[6] F. Riesz and B. Sz-Nagy: *Functional Analysis*, Ungar, New York, 1955.

[7] G.F. Roach: *Greens Functions* (2nd Edn), Cambridge Univ. Press, London, 1970/1982.

[8] G.F. Roach: *An Introduction to Linear and Nonlinear Scattering Theory*, Pitman Monographs and Surveys in Pure and Applied Mathematics, Vol 78, Longman, Essex, 1995.

[9] W. Rudin: *Principles of Mathematical Analysis* (3rd Edn), McGraw-Hill, NewYork, 1976.

[10] I.N. Sneddon: *Fourier Transforms*, McGraw-Hill, New York, 1951.

[11] E.C. Titchmarsh: *Introduction to the Theory of Fourier Integrals*, Oxford Univ. Press, 1937.

5

Two Important Techniques

5.1 Introduction

In this chapter we outline two analytical techniques which are frequently used when discussing scattering problems. The first is centred on spectral theory and how it contributes to the definition and constructive solution of scattering problems. The second uses results from the theory of semigroups to settle questions of existence and uniqueness of solutions to scattering problems.

As in previous chapters the majority of results are simply stated. Proofs are included when it is felt that they might be particularly useful in applications. Full details can be found in the References cited either in the text or in the Commentary.

5.2 Spectral Decomposition Methods

Spectral theory provides mechanisms for decomposing quite complicated problems into a number of simpler problems with properties which are more manageable. We shall illustrate how this can be done in the following sections. We consider a typical, abstract problem first when the underlying space is finite dimensional and then when the space is infinite dimensional. In both cases we will work mainly in a Hilbert space setting and assume that the operator which characterises the problem is self-adjoint. The discussion of these two cases leads quite naturally to a statement of the celebrated spectral theorem. We shall prove this theorem for bounded, self-adjoint operators. The proof for more general operators is discussed in the Commentary.

5.2.1 Basic Concepts

In this chapter we will be concerned with abstract equations having the typical form

$$(A - \lambda I)u = f \tag{5.1}$$

To be more precise let X be a complex, normed linear space and let

$$A: X \supseteq D(A) \to X$$

be a linear operator. In (5.1) I is the identity operator on X and $\lambda \in \mathbf{C}$. We assume that $f \in X$ is a given data element. The aim is to solve (5.1) for the unknown quantity $u \in X$.

Solutions of (5.1) can be written in the form

$$u = (A - \lambda I)^{-1} f =: R_\lambda f \tag{5.2}$$

where $R_\lambda \equiv R_\lambda(A) = (A - \lambda I)^{-1}$ is known as the **resolvent** (operator) of A. Quite how useful the representation (5.2) may be depends crucially on the nature of the resolvent, R_λ. This observation leads naturally to the following notions.

Definition 5.1. A **regular value** of A is a complex number λ such that

(i) $R_\lambda(A)$ exists

(ii) $R_\lambda(A)$ is bounded

(iii) $R_\lambda(A)$ is defined on a dense subset of X.

The set of all regular values of A, denoted $\rho(A)$, is called the **resolvent set** of A.

Definition 5.2. The spectrum of A, denoted $\sigma(A)$, is the complement in the complex plane of the resolvent set $\rho(A)$, that is

$$\sigma(A) = \mathbf{C} \backslash \rho(A)$$

The spectrum of A is partitioned by the following disjoint sets.

(i) The **point spectrum** of A, denoted $\sigma_p(A)$, consists of all those $\lambda \in \mathbf{C}$ such that $R_\lambda(A)$ does *not* exist.

(ii) The **continuous spectrum** of A, denoted $\sigma_c(A)$, consists of all those $\lambda \in \mathbf{C}$ such that $R_\lambda(A)$ exists as an unbounded operator and is defined on a dense subset of X.

(iii) The **residual spectrum** of A, denoted $\sigma_r(A)$ consists of all these $\lambda \in \mathbf{C}$ such that $R_\lambda(A)$ exists as either a bounded or an unbounded operator but in either case is *not* defined on a dense subset of X.

The spectrum of A is the union of these three disjoint sets

$$\sigma(A) = \sigma_p(A) \cup \sigma_c(A) \cup \sigma_R(A)$$

and any $\lambda \in \sigma(A)$ is referred to as a **spectral value** of A.

Before continuing we recall the following properties of linear operators on Banach spaces. The proofs of these various results can be found in the standard texts cited in the Commentary.

Theorem 5.3. *Let X, Y be Banach spaces and let* $A: X \supseteq D(A) \to R(A) \subseteq Y$ *denote a linear operator. Then*

(i) *The* ***inverse operator*** $A^{-1}: R(A) \to D(A)$ *exists if and only if* $Au = \theta_Y$ *implies* $u = \theta_X$ *where* θ_X, θ_Y *are the zero elements in X and Y respectively.*

(ii) *If* A^{-1} *exists then it is a linear operator.*

(iii) *If* $\dim D(A) = n < \infty$ *and* A^{-1} *exists then* $\dim R(A) = \dim D(A)$.

Theorem 5.4. *If a Banach space X is finite dimensional then every linear operator on X is bounded.*

These last two results combine to indicate that in the finite dimensional case

$$\sigma_c(A) = \sigma_r(A) = \phi$$

We thus see that the spectrum of a linear operator on a finite dimensional space consists only of the point spectrum. In this case the operator is said to have a **pure point spectrum**.

The next few results are particularly useful in applications.

Theorem 5.5. *The resolvent set,* $\rho(A)$, *of a bounded linear operator on a Banach space X is an open set. Hence, the spectrum,* $\sigma_c(A)$, *is a closed set.*

Theorem 5.6. *Let X be a Banach space and A a bounded linear operator on X. For all* $\lambda_0 \in \rho(A)$ *the resolvent operator,* $R_\lambda(A)$, *has the representation*

$$R_\lambda(A) = \sum_{k=0}^{\infty} (\lambda - \lambda_0)^k R_{\lambda_0}^{k+1}(A) \tag{5.3}$$

The series is absolutely convergent for every λ *in the open disc given by*

$$|\lambda - \lambda_0| < \|R_{\lambda_0}(A)\|^{-1}$$

in the complex plane. This disc is a subset of $\rho(A)$.

Theorem 5.7. *The spectrum,* $\sigma(A)$, *of a bounded linear operator* $A: X \to X$ *on a Banach space X is compact and lies in the disc given by* $|\lambda| \le \|A\|$. *Hence the resolvent set of A,* $\rho(A)$, *is not empty.*

Definition 5.8. Let X be a Banach space and $A: X \supseteq D(A) \to X$ a linear operator. If $(A - \lambda I)u = \theta$ for some **non-trivial** $u \in D(A)$ then u is an **eigenvector** of A with associated **eigenvalue** λ.

For the remainder of this chapter we confine attention to linear operators on a complex separable Hilbert space H.

The set M_λ consisting of the zero element in H and all eigenvectors of A corresponding to the eigenvalue λ is called the **eigenspace** of A corresponding to the eigenvalue λ.

The eigenspace M_λ is, in fact, a subspace, that is a closed linear manifold, of H. That it is a linear manifold is clear since for any u_1, $u_2 \in M_\lambda$ and α. $\beta \in \mathbf{C}$ we have, by the linearity of A,

$$A(\alpha u_1 + \beta u_2) = \lambda(\alpha u_1 + \beta u_2)$$

To show that M_λ is a **closed** linear manifold let $\{u_k\} \subset M_\lambda$ be a sequence such that $u_k \to u$ as $k \to \infty$. Now consider two cases.

(i) A is a bounded operator.

A is bounded implies that A is continuous. Hence,

$$Au = A \lim u_k = \lim Au_k = \lim \lambda u_k = \lambda u$$

and thus $u \in M_\lambda$ and we conclude that M_λ is closed.

(ii) A is an unbounded, closed operator.

In this case, since $u_k \to u$ as $k \to \infty$ there will exist a w such that

$$w := \lim Au_k = \lim \lambda u_k = \lambda u$$

However, A is a closed operator, which implies that $u \in D(A)$ and $Au = w$. Hence $u \in M_\lambda$ and we can conclude that M_λ is closed. Hence M_λ is a subspace of H.

In this monograph we will be largely concerned with linear operators on a Hilbert space which are either self-adjoint or unitary. Some of the more important properties of such operators are contained in the following.

Theorem 5.9. *The eigenvalues of a self-adjoint operator are real.*

Proof. Let $A: H \to H$ be a bounded, self-adjoint operator and let $\lambda \in \sigma_p(A)$ with associated eigenvector u, then

$$\lambda(u, u) = (\lambda u, u) = (Au, u) = (u, Au) - \bar{\lambda}(u, u)$$

which because u is non-trivial, implies $\lambda = \bar{\lambda}$ and hence $\lambda \in \mathbf{R}$ and so $\sigma_p(A) \subset \mathbf{R}$. ∎

Theorem 5.10. *The eigenvalues of a unitary operator are complex numbers of modulus one.*

Proof. Let $U: H \to H$ be a bounded, unitary operator and let $\mu \in \sigma_p(U)$ with associated eigenvector w. Then

$$(w, w) = (Uw, Uw) = (\mu w, \mu w) = \mu\bar{\mu}(w, w)$$

which, since w is non-trivial, implies $|\mu|^2 = 1$. ■

Theorem 5.11. *The eigenvectors of either a self-adjoint or a unitary operator corresponding to different eigenvalues are orthogonal.*

Proof. Let $A: H \to H$ be a self-adjoint operator and let $\lambda_1, \lambda_2 \in \sigma_p(A)$, $\lambda_1 \neq \lambda_2$, have associated eigenvector u_1, u_2 respectively. Then

$$\begin{aligned}(\lambda_1 - \lambda_2)(u_1, u_2) &= (\lambda_1 u_1, u_2) - (u_1, \lambda_2 u_2) \\ &= (Au_1, u_2) - (u_1, Au_2) \\ &= (Au_1, u_2) - (Au_1, u_2) = 0\end{aligned}$$

which implies $(u_1, u_2) = 0$ because $\lambda_1 \neq \lambda_2$.

Let $U: H \to H$ be a unitary operator and let $\mu_1, \mu_2 \in \sigma_p(A)$, $\mu_1 \neq \mu_2$ have associated eigenvectors w_1, w_2 respectively. Then

$$\mu_1 \mu_2 (w_1, w_2) = (\mu_1 w_1, \mu_2 w_2) = (Uw_1, Uw_2) = (w_1, w_2)$$

the last equality following from the defining property of a unitary operator (see Definition 4.51). Hence we conclude that $(w_1, w_2) = 0$ since $\mu_1 \bar{\mu}_2 \neq 1$. ■

In later chapters we will have occasion to make use of a result of the following form.

Theorem 5.12. *Let H be a complex, separable Hilbert space and let $A, B: H \to H$ be linear operators. If B is bounded and B^{-1} exists then A and BAB^{-1} have the same eigenvalues.*

Proof. If λ is an eigenvalue of A then there exists a *non-trivial* $\varphi \in H$ such that $A\varphi = \lambda\varphi$.

If B^{-1} exists then B must be a 1-1 onto operator (Theorem 4.47). Consequently, $B\varphi$ cannot be zero for all non-trivial φ. Hence

$$BAB^{-1}B\varphi = BA\varphi = B\lambda\varphi = \lambda B\varphi$$

which implies that λ is an eigenvalue of the operator BAB^{-1} with associated eigenvector $B\varphi$.

Conversely, let μ be an eigenvalue of BAB^{-1} then there exists a non-trivial $\psi \in H$ such that $BAB^{-1}\psi = \mu\psi$. Consequently

$$B^{-1}BAB^{-1}\psi = \mu B^{-1}\psi$$

which implies that μ is an eigenvalue of BAB^{-1} with associated eigenvector $B^{-1}\psi$. ■

For self-adjoint and unitary operators on a complex, separable Hilbert space H eigenvectors corresponding to different eigenvalues are orthogonal (Theorem

5.11). Consequently, the eigenspace corresponding to different eigenvalues are orthogonal subspaces of H. This in turn implies that the operators act on the direct sum of the eigenspaces like a diagonal matrix. To see this recall that if λ is an eigenvalue of the linear operator $A:H \to H$ then the associated eigenspace M_λ is defined as

$$M_\lambda = \{\theta \neq \psi \in D(A): A\psi = \lambda\psi\}$$

More fully, let $A:H \to H$ be either a self-adjoint or a unitary operator and let $\lambda_1, \lambda_2, \ldots, \lambda_k, \ldots$ denote its different eigenvalues. For each λ we will write M_k to denote the eigenspace corresponding to λ_k. An orthonormal basis for M_k will be denoted by $\{\varphi_s^k\}_s$.

We remark that whilst the number of eigenvalues may be either finite or infinite nevertheless they are always countable. If this were not so then there would be an uncountable number of different eigenvalues with an associated uncountable number of orthonormal basis vectors. This is impossible in a separable Hilbert space.

We also remark that the dimension of M_k, that is the number of basis elements φ_s^k for M_k, may be either finite or infinite and, furthermore, may be different for different values of k.

Since the eigenvectors for different eigenvalues of A are orthogonal the set of all eigenvectors φ_s^k for different k is orthonormal, that is

$$(\varphi_r^k, \varphi_s^m) = \delta_{rs}\delta_{km}$$

where $(\cdot, \cdot)$ denotes the inner product in H and

$$\delta_{rs} = \begin{cases} 1, & r = s \\ 0, & r \neq s \end{cases}$$

is the Kronecker delta.

To proceed we need the following concept.

Definition 5.13. Let E be a subset of a Hilbert space H and let D denote the set of all **finite** linear combinations of elements of E. The closure of D (in the topology of H) **generates** a subspace $G \subset H$. The subspace G is said to be **spanned** by E.

Consider the subspace

$$H_p := \bigoplus_k M_k \tag{5.4}$$

which consists of all linear combinations of the form $\Sigma_{k,s} a_s^k \varphi_s^k$. We shall refer to H_p as the **point subspace** of H. It is the subspace spanned by all the eigenvectors of A. Evidently the set of vectors φ_s^k, $k, s = 1, 2, \ldots$ is an orthonormal basis for H_p. Thus we have

$$A\varphi_s^k = \lambda_k \varphi_s^k$$

and

$$(\varphi_r^m, A\varphi_s^k) = \lambda_k \delta_{mk} \delta_{rs}$$

Thus on H_p the operator A acts like a diagonal matrix. The off-diagonal terms are all zero whilst the diagonal terms are eigenvalues of A.

Summarising the above we see that for any $\psi \in H_p$ there are scalars a_s^k such that

$$\psi = \sum_{k,s} a_s^k \varphi_s^k \tag{5.5}$$

$$A\psi = \sum_{k,s} \lambda_k a_s^k \varphi_s^k \tag{5.6}$$

The results (5.5), (5.6) provide an example of a **spectral representation** (**decomposition**) of the operator A.

We notice two things.

(i) The spectral representation(5.5), (5.6) is only valid on H_p. For those operators that have a spectrum with more components than just eigenvalues (that is $\sigma_c(A)$ and $\sigma_R(A)$ are not necessarily empty) then (5.5), (5.6) are inadequate; more terms are required.

(ii) On finite dimensional Hilbert spaces the spectrum of a linear operator is a pure point spectrum (Theorem 5.4) in which case (5.5), (5.6) provide a perfectly adequate spectral representation.

5.2.2 Spectral Decompositions

In the introduction to this chapter we said that one of the main reasons for introducing and using spectral theory was that it could provide mechanisms for decomposing quite complicated operators into simpler, more manageable components. In practice a full demonstration of this will involve working through the following stages.

- Determine a characterisation of a given physical problem in terms of an operator $A : H \to H$ where H denotes a complex, separable Hilbert space.
- Determine $\sigma(A)$, the spectrum of A, as a subset of the complex plane **C**.
- Provide a decomposition of **C** into components intimately connected with the nature of $\sigma(A)$.
- Provide a decomposition of H into components, the so-called **spectral components** of H, which are intimately connected with the nature of A and $\sigma(A)$.

- Provide an interpretation of A when it acts on the various spectral components of H. This will introduce the so-called (**spectral**) **parts** of A which are often more manageable than A itself.
- Show how results obtained when dealing with just the parts of A can be combined to provide meaningful and practical results for problems centred on A itself.

5.2.3 Spectral Decompositions on Finite Dimensional Spaces

The only spectral values of operators acting on a finite dimensional space are eigenvalues.

Theorem 5.14. *On a finite dimensional, complex Hilbert space the eigenvectors of either a self-adjoint or a unitary operator span the space.*

Proof. Let H be a finite dimensional, complex Hilbert space and $A:H \to H$ a linear operator which is either self-adjoint or unitary. Let M denote the subspace spanned by the eigenvectors of A and let $P:H \to M$ be the projection operator onto M.

Since linear operators on an n-dimensional space ($n < \infty$) can always be represented in the form of an $n \times n$ matrix the results of matrix algebra indicate that a linear operator on a finite dimensional complex Hilbert space has at least one eigenvalue.

Suppose $M \neq H$ and consider the operator $A(I - P)$ on $M^{\perp}$ (which is clearly finite dimensional). There there must exist a scalar λ and a non-trivial element $v \in M^{\perp}$ such that

$$A(I - P)v = \lambda v$$

Consequently, since $P:H \to M$ implies $Pv = \theta$, $v \in M^{\perp}$ (recall the projection theorem), we have

$$Av = APv + A(I - P)v = \lambda v$$

which implies that v is an eigenvector of A and this contradicts the assumption that M is not the whole space. ∎

This last theorem means that every self-adjoint or unitary operator, A, on an n-dimensional space ($n < \infty$) provides a basis for the space consisting entirely of orthonormal eigenvectors of A. Consequently, let

(i) $\lambda_1, \lambda_2, \ldots, \lambda_m$, denote the different eigenvalues of A

(ii) M_k denote the eigenspace corresponding to the eigenvalue λ_k, $k = 1, 2, \ldots, m$

(iii) $\{\varphi_s^k\}_s$ be an orthonormal basis for M_k, $k = 1, 2, \ldots, m$ where $s = 1, 2, \ldots, s(k)$ and $s(k)$ is a positive integer depending on k.

Then

$$A\varphi_s^k = \lambda_k \varphi_s^k, \quad s = 1, 2, \ldots, s(k), \quad k = 1, 2, \ldots, m$$

and we can conclude that $\{\varphi_s^k\}_s$ is an orthonormal basis for the whole space. The total number of the eigenvectors φ_s^k is n and $m \leqslant n$. The number of orthonormal eigenvectors φ_s^k associated with the eigenvalue λ_k, namely $s(k)$, indicates the dimension of M_k, denoted dim M_k, and, equivalently, is referred to as the **multiplicity** of λ_k. We remark that dim M_k may be different for different values of k.

A closer look at (5.5) suggests that we define the operator

$$P_k : H \to M_k, \quad k = 1, 2, \ldots, m \tag{5.7}$$

$$P_k : \psi \to P_k\psi = \sum_{s=1}^{s(k)} a_s^k \varphi_s^k, \quad \psi \in H \tag{5.8}$$

The representation (5.5) can now be written

$$\psi = \sum_{k=1}^{m} P_k \psi \tag{5.9}$$

which in turn implies the completeness property

$$\sum_{k=1}^{m} P_k = I \tag{5.10}$$

The operator P_k is a projection onto the eigenspace M_k. Since eigenspaces corresponding to different eigenvalues of a self-adjoint or unitary operator are orthogonal it follows that the projections P_k, $k = 1, 2, \ldots, m$ are orthogonal in the sense that

$$P_k P_m = \delta_{km} P_k \tag{5.11}$$

Furthermore, (5.6) can now be written

$$A\psi = \sum_{k,s} \lambda_k a_s^k \varphi_s^k = \sum_{k=1}^{m} \lambda_k P_k \psi$$

which implies

$$A = \sum_{k=1}^{m} \lambda_k P_k \tag{5.12}$$

This is a representation of the operator A in terms of projection operators. It illustrates how the spectrum of A can be used to provide a representation of A in terms of simpler operators. This use of projections seems quite a natural way to obtain the required spectral decompositions.

Unfortunately this particular approach does not generalise to an infinite dimensional space setting. We now describe a slightly different way of obtaining a spectral decomposition of A which *does* generalise to an infinite dimensional space setting where the spectrum of a linear operator can be very much more complicated then just a collection of eigenvalues of the type we have so far been considering.

We consider a self-adjoint operator on a finite dimensional Hilbert space and order its distinct eigenvalues in the form

$$\lambda_1 < \lambda_2 < \ldots < \lambda_{m-1} < \lambda_m$$

For each $\lambda \in \mathbf{R}$ we define an operator-valued function of λ by

$$E_\lambda = \begin{cases} 0, & \lambda < \lambda_1 \\ \sum_{k=1}^{r} P_k, & \lambda_r \leq \lambda < \lambda_{r+1} \\ I, & \lambda \geq \lambda_m \end{cases} \tag{5.13}$$

which we write, more compactly, in the form

$$E_\lambda = \sum_{\lambda_k \leq \lambda} P_k \tag{5.14}$$

Clearly, E_λ is a projection operator onto the subspace of H spanned by all the eigenvectors associated with eigenvalues $\lambda_k = \lambda$.

It follows from (5.13) that

$$E_\mu E_\lambda = E_\lambda E_\mu = E_\mu, \quad \mu \leqslant \lambda \tag{5.15}$$

When (5.15) holds we write

$$E_\mu = E_\lambda, \quad \mu = \lambda \tag{5.16}$$

These various properties indicate that E_λ changes from the zero operator in H to the identity operator on H as λ runs through the spectrum, that is eigenvalues, of A. Furthermore, we notice that E_λ changes by P_k when λ reaches λ_k. With this in mind we define

$$dE_\lambda := E_\lambda - E_{\lambda-\varepsilon}, \quad \varepsilon > 0 \tag{5.17}$$

It now follows that if $\varepsilon > 0$ is small enough to ensure that there is no λ_k such that

$$\lambda - \varepsilon < \lambda_k < \lambda$$

then $dE_\lambda = 0$. Furthermore, if $\lambda = \lambda_k$ then

$$dE_\lambda = dE_{\lambda_k} = P_k \tag{5.18}$$

We are now in the position to indicate a particularly important useful representation of self-adjoint operators.

First, we recall the definition of the Riemann–Stieljes integral of a function g with respect to a function f, namely

$$\int_a^b g(x)df(x) = \lim_{n\to\infty} \sum_{j=1}^{n} g(x_j)|f(x_j) - f(x_{j-1})| \tag{5.19}$$

where $a = x_0 < x_1 < \ldots < x_n = b$ is a partition of the range of integration.

With (5.18) in mind we see that we have

$$\int_{-\infty}^{\infty} dE_\lambda = \lim_{n\to\infty} \sum_{j=1}^{n} 1(E_{x_j} - E_{x_{j-1}})$$

where $\lambda_1 \leq x_0 < x_1 \ldots < x_n = \lambda_m$. Consequently, bearing in mind (5.10) and (5.18) we obtain

$$\int_{-\infty}^{\infty} dE_\lambda = I \tag{5.20}$$

Furthermore with (5.12) in mind and arguing as above we have

$$A - \sum_{k=1}^{m} \lambda_k P_k - \lim_{n\to\infty} \sum_{j=1}^{n} \lambda_k (E_{x_j} - E_{x_{j-1}})$$

which implies

$$A = \int_{-\infty}^{\infty} \lambda dE_\lambda \tag{5.21}$$

The expression (5.21) is the spectral representation of the self-adjoint operator A, which has eigenvalues $\lambda_1 < \lambda_2 < \ldots < \lambda_m$ on an n-dimensional complex Hilbert space H.

For arbitrary $\varphi, \psi \in H$ in the n-dimensional space H the above results lead to

$$(\varphi, \psi) = \int_{-\infty}^{\infty} d(E_\lambda \varphi, \psi) = \int_{-\infty}^{\infty} dw(\lambda) \tag{5.22}$$

$$(A\varphi, \psi) = \int_{-\infty}^{\infty} \lambda d(E_\lambda \varphi, \psi) = \int_{-\infty}^{\infty} \lambda dw(\lambda) \tag{5.23}$$

where $w(\lambda) := (E_\lambda \varphi, \psi)$ defines a **complex-valued function of** λ which changes by $(P_k \varphi, \psi)$ at $\lambda = \lambda_k$.

For a unitary operator $U: H \to H$ which has eigenvalues $\lambda_k = \exp(i\theta_k)$ ordered in the form

$$0 < \theta_1 < \theta_2 < \ldots < \theta_m = 2\pi$$

using similar arguments to those used above we obtain the spectral representation

$$U = \int_0^{2\pi} \exp(i\lambda) dE_\lambda \tag{5.24}$$

which leads to the expression

$$(U\varphi, \psi) = \int_0^{2\pi} \exp(i\lambda) d(E_\lambda \varphi, \psi) \tag{5.25}$$

Similar calculations are possible for compact linear operators on an infinite dimensional space. (See Definition 4.54 and Example 4.55.) For the sake of illustration consider here only a positive, compact operator. All its eigenvalues λ_k are non-negative and we denote them in the form

$$\lambda_1 > \lambda_2 > \ldots > 0$$

with possibly the inclusion of $\lambda_0 = 0$. We can then write

$$A = \sum_{k=1}^{\infty} \lambda_k P_k \tag{5.26}$$

where $\lambda_k \to 0$ as $k \to \infty$.

Denoting by P_0 the projection onto the null space M_0 we then have (compare (5.10))

$$\sum_{k=1}^{\infty} P_k + P_0 = I \tag{5.27}$$

Then as before (compare (5.13), (5.14))

$$E_\lambda = \begin{cases} 0, & \lambda < 0 \\ P_0 + \sum_{j=k}^{\infty} P_j, & \lambda_{k-1} \leq \lambda < \lambda_k \\ I, & \lambda \geq \lambda_1 \end{cases} \tag{5.28}$$

It is a straightforward matter to show, in a similar manner to that used above, that E_λ is a projection operator-valued function of λ. We can also conclude that, just as in the finite dimensional case when we obtained (5.22), (5.23), we again have results of the form

$$(\varphi, \psi) = \int_{-\infty}^{\infty} d(E_\lambda \varphi, \psi) = \int_{-\infty}^{\infty} dw(\lambda) \tag{5.29}$$

$$(A\varphi, \psi) = \int_{-\infty}^{\infty} \lambda d(E_\lambda \varphi, \psi) = \int_{-\infty}^{\infty} \lambda dw(\lambda) \tag{5.30}$$

which implies

$$A = \int_{-\infty}^{\infty} \lambda dE_\lambda \tag{5.31}$$

Consequently, we see that a self-adjoint operator on either a finite or an infinite dimensional space has, in the two special cases considered, an integral representation given by (5.21) and (5.31). In these cases the integral representations are really a means of re-expressing the diagonalisability property. However, for self-adjoint operators which do not belong to the two classes mentioned above this notion of diagonalisability is no longer meaningful. Nevertheless, it might be possible to express any self-adjoint operator in the integral form (5.31) provided the spectral family $\{E_\lambda\}$, $\lambda \in \sigma(A)$ is appropriately defined. This is the content of the celebrated spectral theorem which we will discuss later.

5.2.4 Reducing Subspaces

We introduce this concept in terms of bounded operators. For unbounded operators the following definition and two theorems require a more careful statement which properly takes into account the domains involved. We return to this aspect at the end of this subsection.

Definition 5.15. A **subspace** (closed linear manifold) $M \subseteq H$ reduces a bounded, linear operator $A: H \to H$ if

(i) $A\psi \in M$ for every $\psi \in M$

(ii) $A\varphi \in M$ for every $\varphi \in M^\perp$.

The following two theorems indicate the main properties of reducing subspaces.

Theorem 5.16. *Let*

(i) *H be a complex, separable Hilbert space and $M \subseteq H$ a subspace*
(ii) *$A: H \supseteq D(A) \to H$ be a linear operator*
(iii) *$P: H \to M$ be a projection operator onto M.*

The following statements are equivalent

(a) *M reduces A*
(b) *$PA = AP$*
(c) *$(I - P)A = A(I - P)$.*

Proof. It is obvious that (b) and (c) are equivalent. We shall show that (a) $\Rightarrow$ (b) and that ((b), (c)) $\Rightarrow$ (a).

For any element $\psi \in H$ we have, by the projection theorem,

$$\psi = u + v, \quad u \in M, \quad v \in M^{\perp}$$

Hence $A\psi = Au + Av$ and we conclude that (a)$\Rightarrow$(b) as follows.

(a) $\Rightarrow$ (b):

M reduces $A \Rightarrow Au \in M$ and $Av \in M^{\perp}$.

Therefore

$$PA\psi = Au = AP\psi$$

and hence (a) $\Rightarrow$ (b).

((b), (c)) $\Rightarrow$ (a):

If $u \in M$ and $PA = AP$ then $Pu = u$ and

$$Au = APu = P\,Au$$

Hence $Au \in M$ for all $u \in M$.

If $v \in M^{\perp}$ and $(I - P)A = A(I - P)$ then

$$(I - P)v = v$$

Hence

$$Av = A(I - P)v = (I - P)Av$$

and we conclude that $Av \in M^{\perp}$ for all $v \in M^{\perp}$. Hence ((b), (c)) $\Rightarrow$ (a) ■

This theorem indicates the important practical result

$$A = AP + A(I - P) \tag{5.32}$$

which implies that A is the sum of two parts, namely AP as an operator on M and $A(I - P)$ as an operator on $M^{\perp}$.

Theorem 5.17. *Let A be a bounded, linear operator which is either self-adjoint or unitary on a separable, complex Hilbert space H. Let H_p be the subspace spanned by the eigenvectors of A. Then H_p reduces A.*

The operators induced by A in H_p and $H_p^{\perp}$ are again self-adjoint or unitary.

Proof. (i) Assume A is self-adjoint.

Let $u \in H_p$, then since $Au = \lambda u$ for some $\lambda \in \mathbf{R}$ we can conclude that $Au \in H_p$.

If $v \in H_p^{\perp}$ then

$$(u, Av) = (Au, v) = 0 \quad \text{for any} \quad u \in H_p$$

Hence $Av \in H_p^{\perp}$ and we conclude that H_p reduces A.

Let $P: H \to H$ be a projection. Then there exist operators A_1 in H_p and A_2 in $H_p^{\perp}$ with domains

$$D(A_1) = PH = H_p$$

$$D(A_2) = (I - P)H = P^{\perp}H = H_p^{\perp}$$

such that (see (5.32))

$$AF = A_1(Pf) + A_2(P^{\perp}f), \quad f \in H$$

The operators A_1 and A_2 are the **operators induced** by A in H_p and $H_p^{\perp}$ respectively.

For f, $g \in H_p$ we obtain, recognising this last relation, Theorem 5.16, and the defining properties of the operators

$$(A_1 f, g) = (APf, g) = (PAf, g) = (f, APg) = (f, A_1 g)$$

Hence A_1 is self-adjoint on H_p. Similarly A_2 is self-adjoint on $H_p^{\perp}$.

(ii) Assume that A is unitary.

As before $u \in H_p$ implies $Au \in H_p$.

Let $v \in H_p^{\perp}$ and let w be an eigenvector of A with an associated eigenvalue λ. Then $Aw = \lambda w$ and

$$\lambda(w, Av) = (Aw, Av) = (w, v) = 0$$

the third equality following from the unitarity of A. (See Definition 4.51(ii).) This last result implies $(w, Av) = 0$ because $|\lambda| = 1$ (Theorem 5.10). Thus Av is orthogonal to every element in the orthonormal basis (of eigenvectors of A) for H_p. Therefore $Av \in H_p^\perp$ and hence H_p reduces A.

With A_1, A_2 defined as in part (i) we have for A unitary

$$\|f\| = \|Af\| = \|A_1 f\| \quad \text{for all} \quad f \in H_p$$

Therefore A_1 is an isometric linear operator on H_p (Definition 3.12). Similarly, A_2 is an isometric, linear operator on $H_p^\perp$. To show that A_1 is unitary it remains to show that A_1 maps H_p *onto* H_p. To this end, let $g \in H_p$ be given. Since A is unitary then A maps H onto H and there exists an element $f \in H$ such that $Af = g$.

Since $Af = g \in H_p$ and $A_1 Pf \in H_p$ and $A_2 P^\perp f \in H_p^\perp$ we conclude, from the decomposition of A given in part (i), written in the form $g = A_1(Pf) + A_2(P^\perp f)$ and using Theorem 4.56 that $A_2(P^\perp f) = \theta$ and $\|P^\perp f\| = \|A_2 P^\perp f\| = 0$. Consequently, $f \in H_p$ and

$$g = Af = A_1 f \in H_p$$

■

We shall refer to H_p as the **point subspace** of A. Further, we shall denote $H_p^\perp$, the orthogonal complement of H_p, by H_c, and refer to it as the **subspace of continuity** of A.

We see, from (5.32), that a self-adjoint or a unitary operator splits into two (smaller in some sense) parts. One part of the operator acts on the subspace spanned by eigenvectors and, as such, can be represented by a diagonal matrix of eigenvalues with respect to an orthonormal basis of eigenvectors of the given operator. The other part of the given operator acts on the orthogonal complement of the subspace spanned by the eigenvectors. The simplest form of this decomposition occurs when the eigenvectors of the given operators span the whole space. In this case the associated H_c will be empty.

Example 5.18. Let H be a separable, complex Hilbert space and $N \subseteq H$ a subspace. Let $P: H \to N$ be a projection of H onto N. Assume that there exists a non-trivial $\psi \in H$ such that

$$P\psi = \lambda\psi, \quad \lambda \in \mathbf{C}$$

Then

$$\lambda^2\psi = \lambda P\psi = P^2\psi = P\psi = \lambda\psi$$

which implies that $\lambda = 1$ or 0. We therefore conclude that a projection operator can only have two distinct eigenvalues, namely, $\lambda = 1$ or 0.

If $\psi \in N$ then $P\psi = \psi$ by virtue of the definition of the projection operator P. Also if $\varphi \in N^\perp$ then $P\varphi = \theta$. Therefore, recalling that a projection operator is self-adjoint and that eigenspaces of self-adjoint operators corresponding to different eigenvalues are orthogonal we can infer that

$N =$ eigenspace of P corresponding to the eigenvalue $\lambda = 1$

$N^{\perp} =$ eigenspace of P corresponding to the eigenvalue $\lambda = 0$.

A basis for the whole space is obtained by combining the basis for N and the basis for $N^{\perp}$.

The projection operator P can be represented as a diagonal matrix with respect to this basis. The diagonal elements are one for the basis vectors of N and zero for the basis elements of $N^{\perp}$.

In this example we have completely described a projection operator in terms of its eigenvalues and eigenvectors.

We would emphasise that throughout this subsection we have assumed that all the linear operators are bounded.

For more general operators we can begin the discussion by, as before, assuming a decomposition of a separable, complex Hilbert space H in the form

$$H = M + M^{\perp}$$

where M is a subspace of H. We shall denote by P the projection of H onto M.

Definition 5.19. A, possibly unbounded, operator A, is said to be **decomposed** according to $H = M + M^{\perp}$ if

$$PD(A) \subseteq D(A), \quad APD(A) \subseteq M \quad \text{and} \quad A(I - P)D(A) \in M^{\perp} \tag{5.33}$$

The results (5.33) imply that for any $f \in D(A)$ we have $Pf \in D(A)$ and $APf \in M$ and $A(I - P) \in M^{\perp}$. Hence

$$(I - P)APf = APf - PAPf = APf - APf = \theta$$

$$PA(\mathrm{I} - P)f = \theta$$

and we conclude $(I - P)APf - PA(I - P)f - \theta$. This leads to the conclusion that $APf = PAf$ for $f \in D(A)$. Thus we see that the condition (5.33) is equivalent to the condition that A commutes with the projection P, that is,

$$PA \subseteq AP \tag{5.34}$$

If one of the two equivalent conditions (5.33) and (5.34) is satisfied then the restriction of A to $M \cap D(A)$ can be considered as an operator in the Hilbert space M. This operator is called the **part** of A in H and is frequently denoted A/M.

Definition 5.20. If the operator A is symmetric then the operator A is said to be **reduced** by M if $PD(A) \subseteq D(A)$ and $APD(A) \subseteq M$.

A result involving the reduction of possibly unbounded operators is the following.

Theorem 5.21. *Let H be a separable, complex Hilbert space and $M \subseteq H$ a subspace. Let $A: H \to D(A) \supseteq H$ and let $P: H \to M$ be a projection.*

(i) *If A is a symmetric operator then A is reduced by M if and only if A and P commute.*

(ii) *If A is self-adjoint then A/M is also self-adjoint.*

The proof of this theorem is straightforward and can be found in the texts cited in the Commentary [5], [10].

5.2.5 Spectral Decompositions on Infinite Dimensional Spaces

Spectral studies on infinite dimensional spaces are more complicated than similar studies on finite dimensional spaces. For example, in such spaces there are self-adjoint and unitary operators which have no eigenvalues yet the spectrum of such operators consists of more than the point spectrum (see Definition 5.2). Nevertheless, it is still possible to obtain spectral decompositions in terms of projection operators which have an integral form similar to that already obtained in Subsection 5.2.3.

We shall assume, just as for the finite dimensional case, (see the remarks following Theorem 5.12 and (5.14)) that there exists a non-decreasing family of subspaces $\{M_\lambda\}$ of a complex, separable Hilbert space H. These subspaces depend on a real parameter $\lambda \in (-\infty, \infty)$ such that

(i) The intersection of all the M_λ is θ, the zero element in H.

(ii) The union of all the M_λ is a dense subset of H.

We now introduce a family of projection operators $\{E_\lambda\}$ associated with $\{M_\lambda\}$. First we recall that the family is said to be **non-decreasing** if $M_\mu \subseteq M_\lambda$ for $\mu < \lambda$. Now, bearing in mind (5.13) and the discussion which followed, the following definition is natural.

Definition 5.22. A family of projection operators $\{E_\lambda\}$ depending on the parameter λ is said to be a **spectral family** or a **resolution of the identity** if it has the following properties.

(i) $\{E_\lambda\}$ is non-decreasing in the sense that

$$E_\mu \leq E_\lambda \quad \text{for} \quad \mu < \lambda$$

Equivalently, we have

$$E_\lambda E_\mu = E_\mu E_\lambda = E_{\min(\lambda,\mu)} = E_\mu$$

(ii) If $\varepsilon > 0$ then for any element $\varphi \in H$ and scalar λ

$$E_{\lambda+\varepsilon}\,\varphi \to E_\lambda\varphi \quad \text{as} \quad \varepsilon \to 0$$

(iii) $E_\lambda\psi \to \theta$ as $\lambda \to -\infty$, $E_\lambda\psi \to \psi$ as $\lambda \to +\infty$.
Equivalently, we write

$$s-\lim_{\lambda\to-\infty} E_\lambda = \Theta, \quad s-\lim_{\lambda\to+\infty} E_\lambda = I$$

where Θ denotes the zero operator on H.

With every spectral family $\{E_\lambda\}$ we can associate a self-adjoint or unitary operator. This is a statement of the celebrated spectral theorem which can be quoted in the following form.

Theorem 5.23 (Spectral theorem). *Let H be a complex, separable Hilbert space.*

(i) *For each bounded, self-adjoint operator A on H there exists a unique spectral family $\{E_\lambda\}$ such that*

$$(A\psi,\varphi) = \int_{-\infty}^{\infty} \lambda d(E_\lambda\psi,\varphi) \quad \text{for all } \varphi,\psi \in H$$

Equivalently, we write

$$A = -\int_{-\infty}^{\infty} \lambda dE_\lambda \tag{5.35}$$

(ii) *For each unitary operator U on H there exists a unique spectral family $\{F_\lambda\}$ such that $F_\lambda = \Theta$ for $\lambda \leqslant 0$ and $F_\lambda = I$ for $\lambda \geq 2\pi$ such that*

$$(U\psi,\varphi) = \int_0^{2\pi} e^{i\lambda} d(F_\lambda\psi,\varphi) \quad \text{for all} \quad \varphi,\psi \in D(U) \subset H$$

Equivalently we write

$$U = \int_0^{2\pi} e^{i\lambda} dF_\lambda \tag{5.36}$$

As before, we refer to (5.35) and (5.36) as the **spectral decompositions** of A and U respectively.

The proof of the theorem is quite technical and lengthy. Full details can be found in the texts cited in the Commentary; in this connection we would particularly mention [1], [11], [10], [13]. In defence of this action we recall the sentiments expressed in the introductory chapter. This monograph is not a book on functional analysis or operator theory or spectral theory. Nevertheless, we shall need many results from these three fields. Consequently, we only include proofs when they are needed, either for clarification in the development of material or for their usefulness in applications.

Example 5.24. Let $H := L_2(0, 1)$ and define $A: H \to H$ by

$$(A\psi)(x) = x\psi(x), \quad \psi \in D(A) = H$$

It is an easy exercise to show that A is linear and self-adjoint on H.

Let $\{E_x\}$ denote a family of projections defined by

$$(E_x\psi)(x) = \begin{cases} \psi(z), & z \le x \\ 0, & z > x \end{cases}$$

The following are immediate.

(i) $E_xE_y = E_yE_x = E_x$, $x \le y$.
Equivalently, $E_x \le E_y$, $x \le y$.

(ii) $\|E_{x+\varepsilon}\psi - E_x\psi\|^2 = \int_x^{x+\varepsilon} |\psi(y)|^2 \, dy \to 0 \quad \text{as} \quad \varepsilon \to 0^+$

Hence $E_{x+\varepsilon} \to E_x$ *as* $\varepsilon \to 0^+$.

We assume further that $\psi(x)$ is zero for x outside the interval [0, 1]. In this case we have

$$Ex = \Theta, \quad x < 0, \quad E_x = I, \quad x > 1$$

Consequently, $\{E_x\}$ is a spectral family.

We obtain a spectral decomposition by noticing

$$\begin{aligned} \int_{-\infty}^{\infty} x d(E_x\psi, \varphi) &= \int_{-\infty}^{\infty} x d\left\{\int_0^1 (E_x\psi)(y)\overline{\varphi(y)} dy\right\} \\ &= \int_0^1 x d\left\{\int_0^x \psi(y)\overline{\varphi(y)} dy\right\} \\ &= \int_0^1 x\psi(x)\overline{\varphi(x)} \, dx \\ &= (A\psi, \varphi) \end{aligned}$$

Thus A has the spectral representation

$$A = \int_{-\infty}^{\infty} x dE_x \tag{5.37}$$

We also notice that

$$(E_x\psi, \varphi) - (E_{x-\varepsilon}\psi, \varphi) = \int_{x-\varepsilon}^{x} \psi(y)\overline{\varphi(y)} dy$$

and the right-hand side tends to zero as $\varepsilon \to 0^+$. This implies that in this case $(E_x\psi, \varphi)$, as a function of x, is also continuous from the left. Hence, the relation

$$w(x) = (E_x\psi, \varphi), \quad \varphi, \psi \in H$$

defines a continuous function of x.

Example 5.25. Let $A:H \supseteq D(A) \to H = L_2(\mathbf{R})$ be defined by

$$(A\psi)(x) = x\psi(x), \quad \psi \in D(A) = H$$

Define the spectral family as in Example 5.24. It is then readily shown that in this case A also has a spectral decomposition. However, the spectral family $\{E_x\}$ in this case has the properties that E_x increases over the whole range $-\infty < x < \infty$ with $Ex \to \Theta$ as $x \to -\infty$ and $E_x \to I$ as $x \to \infty$.

We see, from these two examples, that E_λ as a function of λ can increase continuously rather than by a series of jumps (steps) as was the case when working in a finite dimensional setting. This is because A, as defined in each example, has no discrete eigenvalues.

The spectral family associated with a self-adjoint operator on a Hilbert space can provide information about the spectrum of the operator in a relatively simple manner. Indeed, we have seen that in a finite dimensional setting the spectral family is discontinuous at the eigenvalues of the associated operator. This property carries over to an infinite dimensional space setting. In addition, information about the parts of the spectrum, other than the eigenvalue spectrum, which can exist in an infinite dimensional space setting can also be obtained. We collect here, simply as statements, a number of fundamental properties of spectral families.

Theorem 5.26. *Let H be a complex, separable Hilbert space and let $A:H \to H$ be a bounded, linear, self-adjoint operator with an associated spectral family $\{E_\lambda\}$ and spectral decomposition $A = \int_{-\infty}^{\infty} \lambda dE_\lambda$. Then E_λ has a discontinuity at $\lambda = \mu$ if and only if μ is an eigenvalue of A.*

Let $P_\mu : H \to M_\mu$ denote the projection operator onto M_μ the subspace spanned by the eigenvectors of A associated with the eigenvalue μ. Then

(i) $E_\lambda P_\mu = \begin{cases} P_\mu. & \lambda \geq \mu \\ 0. & \lambda < \mu \end{cases}$

(ii) *for* $\varepsilon > 0$

$$E_\mu\psi - E_{\mu-\varepsilon}\psi \to P_\mu\psi$$

as $\varepsilon \to 0$ and for any $\psi \in H$.

For unitary operators the corresponding result is as follows.

Theorem 5.27. *Let H be a complex, separable Hilbert space and let $U:H \to H$ be a linear, unitary operator with an associated spectral family $\{F_\lambda\}$ and spectral decomposition $U = \int_0^{2\pi} e^{i\lambda} dF_\lambda$. Then F_λ has a discontinuity at $\lambda = \mu$ if and only if $e^{i\mu}$ is an eigenvalue of U.*

Let $P_\mu: H \to M_\mu$ denote the projection operator onto M_μ the subspace spanned by the eigenvectors of U associated with the eigenvalue $e^{i\mu}$. Then

(i) $F_\lambda P_\mu = \begin{cases} P_\mu, & \lambda \geq \mu \\ 0, & \lambda < \mu \end{cases}$

(ii) *for* $\varepsilon > 0$

$$F_\mu \psi - F_{\mu-\varepsilon}\psi \to P_\mu \psi$$

as $\varepsilon \to 0$ *and for any* $\psi \in H$.

These two theorems indicate that the jumps in the values of E_λ and F_λ are the same as in the finite dimensional case. However, in the infinite dimensional space setting a continuous increase in E_λ and F_λ is possible (see also Examples 5.24 and 5.25).

The resolvent set of a self-adjoint operator can also be characterised in terms of the associated spectral family. Specifically, the following result can be obtained.

Theorem 5.28. *Let H, A and $\{E_\lambda\}$ be as in Theorem 5.26. A real number μ belongs to $\rho(A)$, the resolvent set of A, if and only if there exists a constant $c > 0$ such that $\{E_\lambda\}$ is constant on the interval $[\mu - c, \mu + c]$.*

The importance of this theorem is that it indicates that $\mu \in \sigma(A)$ if and only if the spectral family $\{E_\lambda\}$ is *not* constant in any neighbourhood of $\mu \in \mathbf{R}$.

We can say more about the spectrum of a self-adjoint operator. First, we need the following important property of self-adjoint operators.

Theorem 5.29. *Let H be a complex, separable Hilbert space and let $A: H \to H$ be a linear, self-adjoint operator. The residual spectrum of A, denoted $\sigma_r(A)$, is empty.*

Proof. Assume $\sigma_r(A)$ is non-empty. By definition, if $\lambda \in \sigma_r(A)$ then the resolvent operator $R_\lambda(A) := (A - \lambda I)^{-1}$ exists as either a bounded or an unbounded operator on $\overline{D(R_\lambda(A))} \neq H$. This implies, by the projection theorem, that there exists a non-trivial element $\varphi \in H$ which is orthogonal to $D((A - \lambda I)^{-1}) = D(R_\lambda(A))$. However, $D(R_\lambda(A))$ is the range of $(A - \lambda I)$. Hence there exists a non-trivial element φ in H such that for all $\psi \in D((A - \lambda I)) = D(A)$ we have

$$0 = ((A - \lambda I)\psi, \varphi) = (A\psi, \varphi) - \lambda(\psi, \varphi) = (\psi, \mathbf{A}^*\varphi) - (\psi, \bar{\lambda}\varphi)$$

which implies that $A^*\varphi = \bar{\lambda}\varphi$, that is $\bar{\lambda} \in \sigma_p(A^*)$. Consequently, since A is self-adjoint we have that

$$\lambda \in \sigma_r(A) \Rightarrow \bar{\lambda} \in \sigma_p(A^*) \Rightarrow \bar{\lambda} \in \sigma_p(A) \Rightarrow \lambda \in \sigma_p(A)$$

This is a contradiction and we can conclude that $\sigma_r(A) = \phi$. ■

This theorem indicates that for a self-adjoint operator A the spectrum of A, denoted $\sigma(A)$, decomposes in the form

$$\sigma(A) = \sigma_p(A) \cup \sigma_c(A)$$

Since points in $\sigma_p(A)$ correspond to discontinuities in $\{E_\lambda\}$, the spectral family of A, the following result follows immediately.

Theorem 5.30. *Let A and $\{E_\lambda\}$ be defined as in Theorem 5.26. A real number μ belongs to $\sigma_c(A)$, the continuous spectrum of A, if and only if $\{E_\lambda\}$ is continuous at μ and is not constant in any neighbourhood of μ.*

The spectral theorem (Theorem 5.23) indicates that a spectral family $\{E_\lambda\}$ determines a self-adjoint operator A according to the relation

$$A = \int_{-\infty}^{\infty} \lambda dE_\lambda$$

Clearly, different spectral families lead to different self-adjoint operators.

In applications we are particularly interested in determining the spectral family $\{E_\lambda\}$ associated with a given self-adjoint operator. This can be achieved by means of the celebrated Stone's formula which relates the spectral family of A and the resolvent of A.

Theorem 5.31 (Stone's formula). *Let H be a complex, separable Hilbert space, and let $A: H \to H$ be a self-adjoint operator. The spectral family, $\{E_\lambda\}$, associated with A and $(A - \lambda I)^{-1}$, the resolvent of A, are related as follows.*

For all $f, g \in H$ and for all $a, b \in \mathbf{R}$

$$([E_b - E_a]f, g) = \lim_{\delta \downarrow 0} \lim_{\varepsilon \downarrow 0} \frac{1}{2\pi i} \int_{a+\delta}^{b+\delta} ([R(t+i\varepsilon) - R(t-i\varepsilon)]f, g)dt \tag{5.38}$$

where $R(t \pm i\varepsilon) = [A - (t + i\varepsilon)I]^{-1}$.

The manner in which this formula is used to provide the required spectral representations is indicated is well illustrated in the References cited in the text and in the Commentary; see, in particular [13].

5.2.6 Functions of an Operator

The spectral theorem (Theorem 5.23) tells us that a self-adjoint operator A has a spectral representation in terms of its associated spectral family $\{E_\lambda\}$ given by (5.35). We will want to form functions of such an operator in later chapters. If $A \in B(H)$ where H, as usual, denotes a complex, separable Hilbert space, then it is easy to see that a natural definition for $\exp(A)$, for example, is obtained by using the familiar expansion for the exponential function, namely

$$\exp(A) = \sum_{n=0}^{\infty} \frac{A^n}{n!}$$

This relation is well defined since the right-hand side of the exp ression converges in the operator norm. However, there are many more complicated operators than the one we have just considered which arise in applications. Nevertheless, the spectral theorem allows us to form a large class of functions of a self-adjoint operator A. This we do in the following manner.

Let φ be a complex-valued, continuous function of the realvariable λ. We can define an operator $\varphi(A)$, where A is a self-adjoint operator on a Hilbert space H, by writing

$$D(\varphi(A)) = \left\{ f \in H : \int_{-\infty}^{\infty} |\varphi(\lambda)|^2 \, d(E_\lambda f, f) < \infty \right\} \tag{5.39}$$

and for $f \in D(\varphi(A))$

$$(\varphi(A)f, g) = \int_{-\infty}^{\infty} \varphi(\lambda) d(E_\lambda f, g), \quad g \in H \tag{5.40}$$

It then follows, formally at least, that we can write

$$\varphi(A) = \int_{-\infty}^{\infty} \varphi(\lambda) dE_\lambda \tag{5.41}$$

For $\varphi(\lambda) = \lambda$ we recover the operator A as expected (see (5.35)).

Some of the basic properties of $\varphi(A)$ are contained in the following exercise.

Exercise 5.32. Let H be a complex, separable Hilbert space and $A : H \to H$ a bounded, self-adjoint operator. Let φ, φ_1, φ_2, be complex-valued, continuous functions defined on the support of $\{E_\lambda\}$. Then the following results are valid.

(i) $\varphi(A)^* = \overline{\varphi}(A)$ where $\overline{\varphi}(\lambda) = \overline{\varphi(\lambda)}$.

(ii) If $\varphi(\lambda) = \varphi_1(\lambda)\varphi_2(\lambda)$ then $\varphi(A) = \varphi_1(A)\varphi_2(A)$.

(iii) If $\varphi(\lambda) = a_1\varphi_1(\lambda) + a_2\varphi_2(\lambda)$ then $\varphi(A) = a_1\varphi_1(A) + a_2\varphi_2(A)$.

(iv) $\varphi(A)$ is normal, that is $\varphi(A)^*\varphi(A) = \varphi(A)\varphi(A)^*$.
(v) $\varphi(A)$ commutes with all bounded operators that commute with A.
(vi) If A is reduced by a projection P then

$$\varphi(A)/PH = \varphi(A/PH)$$

The details are left as an exercise for the reader.

In applications a particularly interesting function of a self-adjoint operator A is its resolvent. The above discussion indicates that if we introduce the function φ_z defined by

$$\varphi_z(\lambda) = (\lambda - z)^{-1}$$

then we can define

$$\varphi_z(A) = R_z(A) = \int_{-\infty}^{\infty} (\lambda - z)^{-1} dE_\lambda \tag{5.42}$$

The next example gives some properties of φ_z.

Example 5.33. Let H be a complex, separable Hilbert space and let $A: H \to H$ be a bounded, self-adjoint operator. Also, let $z \in \mathbf{C}$ be such that $\operatorname{Im} z \neq 0$. Then
(i) $\varphi_z(A) = (A - zI)^{-1} \in B(H)$
(ii) $\| \varphi_z(A) \| \leq 1/\operatorname{Im} z$.

Proof. Since A is self-adjoint, $\sigma(A) \subset \mathbf{R}$. Hence for $\operatorname{Im} z \neq 0$ it follows that $z \in \rho(A)$ and part (*i*) follows by definition of the resolvent.

We now show that $\varphi_z(A)$ is bounded as in (ii). Indeed, for any

$$g \in D((A - zI)^{-1}) = R(A - zI) = (A - zI)D(A)$$

where $R(A - zI)$ denotes the range of $(A - zI)$, we have $f = (A - zI)^{-1} g \in D(A)$. Consequently, writing $z = \alpha + i\beta$ we obtain

$$\begin{aligned} \|g\|^2 &= ((A - zI)f, (A - zI)f) \\ &= \|(A - \alpha I)f\|^2 + |\beta|^2 \|f\|^2 \geqslant |\beta|^2 \|f\|^2 \\ &\qquad = |\beta|^2 \|(A - zI)^{-1} g\|^2 \end{aligned}$$

We conclude that

$$\|(A - zI)^{-1} g\| \leq \frac{\|g\|}{|\beta|} \quad \text{for all} \quad g \in D((A - zI)^{-1})$$

and, on recalling the definition of an operator norm, we establish part (ii). ■

Finally, in this subsection we give some useful consequences of the functional calculus generated by the relations (5.40) and (5.41).

Example 5.34. (i) For $\varphi(x) = x$ we have $\varphi(A) = A$. This follows from (5.37) and the spectral decomposition theorem, Theorem 5.23, since for all $f \in D(A)$, $g \in H$

$$(\varphi(A)f, g) = \int_{-\infty}^{\infty} \varphi(\lambda) d(E_\lambda f, g) = \int_{-\infty}^{\infty} \lambda d(E_\lambda f, g) = (Af, g)$$

Arguing as in (i) the next results follow almost immediately.
(ii) For $\varphi(x) = 1$ we have $\varphi(A) = I$ because

$$(\varphi(A)f, g) = \int_{-\infty}^{\infty} d(E_\lambda f, g) = (f, g)$$

(iii) If f, g are continuous, complex-valued functions of a real variable x and if $(fg)(x) = f(x)g(x)$ then for any $\varphi, \psi \in H$

$$\begin{aligned}
(f(A)g(A)\varphi, \psi) &= \int_{-\infty}^{\infty} f(\lambda) d_\lambda (E_\lambda g(A), \psi) \\
&= \int_{-\infty}^{\infty} f(\lambda) d_\lambda (g(A)\varphi, E_\lambda \psi) \quad \text{since } E_\lambda = E_\lambda^2 = E_\lambda^* \\
&= \int_{-\infty}^{\infty} f(\lambda) d_\lambda \int_{-\infty}^{\infty} g(\mu) d_\mu (E_\mu \varphi, E_\lambda \psi) \quad \text{by (5.37)} \\
&= \int_{-\infty}^{\infty} f(\lambda) d_\lambda \int_{-\infty}^{\lambda} g(\mu) d_\mu (\varphi, E_\mu \psi) \quad \text{since } E_\lambda E_\mu = E_\mu,\ \mu < \lambda \\
&= \int_{-\infty}^{\infty} f(\lambda) g(\lambda) d_\lambda (\varphi, E_\lambda \psi) \\
&= \int_{-\infty}^{\infty} (fg)(\lambda) d_\lambda (E_\lambda \varphi, \psi)
\end{aligned}$$

Hence $(fg)(A) = f(A)g(A)$.
(iv) For any $\varphi \in H$

$$(f(A)\varphi, \varphi) = \int_{-\infty}^{\infty} f(\lambda) d(E_\lambda \varphi, \varphi) = \int_{-\infty}^{\infty} f(\lambda) d(\|E_\lambda \varphi\|^2)$$

5.2.7 Spectral Decompositions of Hilbert Spaces

There are intimate connections betweena self-adjoint operator $A: H \supseteq D(A) \to H$ on a complex, separable Hilbert space H which can be used to provide decompositions into simpler parts of not only a given operator such as A but also the underlying Hilbert space, the associated spectrum of A and other related quantities. As an illustration recall that for any self-adjoint operator $A: H \to H$ the point spectrum, $\sigma_p(A)$, of A is (Definition 5.2)

$$\sigma_p(A) := \{\lambda \in \boldsymbol{R} : \exists\ \theta \neq u \in H \text{ s.t. } Au = \lambda u\}$$

We define (see Definition 5.13) $H_p(A)$ to be the linear span of all eigenfunctions of A. As we have seen $H_p(A)$ is a subspace of H and is called the **point subspace** of H with respect to A. Hence by means of the projection theorem we can write

$$H = H_p(A) \oplus H_c(A) \quad \text{where} \quad H_c(A) = H_p^{\perp}(A) \tag{5.43}$$

We refer to $H_c(A)$ as the **subspace of continuity** of H with respect to A.

Let

$$P_p : H \to H_p(A)$$

denote the projection onto $H_p(A)$. Then for any $f \in H$

$$f = P_p f + (I - P_p) f = P_p f + P_c f \tag{5.44}$$

where

$$P_c := (I - P_p) : H \to H_c(A)$$

is a projection orthogonal to P_p.

Thus in (5.43) and (5.44) we have a decomposition of H with respect to A. Furthermore, a decomposition of A is also available. Since, on using, (5.44) we obtain

$$Af = AP_p f + AP_c f =: A_p f + A_c f \tag{5.45}$$

where A_p, regarded as an operator on $H_p(A)$, is called the **discontinuous part** of A whilst A_c, regarded as an operator on $H_c(A)$, is called the **continuous part** of A.

Also, using (5.44) together with the associated spectral family $\{E_\lambda\}$, we obtain

$$(E_\lambda f, f) = (E_\lambda P_p f, f) + (E_\lambda P_c f, f) \tag{5.46}$$

This result provides a means of decomposing integrals such as those appearing in (5.21) to (5.23). To see how this can be achieved we introduce the notion of **spectral measure**.

Let H be a complex separable Hilbert space and $A : H \to H$ a bounded, self-adjoint operator with associated spectral family $\{E_\lambda\}$.

For any interval $\Delta := (\lambda', \lambda''] \subset \mathbf{R}$ we define

$$E_\Delta = E_{\lambda''} - E_{\lambda'} \tag{5.47}$$

The definition of the spectral family (Definition 5.22) together with (5.47) indicate that

(i) E_Δ is a projection onto the subspace $M_{\lambda''} \ominus M_{\lambda'}$, that is onto the orthogonal complement of $M_{\lambda'}$ in $M_{\lambda''}$.

(ii) If Δ_1, Δ_2 are disjoint intervals on $\mathbf{R}$ then

$$E_{\Delta_1} E_{\Delta_2} = E_{\Delta_2} E_{\Delta_1} = 0$$

that is, the ranges of the E_{Δ_1} and E_{Δ_2} are orthogonal. This follows directly by using (5.47), writing the various products out in full and using Definition 5.22.

(iii) $E_{\Delta_1} E_{\Delta_2} = E_{\Delta_1 \cap \Delta_2}$. Again as in (ii) this follows by direct calculation.

The family $\{E_\lambda\}$ defined in this manner is called a **spectral measure** on the class of all sets Δ of the form indicated above.

The definition of E_Δ in (5.47) can be extended to closed and open intervals. To do this we first recall the notation.

$$E_{\lambda\pm 0} = s - \lim_{\eta \to 0^+} E_{\lambda \pm \eta} \tag{5.48}$$

A spectral family$\{E_\lambda\}$ is said to be **right continuous** if $E_{\lambda+0} = E_\lambda$ and **left continuous** if $E_{\lambda-0} = E_\lambda$. We shall assume in the following that spectral families are right continuous.

We now define

$$E\{\lambda\} = E_\lambda - E_{\lambda-0} \tag{5.49}$$

The required extensions of (5.48) are given as follows.

(a) $\Delta_1 = [\lambda', \lambda'']$ then we set

$$E_{\Delta_1} = E_{d_1} + E\{\lambda'\}, \quad d_1 = (\lambda', \lambda'']$$

(b) $\Delta_2 = (\lambda', \lambda'')$ then we set

$$E_{\Delta_2} = s - \lim_{n \to \infty} E_{d_2}, \quad d_2 = \left(\lambda', \lambda'' - \frac{1}{n}\right]$$

This is meaningful since $\Delta_2 = \bigcup_n d_2$.

A more general introduction and treatment of spectral measures can be found in the texts cited in the Commentary. We would mention in particular [1], [11]. In this monograph we shall be mainly concerned with integration with respect to numerical measures generated by a function of the form

$$w : \lambda \to w(\lambda) = (E_\lambda \varphi, \psi) \ \forall \varphi, \quad \psi \in H, \quad \lambda \in R$$

In this connection the following theorem is instructive [10], [13], [11].

Theorem 5.35. *Let H be a complex separable Hilbert space and $A : H \supseteq D(A) \to H$ a self-adjoint operator with spectral family $\{E_\lambda\}$. For any φ, $\psi \in H$ the complex valued function $\lambda \to (E_\lambda \varphi, \psi)$ is of bounded variation.*

Using standard methods [11] we can define (Riemann–Stieltjes) integrals over any finite interval with respect to the measure generated by the numerical value function w defined by

$$w : \lambda \to w(\lambda) = (E_\lambda \varphi, \psi), \quad \varphi, \psi \in H$$

As a consequence if f is a complex-valued, continuous function of λ then the integral

$$\int_a^b f(\lambda) d(E_\lambda \varphi, \psi)$$

defined as the limit of Riemann–Stieltjes sums, exists for every finite a, b and for all $\varphi, \psi \in H$. Improper integrals over $\mathbf{R}$ are defined as in the case of Riemann integrals as

$$\int_{-\infty}^{\infty} f(\lambda) d(E_\lambda \varphi, \psi) = \lim \int_a^b f(\lambda) d(E_\lambda \varphi, \psi) \tag{5.50}$$

as $a \to -\infty$ and $b \to \infty$ whenever they exist.

We re-introduce here, in a slightly different way, some already familiar concepts and notation. This will have advantages later.

Let H be a complex, separable Hilbert space and $A : H \supseteq D(A) \to H$ a self-adjoint operator with associated spectral family $\{E_\lambda\}$. We have seen (Theorems 5.26 and 5.27 and the notation introduced in (5.48)) that $P_\lambda = E_\lambda - E_{\lambda-0}$ is nonzero if and only if λ is an eigenvalue of A and P_λ is the orthogonal projection onto the associate eigenspace. The set of all eigenvalues of A we have denoted by $\sigma_p(A)$ and referred to it as the **point spectrum** of A. Let H_p denote the subspaces spanned by all the eigenvectors of A, that is, spanned by all the eigenvectors $P_\lambda H$. If $H_p = H$ then A is said to have a **pure point spectrum**.

In general $\sigma_p(A)$ is *not* a closed set. To see that this is the case consider an operator A that has a pure point spectrum. The spectrum of A, denoted $\sigma(A)$, is the point spectrum of A together with all its points of accumulation, that is $\sigma(A) := \overline{\sigma_p(A)}$. Furthermore, H_p reduces A since $P_\lambda H$ does and A_p, the part of A in H_p, has pure point spectrum. The proof of these statements is left as an exercise.

In the case when A has no eigenvalues then $H_p = \{\theta\}$ and A is said to have a **purely continuous spectrum**. We define $H_c = H_p^\perp$. In general the part A_c of A defined in H_c has a purely continuous spectrum which we denote by $\sigma(A_c)$. The spectrum of A_c, denoted $\sigma(A_c)$, is called the **continuous spectrum** of A and will be denoted in future by $\sigma_c(A)$.

In the above we have offered a decomposition of $\sigma(A)$, the spectrum of A, in terms of a decomposition of H rather than a decomposition of $\mathbf{C}$ as previously.

The subspaces $H_p \equiv H_p(A)$ and $H_c \equiv H_c(A)$ are called the **subspace of discontinuity** and the **subspace of continuity** respectively. When there is no danger of confusion we shall simply write H_p and H_c.

The following result characterises H_c.

Theorem 5.36. *$f \in H_c$ if and only if $(E_\lambda f, f)$ is a continuous function of λ.*

Proof. Let $w(\lambda) = (E_\lambda f, f)$. If w is a continuous function of λ then $w(\lambda) \to w(\mu)$ as $\lambda \to \mu$. This implies that $((E_\lambda - E_\mu)f, f) \to 0$ as $\lambda \to \mu$. Hence, recalling (5.49), we can conclude that $(P_\lambda f, f) = 0$ for all $\lambda \in \mathbf{R}$.

Since P_λ is a projection and recalling the Schwarz inequality $|(a, b)| \leq \|a\| \|b\|$, we obtain, for all $g \in H$

$$|(E_\lambda g, f)|^2 = |(E_\lambda g,\ E_\lambda f)|^2 \leq (E_\lambda g, g)(E_\lambda f, f) = 0$$

Hence f is orthogonal to the ranges of all P_λ and therefore to H_p. Hence $f \in H_c$.

Conversely, if $f \in H_c = H_p^\perp$ then f is orthogonal to P_λ for all λ so that, again recalling (5.49), $(E_\lambda f, f)$ is continuous. ■

We shall find it useful when developing a scattering theory to further subdivide (decompose) H_c.

We have seen above that the spectral family $\{E_\lambda\}$ generates a spectral measure E_Δ. Thus, for any fixed $f \in H$ we can construct a non-negative measure by defining

$$m_f(\Delta) = (E_\Delta f, f) = \|E_\Delta f\|^2 \tag{5.51}$$

We now introduce two further subspaces of H.

Definition 5.37. An element $f \in H$ is said to be **absolutely continuous with respect to** A if m_f is absolutely continuous with respect to the Lebesgue measure, $|\cdot|$, on $\mathbf{R}$. That is, $|\Delta| = 0$ implies $m_f(\Delta) = \|E_\Delta f\|^2 = 0$.

Definition 5.38. A measure m_f is **singular** with respect to Lebesgue measure on $\mathbf{R}$ if there is a set Δ_0 with $|\Delta_0| = 0$ such that $m_f(\Delta) = m_f(\Delta \cap \Delta_0)$ for all sets $\Delta \subseteq \mathbf{R}$. In which case f is said to be **singular with respect to** A.

The set of all elements in H which are absolutely continuous (singular) with respect to A denoted $H_{ac}(H_s)$ is called the **subspace of absolute continuity (singularity)** with respect to A.

Theorem 5.39. *H_{ac} and H_s are subspaces of H, are orthogonal complements of each other and reduce A.*

A proof of this theorem can be found in [6] and [10].

Since the point set $\{\lambda\}$ has Lebesgue measure zero we have $(E_\lambda f, f) = (E_{\lambda-0} f, f)$ for all $f \in H_{ac}$ and all $\lambda \in \mathbf{R}$. Therefore, by Theorem 4.40, we have $H_{ac} \subseteq H_c$ and $H_p \subseteq H_s$. If we now set

$$H_{sc} := H_c \ominus H_{ac} \tag{5.52}$$

then we obtain, for each self-adjoint operator $A : H \to H$ the following decompositions of H.

$$H = H_{ac} \oplus H_s = H_{ac} \oplus H_{sc} \oplus H_p \tag{5.53}$$

If $H_{ac} = H$ then A is said to be **(spectrally) absolutely continuous**. If $H_s = H$ then A is said to be **(spectrally) singularly continuous**.

The parts of A on these various subspaces are denoted A_{ac}, A_s, A_{sc} respectively. We write $\sigma_{ac}(A)$, $\sigma_s(A)$, $\sigma_{sc}(A)$ to denote the absolutely continuous, the singular and the singularly continuous spectrum of A respectively. The associated components of the spectrum of A are given by $\sigma(A_{ac})$, $\sigma(A_s)$ and $\sigma(A_{sc})$ respectively.

The physical relevance of the decomposition (5.53) can be considerable, particularly in some areas of quantum scattering [10]. For the moment we simply state that $H_p(A)$ usually contains the **bound states** of A whilst $H_c(A)$, and more especially $H_{ac}(A)$ consists of **scattering states** of A. For most self-adjoint operators arising in applications it turns out that $H_{sc} = \{\theta\}$ which implies that $H_c = H_{ac}$.

Further detailed discussions along these lines can be found in the texts cited in the Commentary.

Finally in this subsection, we mention yet another way of decomposing the spectrum of a self-adjoint operator $A : H \to H$ where H is a Hilbert space.

Definition 5.40. The set of all $\lambda \in \sigma(A)$ with the range of P_λ finite dimensional forms $\sigma_d(A)$ the **discrete spectrum** of A.

The set complementary to $\sigma_d(A)$ in $\sigma(A)$ constitutes the **essential spectrum** of A and is denoted $\sigma_e(A)$.

The sets $\sigma_d(A)$ and $\sigma_e(A)$ are disjoint. The set $\sigma_e(A)$ consists of the continuous spectrum of A, the accumulation points of the point spectrum of A and eigenvalues of A of infinite multiplicities.

Finally, we would remark that in this chapter we have not worked explicitly with unbounded operators. If we would wish to do this then more care must be taken when handling the domain of the operator. In this connection see Chapter 3 and [5], [6].

5.3 Semigroup Methods

In Chapter 1 we saw that in the acoustic case an IBVP could be reduced, formally at least, to an IVP. The IBVP was defined in $\mathbf{R}^n \times \mathbf{R}$ in terms of a partial differential equation which was of second order in time whilst the IVP was a Cauchy problem for a system of ordinary differential equations which was first order in time and defined on an appropriate energy space. It turns out that for these first order equations results concerning existence, uniqueness and stability of solutions can be obtained in an efficient and elegant manner using results from the theory of semigroups and the theory of Volterra integral equations. This approach will be seen to offer good prospects for developing constructive methods of solution.

We would again point out that, in keeping with the spirit of this monograph, most of the results offered in this chapter are simply stated without proof. This we have done because in most practical problems the main aim is centred on making progress by the application of analytical mathematical results rather than by working through their proofs, which are often quite lengthy. However, as always, References will be given either in the text or in the Commentary indicating where detailed workings can be found.

We remark that we shall work here in a Hilbert space setting. However, a similar analysis can be conducted in a more general Banach space setting [13], [7].

Let H denote a Hilbert space and consider the IVP

$$\left\{\frac{d}{dt}-G\right\}w(t)=0,\quad t\in\mathbf{R}^{+}=:(0,\infty),\quad w(0)=w_0 \tag{5.54}$$

where $w\in C(\mathbf{R}^{+},H)$ and $G:H\supseteq D(G)\to H$. We remark that any boundary conditions which are imposed on the originating problem are accommodated in the definition of $D(G)$, the domain of G.

An IVP of the form (5.54) governs the manner in which a system evolves from an initial state $w(0)=w_0$ to another state, $w(t)$, at some other time $t\neq 0$. The operator G characterises the particular class of problem being considered. The presentation in this section will allow a wide range of specific forms to be accommodated.

When faced with a problem such as (5.54) the first requirement is to clarify the meaning of the defining equation. The definition of an ordinary derivative indicates that (5.54) should be interpreted to mean that

(i) $w(t)\in D(G)$

(ii) $\lim_{h\to 0}\left\|h^{-1}\{w(t+h)-w(t)\}-Gw(t)\right\|=0$

where $\|\cdot\|$ denotes the norm in H.

When the problem (5.54) models a physical, evolutionary system then ideally the problem (5.54) should be well-posed in the following sense.

Definition 5.41. A problem is said to be **well-posed** if it has a unique solution which depends continuously on the given data.

This definition implies that small changes in the given data produce only small changes in the solution.

Let the evolutionary problem (5.54) be well-posed and let $U(t)$ denote the transformation which maps $w(s)$, the solution at time s, onto $w(s+t)$, the solution at time $(s+t)$, that is

$$w(s+t)=U(t)w(s)$$

In particular we have

$$w(t)=U(t)w(0)=U(t)w_0$$

Therefore, since (5.54) is assumed to be well-posed, a solution of (5.54) is unique and we have

$$U(s+t)w_0 = w(s+t) = U(t)w(s) = U(t)U(s)w_0$$

which implies the so-called **semigroup properties**

$$U(s+t) = U(s)U(t), \quad s, t \in \mathbf{R}^+, \quad U(0) = I \tag{5.55}$$

We are thus led to a consideration of a family of operators $\{U(t)\}_{t\geq 0}$. Our first aim is to determine the family $\{U(t)\}_{t\geq 0}$. With this in mind we notice that (5.55) is reminiscent of the properties of the familiar exponential function. In support of this remark we recall the following result which is obtained when investigating **Cauchy's functional equation**. Specifically, if $f: [0, \infty) \to \mathbf{R}$ is such that

(i) $f(s+t) = f(s)f(t)$ for all $s, t \geq 0$

(ii) $f(0) = 1$

(iii) f is continuous on $[0, \infty)$ (on the right at the origin) then f is defined by

$$f(t) = \exp\{tA\}$$

for some **constant** $A \in \mathbf{R}$.

Furthermore, if we apply an integrating factor technique to (5.54) then, formally at least, we have

$$w(t) = \exp\{tG\}w_0$$

provided that G is regarded as a constant. With these results in mind it is natural to conjecture that

$$U(t) = \exp\{tG\} \tag{5.56}$$

for some **operator** G. Of course, this conjecture has to be proved if it is to be of any use. In this connection we introduce the following notion.

Definition 5.42. A family $U := \{U(t)\}_{t\geq 0}$ of **bounded**, linear operators on a Hilbert space into itself is called a **strongly continuous, one-parameter semigroup**, denoted a C_0-semigroup, provided

(i) $U(s+t)f = U(s)U(t)f = U(t)U(s)f$ for all $f \in H$ and $s, t \geq 0$

(ii) $U(0)f = f$ for all $f \in H$

(iii) the mapping $t \to U(t)$ is continuous for $t \geq 0$ and for all $f \in H$. If, in addition we have

(iv) $\|U(t)f\| \leq \|f\|$ for all $f \in H$ then U is called a **C_0-contraction semigroup**.

We remark that the restriction $U(t) \in B(H)$ enable us to compare different solutions by means of the relation

$$\|U(t)u_0 - U(t)v_0\| \leq \|U(t)\| \, \|u_0 - v_0\|$$

which only makes sense if $U(t) \in B(H)$.

Once we have introduced the notion of a C_0-semigroup three questions are immediate.

Q1: Given a C_0-semigroup $\{U(t)\}_{t\geq 0}$ how can we obtain the operator G whose existence was conjectured in (5.56)?

Q2: What types of operators, G, can appear in (5.56)?

Q3: Given a suitable operator G how can we construct an associated semigroup $\{\exp(tG)\}_{t\geq 0}$?

These three questions are investigated in great detail in the general theory of C_0-semigroups [7], [9]. Whilst we will always be aware of Q1 and Q2 our main interest in this monograph is centred on Q3.

If we recall the interpretation given to (5.54) then the following definition appears quite natural.

Definition 5.43. The **infinitesimal generator** of the C_0-semigroup $\{U(t)\}_{t\geq 0}$ is the linear operator

$$G : H \supseteq D(G) \to H$$

defined by

$$\begin{aligned} Gf &:= \lim_{h\to 0}\{h^{-1}(U(t)f - f)\}, && \text{for } f \in D(G) \\ D(G) &= f \in H : \lim_{h\to 0}\{h^{-1}(U(t)f - f)\} && \text{exists in } H \end{aligned}$$

Example 5.44. The defining properties of a C_0-semigroup suggest that, formally at least, for the family $U = \{U(t)\}_{t\geq 0}$ defined by

$$U(t) = \exp(tG)$$

we have

$$G = \left.\frac{dU(t)}{dt}\right|_{t=0} = U'(0)$$

which indicates that U is the semigroup generated by the operator G. Furthermore, the familiar integrating factor technique indicates that the solution of (5.54) can be expressed in the form

$$w(t) = \exp\{tG\}\cdot w_0 = U(t)w_0$$

provided $w_0 \in D(G)$. To see this last point notice that (5.54) implies that we must have

$$Gw_0 - \left.\frac{dU(t)w}{dt}\right|_{t=0} = Gw_0 - Gw_0 = 0$$

and this result is only valid if $w_0 \in D(G)$.

We now collect some well-known facts from the theory of semigroups. The presentation is essentially informal and is intended to provide a reference source rather than a comprehensive, self-contained account. More details can be found in the references cited here and in the Commentary.

Theorem 5.45 ([7], [13]). *Let $G \in B(H)$. The family $\{U(t)\}_{t\geq 0}$ defined by*

$$U := \left\{ U(t) = \exp(tG) = \sum_{n=0}^{\infty} \frac{(tG)^n}{n!} : t \in \mathbf{R}^+ \right\}$$

is a C_0-semigroup which satisfies

$$\|U(t) - I\| \to 0 \quad as \quad t \to 0$$

Moreover, G is the generator of U.

Conversely, if U is a C_0-semigroup satisfying the above relation then the generator of U is an element $G \in B(H)$.

Corollary 5.46. *If $U(t) = \exp(tG)$, defined as Theorem 5.45, then*

(i) $\|U(t)\| = \|\exp(tG)\| \leq \exp\{t \, \|G\|\}$, $t \in \mathbf{R}^+$

(ii) $U(t) : \mathbf{R}^+ \to H$ *continuously for all* $t \in \mathbf{R}^+$

(iii) $\dfrac{d^n}{dt^n}\{U(t)\} = G^n U(t) = U(t) G^n.$

In many cases of practical interest the operator G appearing in (5.54) could be unbounded. For such an operator some of the quantities used above such as $\|G\|$ and $\exp(tG)$, defined as in Theorem 5.45, are meaningless. Consequently, results such as Theorem 5.45 have to be modified. With this in mind the following results are available [7], [13].

Theorem 5.47. *Let H be a Hilbert space and $U := \{U(t),\ t \geq 0\}$ a C_0-semigroup with generator G. Then*

(i) $U(t)U(s) = U(s)U(t)$ *for all* $t, s \geq 0$

(ii) *U is exponentially bounded in the sense that there exist constants $M > 0$ and $\omega \in \mathbf{R}$ such that*

$$\|U(t)\| \leq M \exp(\omega t)$$

(iii) $\overline{D(G)} = H$ *and the operator G is closed*

(iv) $\frac{d}{dt}\{U(t)f\} = U(t)Gf = GU(t)f$ *for all* $f \in D(G)$

(v) *for all* $\lambda \in C$ *such that* $Re\{\lambda\} > \omega$ *there exists* $R\{G, \lambda\} := (\lambda I - G)^{-1}$ *and*

$$R\{G, \lambda\}f = \int_0^\infty e^{-\lambda t}U(t)f, \quad f \in H$$

In the next subsections we give conditions which ensure that problems of the form (5.54) are well-posed. We indicate when the operator G in (5.54) actually generates a C_0-semigroup suitable for ensuring that (5.54) is well-posed.

5.3.1 Well-posedness of Problems

Let H be a Hilbert space and G a densely defined operator on H. Consider the following IVP

$$\left\{\frac{d}{dt} - G\right\}w(t) = 0, \quad t \in \mathbf{R}^+, \quad w(0) = w_0 \tag{5.57}$$

A more precise definition than that given earlier of the well-posedness of problems such as (5.57) is as follows.

Definition 5.48. The problem (5.57) is **well-posed** if the resolvent set $\rho(G) \neq \phi$ and if for all $w_0 \in D(G)$ there exists a unique solution $w: \mathbf{R}^+ \to D(G)$ of (5.57) with $w \in C^1((0, \infty), H) \cap C([0, \infty], H)$.

Results along the following lines can now be established.

Theorem 5.49. *The problem* (5.57) *is well-posed if G generates a* C_0*-semigroup U on H. In this case the solution of* (5.57) *is given by* $w(t) = U(t)w_0$, $t \in \boldsymbol{R}^+$.

Proof. Let G generate a C_0-semigroup $U = \{U(t): t \in \mathbf{R}^+\}$. If $w_0 \in D(G)$ then by Theorem 5.47 we see that $w(\cdot) = U(\cdot)w_0 \in C^1(\mathbf{R}^+, H)$ is $D(G)$-valued and (5.57) holds. To prove well-posedness it remains to establish uniqueness. To this end let φ be any solution of (5.57) Then, for $0 \leq s \leq t < \infty$ we have

$$\frac{d}{ds}\{U(t-s)\varphi(s)\} = U(t-s)G\varphi(s) - U(t-s)G\varphi(s) = 0$$

Hence $U(t - s)\varphi(s)$ is independent of s. Consequently, since $U(0) = I$ this independence allows us to write

$$\varphi(t) = U(t - s)\varphi(s) = U(t)\varphi(0) = U(t)w_0$$

The last equality follows from the assumption that $\varphi(t)$ is any solution of (5.57) and must therefore satisfy the imposed initial condition. The required uniqueness now follows since the right-hand side is $w(t)$. ■

The converse of this theorem also holds. The details can be found in [7] and titles cited in the Commentary.

We will also be interested in non-homogeneous forms of the problem (5.54). Specifically, we will want to discuss problems of the form

$$\left\{\frac{d}{dt}-G\right\}v(t)=f(t), \quad t\in\mathbf{R}^{+}, \quad v(0)=v_0 \tag{5.58}$$

where f and v_0 are given data functions. In this connection the following result holds.

Theorem 5.50. *Let H be a Hilbert space and $G:H\supseteq D(G)\rightarrow H$ be the generator of a C_0-semigroup $U=\{U(t):t\geq 0\}\subseteq B(H)$. If $v_0\in D(G)$ and $f\in C^1(\boldsymbol{R}^+, H)$ then (5.58) has a unique solution $v\in C^1(\boldsymbol{R}^+, H)$ with values in $D(G)$.*

A proof of this theorem can be obtained by first noticing that a formal application to (5.58) of the familiar integrating factor technique yields

$$v(t)=U(t)v_0+\int_0^t U(t-s)f(s)ds \tag{5.59}$$

It now remains to prove that (5.59) is indeed a solution of (5.48) and moreover that v has all the properties indicated in the statement of Theorem 5.50. This is a straightforward matter. The details are left as an exercise but can be found in [13] if required.

5.3.2 Generators of Semigroups

We have seen that the well-posedness of an IVP can be established provided there is associated with the IVP a C_0-semigroup. The following results help to characterise these linear operators which actually generate C_0-semigroups. The results are simply listed for our convenience in this monograph. Detailed proofs can be found in the texts cited in the Commentary. We particularly mention [7], [13] as starting texts. We remark that now the linear operators we will be dealing with are not necessarily bounded.

Theorem 5.51. *Let $U=\{U(t):t\geq 0\}\supseteq B(H)$ be a C_0-semigroup with generator G. Then $D(G)$ is dense in H. Furthermore, $G:H\supseteq D(G)\rightarrow H$ is a closed, linear operator.*

Theorem 5.52. *A C_0-semigroup is uniquely determined by its generator.*

Theorem 5.51 is a valuable result whenever we know that G is the generator of a C_0-semigroup. We also want to know when it is that an operator G is indeed the generator of a C_0-semigroup. The answer is given by the celebrated Hille–Yosida theorem [6], [7].

It will be convenient at this stage to introduce the following notation.

For real numbers $M > 0$ and $\omega > 0$ the set of all generators of C_0-semigroups $\{U(t)\}_{t\geq 0}$ which satisfy on a Hilbert space H the relation

$$\|U(t)\| \leq M \exp(\omega t)$$

will be denoted $\mathcal{G}(M, \omega, H)$. We remark that this notation is also used in general Banach spaces.

Necessary and sufficient conditions for an operator G to belong to $\mathcal{G}(M, \omega, H)$ are provide by the following theorem.

Theorem 5.53 (Hille–Yosida theorem). *A linear operator $G : H \supseteq D(G) \to H$ is an element of $\mathcal{G}(M, \omega, H)$ if and only if*

(i) *G is a closed linear operator with domain $D(G)$ dense in H*

(ii) *a real number $\lambda > \omega$ is such that $\lambda \in \rho(G)$ the resolvent set of G*

(iii) *$R(G, \lambda) := (\lambda I - G)^{-1}$ is such that*

$$\left\|[R(G,\lambda)]^n\right\| \leq \frac{M}{(\lambda-\omega)^n}, \quad n = 1, 2, \ldots$$

We now turn to Q3. Specifically, if a given operator G satisfies the conditions of the Hille–Yosida theorem then how can the C_0-semigroup $\{U(t)\}_{t\geq 0}$ generated by G be constructed?

We have seen, in Theorem 5.45, that when $G \in B(H)$ the semigroup $\{U(t)\}_{t\geq 0}$ generated by G is defined by

$$U := \left\{ U(t) = \exp(tG) = \sum_{n=0}^{\infty} \frac{(tG)^n}{n!} : t \in \mathbf{R}^+ \right\} \tag{5.60}$$

However in many applications the given operator G is not necessarily bounded. Consequently, for a not necessarily bounded operator we try to find a family of *bounded* operators which approximate G in some sense. As a first attempt in this direction let $\{G_k\}_{k=1}^{\infty}$ be a family of bounded operators the elements of which generate the associated semigroups $U_k := \{U_k(t)\}_{t\geq 0}$ of the form given in (5.60). We would like to have a result of the form

$$\exp\{tG\} = \lim_{k\to\infty}(\exp\{tG_k\})$$

This would then enable us to write

$$U(t)g = \lim_{k\to\infty} U_k(t)g, \quad t \geq 0, \quad g \in H$$

Theorem 5.54. *Let G be the generator of a C_0-semigroup $\{U(t)\}_{t\geq 0} \subset B(H)$ then*

$$U(t)g = \lim_{h\to 0^+} (\exp\{tG_h\})g, \quad t \geq 0, \quad g \in H$$

where G_h is defined by

$$G_h g = \{U(h)g - g\}/h, \quad g \in H$$

and the convergence is uniform with respect to t for $0 \leq t \leq t_0$ with $t_0 \geq 0$ arbitrary [7], [9].

There is a major practical difficulty associated with this result which is centred on the approximations $G_h = \{U(h) - I\}/h$. These quantities are only known if $U(h)$ is known and this is what we are trying to find!

A way around the above difficulty can be obtained by first recalling that when G is a real or complex number then

$$\exp\{tG\} = [EXP\{-tG\}]^{-1} = \lim_{n\to\infty} \{[I - tG/n]^n\}^{-1} = \lim_{n\to\infty} \{[I - tG/n]^{-1}\}^n$$

When G is an operator in H then the analogue of the result would seem to be given by

$$(\exp\{tG\})g = \left(\lim_{n\to\infty} \{[I - tG/n]^{-1}\}^n\right) g =: \lim_{n\to\infty} \{[V(t/n)]^n g\}$$

This indeed proves to be the case as the following result can be obtained [7], [9].

Theorem 5.55. *Let G be the generator of a C_0-semigroup $\{U(t)\}_{t\geq 0} \subset B(H)$. Then for all $g \in H$*

$$U(t)g = \lim_{n\to\infty} \{[V(t/n)]^n g\}, \quad t \geq 0, \quad \text{for all } g \in H$$

where

$$V(t/n) = [I - tG/n]^{-1}$$

The convergence is uniform with respect to t for $0 \leq t \leq t_0$.

Finally, in this subsection we give Stone's theorem which will be particularly useful when developing scattering theories. In order to do this we require some preparation.

Definition 5.56. A **C_0-group** on a Hilbert space H is a family of operators $U := \{U(t) : t \in \mathbf{R}\} \subset B(H)$ satisfying all the requirements of Definition 5.42 but with $s, t \in \mathbf{R}$. The generator, G, of a C_0-group U on H is defined by

$$Gf = \lim_{h \to 0} h^{-1}\{U(h)f - f\}$$

where $D(G)$ is the domain of definition G which is the set of all $f \in H$ for which the above limit exists. This limit is two-sided in the sense $t \to 0$ and not just $t \to 0^+$.

We would point out that G is the generator of a C_0-group, U, if and only if $G_{\pm}$, defined as above but with $t \to 0^{\pm}$ respectively, generate C_0-semigroups $U_{\pm}$ where

$$U(t) = \begin{cases} U_+(t), & t \geq 0 \\ U_-(t), & t \leq 0 \end{cases}$$

Definition 5.57. Let H be a Hilbert space with structure $(\cdot, \cdot)$ and $\|\cdot\|$.

(i) An operator $A : H \supseteq D(A) \to H$ is a **symmetric** operator on H if it is densely defined on H and if

$$(Af, g) = (f, Ag) \quad \textit{for all} \quad f, g \in D(A)$$

(ii) The operator A is **skew-symmetric** if $A \subset -A^*$.

(iii) The operator A is **self-adjoint** if $A = A^*$.

(iv) The operator A is **skew-adjoint** if $A = -A^*$.

(v) When H is a complex Hilbert space then A is skew-adjoint if and only if (iA) is self-adjoint.

An instructive exercise is to prove the following.

Theorem 5.58. *Let H be a Hilbert space and let $G : H \supseteq D(G) \to H$ generate a C_0-semigroup $U := \{U(t) : t \in \mathbf{R}^+\} \subset B(H)$. Then $U^* := \{U^*(t) : t \in \mathbf{R}^+\}$ is a semigroup with generator G^*.*

The semigroup U is a **self-adjoint** C_0-semigroup, that is, $U(t)$ is self-adjoint for all $t \in \mathbf{R}^+$ *if* and only if its generator is self-adjoint.

Definition 5.59. A **C_0-unitary group** is a C_0-group of unitary operators.

We now state

Theorem 5.60 (Stone's theorem) [3]. *Let H be a Hilbert space. An operator $G : H \supseteq D(G) \to H$ is the generator of a C_0-unitary group U on H if and only if G is skew-adjoint.*

References and Further Reading

[1] N.I. Akheizer and L.M. Glazman: *Theory of Linear Operators in Hilbert Space*, Pitman-Longman, London, 1981.

[2] H. Amann: *Ordinary Differential Equations, An Introduction to Nonlinear Analysis*, W. de Gruyter, Berlin, 1990.

[3] J.A. Goldstein: *Semigroups of Linear Operators* and *Applications*, Oxford University Press, Oxford, 1986.

[4] E. Goursat: *Cours d'analyse mathématique* **3**, Paris, 1927.

[5] G. Helmberg: *Introduction to Spectral Theory in Hilbert Space*, Elsevier, New York, 1969.

[6] T. Kato: *Perturbation Theory for Linear Operators*, Springer, New York, 1966.

[7] A.C. McBride: *Semigroups of Linear Operators: An Introduction*, Pitman Research Notes in Mathematics Series, Vol. 156, Longman, Essex, 1987.

[8] S.G. Mikhlin: *Integral Equations and Their Application to Certain Problems in Mechanics Physics and Technology*, Pergammon Press, Oxford, 1957.

[9] A. Pazy: *Semigroups of Linear Operators* and *Applications to Partial Differential Equations*, Springer, New York, 1983.

[10] D.B. Pearson: *Quantum Scattering and Spectral Theory*, Academic Press, London, 1988.

[11] F. Riesz and B. Sz-Nagy: *Functional Analysis*, Ungar, New York, 1981.

[12] G.F. Roach: *Greens Functions* (2nd Edn), Cambridge Univ. Press, London, 1970/1982.

[13] G.F. Roach: *An Introduction to Linear and Nonlinear Scattering Theory*, Pitman Monographs and Surveys in Pure and Applied Mathematics, Vol. 78, Longman, Essex, 1995.

[14] P.E. Sobolevskii: Equations of parabolic type in a Banach space, *Amer. Math. Soc. Transl.* **49**, 1–62, 1966.

[15] H. Tanabe: On the equation of evolution in a Banach space, *Osaka Math* J. **12**, 363–376, 1960.

[16] H. Tanabe: *Evolution Equations*, Pitman Monographs and Studies in Mathematics, Vol. 6, Pitman, London, 1979.

[17] F. Tricomi: *Integral Equations*, Interscience, New York, 1957.

6

A Scattering Theory Strategy

6.1 Introduction

In this chapter we gather together a strategy for investigating wave phenomena in a time domain setting and for developing an associated scattering theory. In the course of this we make more precise many of the statements found in Chapter 1.

In contrast to other strategies in this connection (for example see [9]) the approach adopted here is centred on the eigenfunction expansion method mentioned in Chapter 5. This will be seen to lead, quite readily, to a mathematical description of what is meant by scattering processes and associated scattering states. Furthermore, eigenfunction expansions methods associated with the FP and the PP will be seen to offer good prospects for the immediate and practical construction of solutions and associated wave operators.

A scattering process describes the effects of a perturbation on a system about which everything is known in the absence of the perturbation. Such a process can be conveniently characterised in terms of three main features; generation, interaction and measurement. In the generation stage an incident wave, a signal, is generated, far away in both space and time, from any perturbation which might have to be considered, for example, a target body or some potential. At this stage the interaction between an incident wave and the perturbation is negligible and the system evolves as though it were a free system, that is, a system in which there are no perturbations. Eventually, the incident wave and the perturbation interact and exert considerable influences on each other. The resulting effects, that is, the scattered waves, often have a very complicated structure. After the interaction, during which the scattering has occurred, the now scattered wave and the perturbation can once more become quite distant from each other and the interaction effects again become negligible. Consequently, any measurement of the scattered wave at this stage would indicate that the system is, once again, evolving as a free system, but not necessarily the same free system as that considered originally.

In practical situations measurements of a wave far away from any perturbation are really the only data available. This suggests that one of the fundamental questions to be addressed when investigating scattering processes is of the following type. If an observer, far distant from any perturbation measures the scattered wave (signal) then what was the incident wave (signal)? We would like to be able to answer this question without having to investigate, in too much detail, the actual interaction stage. Consequently, the asymptotic behaviour of solutions to wave equations and especially in the asymptotic equality of solutions of the associated free and perturbed systems becomes of particular interest.

Even more basic than the above question is the assumption we have made in Chapter 1, namely, that the scattered wave can indeed be characterised in terms of quantities associated with some free system. This leads to the so-called **asymptotic condition** and the notion of **asymptotic completeness**. We shall discuss these concepts later. Our first concern is to determine whether or not the systems of interest actually have solutions which produce propagating waves.

6.2 Propagation Aspects

We consider the IVPs

$$\{\partial_t^2 + A_j\}u_j(x, t) = 0, \quad (x, t) \in \mathbf{R}^n \times \mathbf{R}, \quad j = 0, 1 \tag{6.1}$$

$$u_j(x, 0) = \varphi_j(x), \quad u_{jt}(x, 0) = \psi_j(x), \quad j = 0, 1 \tag{6.2}$$

where $j = 0$ represents an FP and $j = 1$ a PP. We shall assume that

$$A_j : H(\mathbf{R}^n) \to H(\mathbf{R}^n) \equiv L_2(\mathbf{R}^n), \quad j = 0, 1$$

and that $H(\mathbf{R}^n)$ is a Hilbert space. We remark that here, for ease of presentation, we have assumed that both IVPs are defined in the same space. It should be noticed that we will not always be able to assume this. This assumption could well hold when A_0 is perturbed by additional terms, as in potential scattering. It is unlikely to hold, without further assumptions, when $D(A_0)$ is perturbed, as would be the case for target scattering. Furthermore, for the sake of illustration we shall assume here that A_0 is a realisation in $H(\mathbf{R}^n)$ of the *negative* Laplacian and A_1 is some perturbation of A_0.

An analysis of the given IVPs (6.1), (6.2) can begin by interpreting them as IVPs for ordinary differential equations rather than for partial differential equations. This can be achieved in the following manner. Let X be a Hilbert space. Furthermore, let $\Lambda \subseteq \mathbf{R}$ be a (Lebesgue measurable) subset of $\mathbf{R}$ and let f denote a function of $x \in \mathbf{R}^n$ and $t \in \mathbf{R}$ which has the action

$$f \equiv f(\cdot, \cdot): t \to f(\cdot, t) =: f(t) \in X, \quad t \in \Lambda \tag{6.3}$$

that is, f is interpreted as an X-valued function of $t \in \Lambda$.

We shall denote by $L_2(\Lambda, X) =: H$ the set of all equivalence classes of measurable functions defined on Λ with values in X satisfying

$$\|f\|_H^2 := \int_\Lambda \|f(t)\|_X^2 \quad dt. < \infty \tag{6.4}$$

where $\|\cdot\|_H$ denotes a norm on H and $\|\cdot\|_X$ the norm on X. It is an easy matter to show that H is a Hilbert space with inner product

$$(f,g) := \int_\Lambda (f(t), g(t))_X \quad dt \tag{6.5}$$

Therefore, with this notation and understanding we can interpret u in (6.1), (6.2) as

$$u \equiv u(\cdot, \cdot): t \to u(\cdot, t) =: u(t) \in X$$

The IVP (6.1), (6.2) can now be realised as an IVP for an ordinary differential equation, *defined in* H, of the form

$$\{d_t^2 + A_j\}u_j(t) = 0, \quad u_j(0) = \varphi_j, \quad u_{jt}(0) = \psi_j, \quad j = 0, 1 \tag{6.6}$$

When these IVPs are known to have solutions then they can be represented in the form

$$u_j(t) = (\cos(tA_j^{1/2}))\varphi_j + A_j^{-1/2}(\sin(tA_j^{1/2}))\psi_j, \quad j = 0, 1 \tag{6.7}$$

Hence, the solution of the given problem, (6.1), (6.2), can be written in the form

$$uj(x, t) = (\cos(tA_j^{1/2}))\varphi_j(x) + A_j^{-1/2}(\sin(tA_j^{1/2}))\psi_j(x), \quad j = 0, 1 \tag{6.8}$$

From (6.8), provided that the spectral theorem is available, it would then follow that, for $j = 0, 1$.

$$u_j(x,t) = \int_{\sigma(A_j)} \{\cos(t\sqrt{\lambda})\} dE_j(\lambda)\varphi_j(x) + \int_{\sigma(A_j)} \left\{\frac{\sin(t\sqrt{\lambda})}{\sqrt{\lambda}}\right\} dE_j(\lambda)\Psi_j(x) \tag{6.9}$$

where $\sigma(A_j)$ denotes the spectrum of A_j and $\{E_j(\lambda)\}_{\lambda \in \sigma(A_j)}$ is the spectral family of A_j.

The representation (6.8) is only meaningful if we know that the problems (6.1), (6.2) actually have solutions which, moreover, are known to be unique. Furthermore, the practical usefulness of the representation (6.9) depends crucially on how readily the spectral family $\{E_j(\lambda)\}_{\lambda \in \sigma(A_j)}$ can be determined.

An alternative approach frequently adopted when discussing wave motions governed by an IVP of the generic form (6.1), (6.2) is to replace the given IVP by an equivalent *system* of equations which are of first order in time. We have already given an indication of how this can be done in Chapter 1 and Chapter 5. This approach has a number of advantages since it can provide a straightforward means of including energy considerations. Results governing the existence and uniqueness of solutions with finite energy can then be quite readily obtained. To develop this approach we introduce an "energy space" $H_E(\mathbf{R}^n)$ which is defined as the completion of $C_0^\infty(\mathbf{R}^n) \times C_0^\infty(\mathbf{R}^n)$ with respect to the energy norm $\|\cdot\|_E$ where for

$$\mathbf{f} = \begin{bmatrix} f_1 \\ f_2 \end{bmatrix} =: \langle f_1, f_2 \rangle \in C_0^\infty(\mathbf{R}^n) \times C_0^\infty(\mathbf{R}^n) \tag{6.10}$$

we define

$$\|\mathbf{f}\|_E^2 := \int_{\mathbf{R}^n} \{|\nabla f_1(x)|^2 + |f_2(x)|^2\} dx \tag{6.11}$$

We notice that $H_E(\mathbf{R}^n)$ has the decomposition

$$H_E(\mathbf{R}^n) = H_D(\mathbf{R}^n) \oplus L_2(\mathbf{R}^n) \tag{6.12}$$

where $H_D(\mathbf{R}^n)$ is the completion of $C_0^\infty(\mathbf{R}^n)$ with respect to the norm defined by

$$\|f\|_E^2 := \int_{\mathbf{R}^n} \left\{|\nabla f(x)|^2\right\} dx, \quad f \in C_0^\infty(\mathbf{R}^n) \tag{6.13}$$

Furthermore, $H_E(\mathbf{R}^n)$ is readily seen to be a Hilbert space with respect to the inner product $(\cdot, \cdot)_E$ defined by

$$(\mathbf{f}, \mathbf{g})_E := (\nabla f_1, \nabla g_1) + (f_2, g_2) \tag{6.14}$$

where $\mathbf{f} =: \langle f_1, f_2 \rangle$, $\mathbf{g} =: \langle g_1, g_2 \rangle$ are elements of $H_E(\mathbf{R}^n)$ and $(\cdot\!: \cdot)$ denotes the usual $L_2(\mathbf{R}^n)$ inner product.

We now write the IVP (6.1), (6.2) in the form

$$\begin{bmatrix} u_j \\ u_{jt} \end{bmatrix}_t (x,t) + \begin{bmatrix} 0 & -I \\ A_j & 0 \end{bmatrix} \begin{bmatrix} u_j \\ u_{jt} \end{bmatrix} (x,t) = \begin{bmatrix} 0 \\ 0 \end{bmatrix}, \quad j = 0, 1 \tag{6.15}$$

$$\begin{bmatrix} u_j \\ u_{jt} \end{bmatrix} (x,0) = \begin{bmatrix} \varphi_j \\ \Psi_j \end{bmatrix} (x), \quad j = 0, 1 \tag{6.16}$$

This array can be written compactly in the form

$$(\partial_t - i\mathbf{G}_j)\mathbf{u}_j(x,t) = 0, \quad \mathbf{u}_j(x,0) = \mathbf{u}_j^0(x), \quad j = 0, 1 \tag{6.17}$$

where

$$\mathbf{u}_j(x,t) = \begin{bmatrix} u_j \\ u_{jt} \end{bmatrix}(x,t), \quad \mathbf{u}_j^0(x) = \begin{bmatrix} \varphi_j \\ \psi_j \end{bmatrix}(x) \tag{6.18}$$

$$-i\mathbf{G}_j = \begin{bmatrix} 0 & -I \\ A_j & 0 \end{bmatrix} \tag{6.19}$$

We now interpret $\mathbf{u}_j$, $j = 0, 1$, as H_E-valued functions of t in the sense that

$$\mathbf{u}_j \equiv \mathbf{u}_j(\cdot, \cdot): t \to \mathbf{u}_j(\cdot : t) =: \mathrm{u}_j(t) \in H_E(\mathbf{R}^n), \quad j = 0, 1 \tag{6.20}$$

In this case (6.17) can be reformulated in $H_E(\mathbf{R}^n)$ as an IVP for an ordinary differential equation of the form

$$\{d_t - i\mathbf{G}_j\}\mathbf{u}_j(t) = 0, \quad \mathbf{u}_j(0) = \mathbf{u}_j^0 \tag{6.21}$$

where for $j = 0, 1$

$$\mathbf{G}_j : H_E(\mathbf{R}^n) \supseteq D(\mathbf{G}_j) \to H_E(\mathbf{R}^n)$$

$$G_j\xi = i\begin{bmatrix} 0 & -I \\ A_j & 0 \end{bmatrix}\begin{bmatrix} \xi_1 \\ \xi_2 \end{bmatrix}, \quad \xi = \langle \xi_1, \xi_2 \rangle \in D(\mathbf{G}_j)$$

$$D(\mathbf{G}_j) = \{\xi = \langle \xi_1, \xi_2 \rangle \in H_E(\mathbf{R}^n): A_j\xi_1 \in L_2(\mathbf{R}^n),\ \xi_2 \in H_D(\mathbf{R}^n)\}$$

Once we have obtained the representations (6.21) of the given IVPs (6.1), (6.2) then the following questions are immediate.

*Q*1: Are the problems (6.21) and (6.1), (6.2) well-posed?

*Q*2: How can the solutions of the problems (6.21) be represented whenever they exist?

*Q*3: How can solutions of (6.21) yield the required solutions to (6.1), (6.2)?

It is clear that if the problems (6.21) are well-posed then it will follow that the problems (6.1), (6.2) are also well-posed. To establish the wellposedness of (6.21) we use results from the theory of semigroups introduced in Chapter 5. For our later convenience we gather together here the relevant results. For ease of presentation we shall consider, for the moment, the IVP

$$\left\{\frac{d}{dt} - B\right\}w(t) = 0, \quad t \in \mathbf{R}^+, \quad w(0) = w_0 \tag{6.22}$$

Theorem 5.49. *The problem (6.22) is well-posed if B generates a C_0-semigroup U on H. In this case the solution of (6.22) is given by $w(t) = U(t)w_0$, $t \in \mathbf{R}^+$.*

We will also want to discuss non-homogeneous problems of the form

$$\left\{\frac{d}{dt}-B\right\}v(t)=f(t),\quad t\in\mathbf{R}^+,\quad v(0)=v_0 \tag{6.23}$$

where f and v_0 are given data functions. In this connection the following result holds.

Theorem 5.50. *Let H be a Hilbert space and B: $H \supseteq D(B) \to H$ be the generator of a C_0-semigroup $U = \{U(t)\colon t \geqslant 0\} \subseteq B(H)$. If $v_0 \in D(B)$ and $f \in C^1(\mathbf{R}^+, H)$ then (6.23) has a unique solution $v \in C^1(\mathbf{R}^+, H)$ with values in $D(B)$.*

A formal application to (6.23) of the familiar integrating factor technique yields the solution form

$$v(t)=U(t)v_0+\int_0^t U(t-s)f(s)ds$$

These results settle the wellposedness of the IVPs concerned provided we can show that B is the generator of a suitable semigroup. With this in mind, we recall.

Theorem 5.60 *(Stone's theorem) [6] Let H be a Hilbert space. An operator B: $H \supseteq D(B) \to H$ is the generator of a C_0-unitary group U on H if and only if B is skew-adjoint.*

Returning now to our original notation we remark that in most cases of practical interest it can be shown [14] that the $\mathbf{G}_j$, $j = 0, 1$ are positive, self-adjoint operators on $H_E(\mathbf{R}^n)$. Furthermore, since the wellposedness of the problem (6.22) will imply the wellposedness of the problem (6.1), (6.2) for each $j = 0, 1$ we can summarise the use of the above results as follows.

Theorem 6.1 *Let H be a Hilbert space and A_j: $H \subseteq D(A_j) \to H$, $j = 0, 1$ be positive, self-adjoint operators on H. Let H_E denote an energy space associated with H. If, for $j = 0, 1$, the operators $\mathbf{G}_j$: $H_E \subseteq D(\mathbf{G}_j) \to H_E$ of the form $iG_i = \begin{bmatrix} 0 & 1 \\ -A_j & 0 \end{bmatrix}$ are self-adjoint on H_E then $(i\mathbf{G}_j)$ generates a C_0-group $\{\mathbf{U}_j(t),\ t \in \mathbf{R}\}$ defined by*

$$\mathbf{U}_j(t)=\exp(it\mathbf{G}_j)=\cos\left(tA_j^{1/2}\right)\begin{bmatrix} I & 0 \\ 0 & I \end{bmatrix}-A_j^{-1/2}\sin\left(tA_j^{1/2}\right)\begin{bmatrix} 0 & -I \\ A_j & 0 \end{bmatrix} \tag{6.24}$$

$$=\begin{bmatrix} \cos\left(tA_j^{1/2}\right) & A_j^{-1/2}\sin\left(tA_j^{1/2}\right) \\ -A_j^{1/2}\sin\left(tA_j^{1/2}\right) & \cos\left(tA_j^{1/2}\right) \end{bmatrix} \tag{6.25}$$

Therefore, recalling the material in Chapter 5, it follows that the IVPs

$$\{d_t^2 + A_j\}u_j(t) = 0 \tag{6.26}$$

$$u_j(0) = \varphi_j \in D(A_j),\quad u_{jt}(0) = \psi_j \in D(A_j^{1/2}) \tag{6.27}$$

are also well-posed.

The solution of (6.21) can be obtained, using an integrating factor technique, in the form

$$\mathbf{u}_j(t) = \exp\{it\mathbf{G}_j\}\mathbf{u}_j^0 = \mathbf{U}_j(t)\mathbf{u}_j^0 \tag{6.28}$$

Consequently, provided we ensure that the $(i\mathbf{G}_j), j = 0, 1$ generate C_0-groups and that they are of the form (6.24) then it is clear that (6.1), (6.2) are well-posed problems and, moreover, *the first component of* (6.28) *yields the* ***same*** *solution as* (6.8). These observations will enable us to settle propagation problems associated with (6.1), (6.2).

However, we recall that a practical interpretation of these solution forms relies on a detailed knowledge of the spectra, $\sigma(A_j)$, $j = 0, 1$ and the spectral families. $\{E_j(\lambda)\}_{\lambda \in \sigma(A_j)}, j = 0, 1$. The spectral families can be determined by means of Stone's formula [10], [14] which for $j = 0, 1$ has the form

$$((E_j(\lambda) - E_j(\mu))f, g) = \lim_{\delta\downarrow 0, \varepsilon\downarrow 0} \int_{\mu+\delta}^{\lambda+\delta} ([R_j(t+i\varepsilon) - R_j(t-i\varepsilon)]f, g)dt \tag{6.29}$$

where

$$R_j(t \pm i\varepsilon) = (A_j - (t \pm i\varepsilon))^{-1} \tag{6.30}$$

Hence. for $j = 0, 1$ the spectral families, $\{E_j(\lambda)\}_{\lambda \in \sigma(A_j)}, j = 0, 1$, can be obtained via an investigation of the resolvent, $R_j(\lambda)$ of A_j. This in turn yields details of the underlying spectral properties of A_j.

From a practical point of view the determination of the spectral families is quite demanding and detailed investigations are often left to specific cases. However, since an investigation of $R_j(\lambda)$ is always required it would seem that an alternative approach based directly on the theory of eigenfunction expansions could offer good prospects for developing constructive methods. We shall tend to concentrate on this approach in the following chapters.

6.3 Solutions with Finite Energy and Scattering States

Bearing in mind (6.12) and the notion of energy in a wave [3] we introduce

$$E(u,t) = \int_{R^n} \{|\nabla u(x,t)|^2 + |u_t(x,t)|^2\}dx = \|u(t)\|_E^2 \tag{6.31}$$

and

$$E(B,u,t) = \int_{B} \{|\nabla u(x,t)|^2 + |u_t(x,t)|^2\}dx = \|u(t)\|_{B,E}^2 \tag{6.32}$$

where $B \subseteq \mathbf{R}$ is any bounded set. $E(u, t)$ denotes the global energy of the wave at time t whilst $E(B, u, t)$ denotes the energy of the wave in B at time t.

In this section we shall only be interested in those systems in which the global energy is conserved, that is

$$E(u, t) = E(u, 0) = \text{constant} \tag{6.33}$$

Therefore, if we are dealing with wave equation problems of the typical form (6.1), (6.2) then the energy integrals (6.31) associated with these problems have the form

$$E(u_j, t) = \int_{R^n} \{|\nabla \varphi_j(x)|^2 + |\varphi_j(x)|^2\} dx, \quad j = 0, 1 \tag{6.34}$$

For the FP, that is for the case $j = 0$, the following result can be obtained [17].

Theorem 6.2 (i) A_0 *is a self-adjoint, non-negative operator on* $L_2(\mathbf{R}^n)$.
(ii) A_0 *has a unique, non-negative square root* $A_0^{1/2}$ *with domain*

$$D(A_0^{1/2}) = \{u \in L_2(\mathbf{R}^n) \colon D_u^\alpha \in L_2(\mathbf{R}^n),\ |\alpha| \le 1\} =: L_2^1(\mathbf{R}^n)$$

where α *is a multi-index of the form* $\alpha = (\alpha_1, \alpha_2, \ldots, \alpha_n)$ *and the* α_k *are non-negative integers for* $k = 1, 2, \ldots, n$ *and* $|\alpha| = \Sigma_{k=1}^n \alpha_k$. *Further, we define* $D^\alpha := D_{1_1}^\alpha D_{2_2}^\alpha \ldots D_{n_n}^\alpha$, *where* $\partial/\partial x_k$, $k = 1, 2, \ldots, n$.

We remark that $L_2^m(\mathbf{R}^n)$, $m = 0, 1, \ldots$ are the usual Sobolev Hilbert spaces [1].

Consequently, using (6.12) we see that if $\varphi_0 \in D(A_0^{1/2} = L_2^1(\mathbf{R}^n)$ and $\psi_0 \in L_2(\mathbf{R}^n)$ then the representation

$$u_0(t) = (\cos(tA_0^{1/2}))\varphi_0 = A_0^{-1/2}(\sin(tA_0^{1/2}))\psi_0 \tag{6.35}$$

implies that $u_0(t) \in L_2^1(\mathbf{R}^n)$ and $u_{0t}(t) \in L_2(\mathbf{R}^n)$. In this case the energy integral $E(u_0, t)$ is finite and u_0 is called a **solution with finite energy** (wfe).

When we come to deal with the PP then we will require a similar result to Theorem 6.2 for the operator A_1 in (6.1).

Although we have assumed that the global wave energies $E(u_j, t)$, $j = 0, 1$ remain constant this is not necessarily the case for the local energies $E(B, u_j, t)$, $j = 0, 1$. As a consequence it is natural to say that the $u_j(x, t)$, $j = 0, 1$, represent **scattering waves** if for every bounded, measurable set $B \subseteq \mathbf{R}^n$

$$\lim_{t \to \infty} E(B, u_j, t) = 0 \tag{6.36}$$

If we assume that φ_0 and ψ_0 are real-valued functions such that $\varphi_0 \in H(\mathbf{R}^n) =: L_2(\mathbf{R}^n)$ and $\psi_0 \in D(A_0^{-1/2})$ and if we define

$$h_0 = \varphi_0 + i_0^{-1/2}\psi_0 \tag{6.37}$$

then (6.35) can be expressed in the form

$$u_0(t) = u_0(\cdot\cdot\ t) = \mathrm{Re}(v_0(\cdot, t)) \tag{6.38}$$

where

$$v_0(t) \equiv v_0(\cdot, t) = \exp\{-itA_0^{1/2}\}h_0 =: U_0(t)h_0 \tag{6.39}$$

is the complex-valued solution in $H(\mathbf{R}^n)$ of (6.1), (6.2) with $j = 0$. The representation (6.38), (6.39) implies that the evolution and asymptotic behaviour of $u_0(x, t)$ is determined by that of $v_0(x, t)$.

If, with (6.39) in mind, a wave system of interest evolves according to

$$v(x, t) = U_0(t)h(x) \tag{6.40}$$

then it is natural to say that $h \in H(\mathbf{R}^n)$ is a **scattering state** if and only if (6.36) holds.

If we introduce the mapping

$$Q_q\colon H_E(\mathbf{R}^n) \to H_E(\mathbf{R}^n) \tag{6.41}$$

where

$$Q_q w(x) = \chi_q(x) w(x) \quad \text{for all} \quad x \in \mathbf{R}^n$$

and χ_q is the characteristic function for

$$B(q) := \{x \in \mathbf{R}^n\colon\ |x| \leq q\}$$

then we notice that

$$E(B(q), v_0, t) = \|Q_q U_0(t) h_0\|_E^2$$

Hence, (6.36) is equivalent to

$$\lim_{t\to\infty} \|Q_q U_0(t) h_0\|_E = 0 \quad \text{for every} \quad q > 0 \tag{6.42}$$

It is a straightforward matter to verify, for $0 \leq q < \infty$, that Q_q is an orthogonal projection on H_E and that

$$s - \lim_{q\to\infty} Q_q = I$$

This leads to the following definition [18].

Definition 6.3. (i) A family of orthogonal projections on H_E, denoted by $\{Q_q: 0 \leq q < \infty\}$ is called a family of **localising operators** on H_E if $Q_q: H_E \to H_E$ satisfies $s - \lim_{q\to\infty} Q_q - I$.

(ii) An element $h_0 \in H_E$ is a **scattering state** for A_0 and $\{Q_q\}$ if and only if (6.42) holds. The set of all such scattering states will be denoted by H^s.

6.4 Construction of Solutions

Once questions of existence and uniqueness of solution have been settled then we can turn our attention to methods for actually determining such solutions.

We consider the IVP (6.1), (6.2) for the case $j = 0, 1$ and, with future applications in mind, we shall take $n = 3$.

We first notice that Theorem 6.2 indicates that the spectral theorem is available for interpreting the solution forms (6.35) and (6.39). Specifically, if $\{E_0(\lambda)\}_{\lambda\in\sigma(A_0)}$ denotes the spectral family of A_0 then we have the spectral representations

$$A_0 = \int_0^\infty \lambda dE_0(\lambda) \tag{6.43}$$

$$\Phi(A_0) = \int_0^\infty \Phi(\lambda) dE_0(\lambda) \tag{6.44}$$

where Φ is a bounded, Lebesgue measurable function of λ. However, as we have already mentioned, a difficulty associated with the results (6.43) and (6.44) concerns the practical determination of the spectral family $\{E_0(\lambda)\}_{\lambda\in\sigma(A_0)}$. For the case of the FP that we are concerned with the situation can be eased by introducing results for Fourier transforms in $L_2(\mathbf{R}^3) := H(\mathbf{R}^3)$. The Plancherel theory indicates that for any $f \in H(\mathbf{R}^3)$ the following limits exist.

$$(F_0 f)(p) = \hat{f}(p) := \lim_{r\to\infty} \frac{1}{(2\pi)^{3/2}} \int_{|x|\leq r} \exp(-ix\cdot p) f(x) dx \tag{6.45}$$

$$f(x) = (F_0^* \hat{f})(x) := \lim_{r\to\infty} \frac{1}{(2\pi)^{3/2}} \int_{|p|\leq r} \exp(ix\cdot p) \hat{f}(p) dp \tag{6.46}$$

where $x, p \in \mathbf{R}^3$. It can also be shown that for any bounded, Lebesgue measurable function Φ we have

$$(\Phi(A_0)f)(x) = \lim_{r\to\infty} \frac{1}{(2\pi)^{3/2}} \int_{|p|\leq r} \exp(ox\cdot p) \Phi(|p|^2 \hat{f}(p) dp \tag{6.47}$$

We would emphasise that the limits in (6.45) to (6.47) have to be taken in the $H(\mathbf{R}^3)$ sense. Furthermore, the theory of Fourier transform indicates that $F_0: H(\mathbf{R}^3) \to H(\mathbf{R}^3)$ and, moreover, that it is a unitary operator. Consequently, we have $F_0^{-1} = F_0^*$.

We notice that

$$w_0(x,p) = \frac{1}{(2\pi)^{3/2}} \exp(ix \cdot p), \quad x, p \in R^3 \tag{6.48}$$

satisfies the Helmholtz equation

$$(\Delta + |p|^2) w_0(x,p) = 0, \quad x, p \in \mathbf{R}^3 \tag{6.49}$$

Thus, w_0 might be thought to be an eigenfunction of $A_0 = -\Delta$ with associated eigenvalue $|p|^2$. However, a direct calculation shows that $w_0 \notin H(\mathbf{R}^3)$ and so w_0 must be a generalised eigenfunction of A_0. Nevertheless, the Fourier Plancherel theory, *which has been developed independently of any scattering aspects*, indicates that all the limits (6.45), (6.46) and (6.47) exist. Consequently, the spectral decomposition of A_0 can be written as a generalised eigenfunction expansion in the form

$$(F_0 f)(p) = \hat{f}(p) := \lim_{r \to \infty} \int_{|x| \le r} \overline{w_0(x,p)} f(x) dx \tag{6.50}$$

$$f(x) = (F_0^* \hat{f})(x) := \lim_{r \to \infty} \int_{|p| \le r} w_0(x,p) \hat{f}(p) dp \tag{6.51}$$

$$(\Phi(A_0) f)(x) = \lim_{r \to \infty} \int_{|p| \le r} w_0(x,p) \Phi(|p|^2) \hat{f}(p) dp \tag{6.52}$$

where as before all limits are taken in the $H(\mathbf{R}^3)$ sense. It will be useful later on to bear in mind that (6.52) can also be written in the form

$$F(\Phi(A_0) f)(p) = \Phi(|p|^2) \hat{f}(p) \tag{6.53}$$

These various results imply that the wave function v_0, introduced in (6.39) can be interpreted in the form

$$v_0(x,t) \int_{\mathbf{R}^3} w_0(x,p) \exp(-it|p|) \hat{h}_0(p) dp \tag{6.54}$$

We remark that the improper integral in (6.54) must be interpreted in the $H(\mathbf{R}^3)$ limit sense as in (6.50) to (6.52). With this understanding we should also notice that in (6.54)

$$w_0(x,p) \exp(-it|p|) = \frac{1}{(2\pi)^{3/2}} \exp(i(x \cdot p - t|p|)) \tag{6.55}$$

are solutions of (6.1) with $j = 0$ and as such represent plane waves propagating in the direction of the vector p. Therefore, the wave function given by (6.54) is a representation of a wave (acoustic) in terms of elementary plane waves (6.55).

We now turn our attention to the PP given by (6.1), (6.2) with $j = 1$. As we have already mentioned, for ease of presentation at this stage we shall assume that the FP and the PP are both defined in the same Hilbert space. We have seen that the complex-valued solution of the FP is given by (6.39). Consequently, arguing as for the FP we find that, provided $\varphi_1 \in H(\mathbf{R}^3)$ and $\psi_1 \in D(A_1^{-1/2})$, the complex-valued solution of the PP is given by

$$v_1(x, t) \equiv v_1(x, t) = \exp\{-itA_1^{-1/2}\}h_1(x) =: U_1(t)h_1(x) \tag{6.56}$$

where

$$h_1 = \varphi_1 + iA_1^{-1/2} \tag{6.57}$$

For the FP the Fourier Plancherel theory provides us with the generalised eigenfunction expansion (6.50) to (6.52). As a consequence we could interpret (6.39) in the form (6.54). We would like to have a similar result for the PP. Specifically, associated with A_1 we want a generalised eigenfunction expansion theorem of the form

$$(F_1 f)(p) = \tilde{f}(p) := \lim_{r\to\infty} \int_{|x|\le r} \overline{w_1(x, p)} f(x) dx \tag{6.58}$$

$$f(x) = (F_1^* \tilde{f})(x) := \lim_{r\to\infty} \int_{|p|\le r} w_1(x, p) \tilde{f}(p) dp \tag{6.59}$$

$$(\Phi(A_1) f)(x) = \lim_{r\to\infty} \int_{|p|\le r} w_1(x, p) \Phi(|p|^2) \tilde{f}(p) dp \tag{6.60}$$

where, as previously, the above limits have to be taken in the $H(\mathbf{R}^3)$ sense. The kernels $w_1(x, p)$ are taken to be solutions of

$$(A_1 - |p|^2) w_1(x, p) = 0, \quad x, p \in \mathbf{R}^3 \tag{6.61}$$

and as such are to be generalised eigenfunctions of A_1.

We would emphasise that for any specific perturbed problem it has to be *proved* that a generalised eigenfunction expansion (spectral decomposition) such as (6.58) to (6.60) is indeed available for use. For specific physical problems this can often involve a great deal of work. A full spectral analysis of A_1 is required and functions such as w_1, which are intimately connected with the particular problem being considered, have to be determined. We shall return to these various aspects in later chapters when we come to deal with specific scattering problems. For the remainder of this chapter we shall assume that such generalised eigenfunction expansions are available. Consequently, we will then be able to write (6.56) in the following form.

$$v_1(x, t) = \int_{\mathbf{R}^3} w_1(x, p) \exp(-it|p|) \tilde{f}_1(p) dp \tag{6.62}$$

which is interpreted in the same way as (6.54).

We remark that in (6.54) and (6.62) the p need not be the same for both. It is associated with eigenvalues of A_0 in (6.54) and with eigenvalues of A_1 in (6.62).

From (6.54) and (6.62) it is a straightforward matter to obtain the representations

$$u_0(x,t)=\int_{R^2} w_0(x,p)\left\{\tilde{\varphi}_0(p)\cos(t|p|)+\tilde{\psi}_0(p)\frac{\sin(t|p|)}{|p|}\right\}dp \tag{6.63}$$

$$u_1(x,t)=\int_{R^3} w_1(x,p)\left\{\tilde{\varphi}_1(p)\cos(t|p|)+\tilde{\psi}_1(p)\frac{\sin(t|p|)}{|p|}\right\}dp \tag{6.64}$$

Hence, provided we can establish an eigenfunction expansion theorem of the form (6.58) to (6.60) then, since all the terms in (6.63) and (6.64) are computable, we have available, in (6.63) and (6.64), a practical means of constructing solutions to the FP and PP respectively.

For the purpose of developing a scattering theory it remains to investigate whether or not these solutions can be considered as being asymptotically equal, in some sense, as $t \to \pm\infty$. We shall begin to investigate this aspect in the next section.

6.4.1 Wave Operators and Their Construction

In Chapter 1 we introduced the notions of Asymptotic Equality (AE) and Wave Operator (WO). Specifically, we say that $v_j, j=0,1$ the complex solutions of (6.1), (6.2), are AE as $t \to \pm\infty$ if

$$\lim_{t\to\pm\infty} \|v_1(t)-v_0(t)\|=0 \tag{6.65}$$

where $\|\cdot\|$ denotes the norm on $H(\mathbf{R}^3)$.

Using (6.39) and (6.56) we find

$$\begin{aligned}\|v_1(t)-v_0(t)\| &= \|U_1(t)h_1-U_0(t)h_0\|\\ &=\|U_0^*(t)U_1(t)h_1-h_0\|\\ &=: \|W(t)h_1-h_0\|\end{aligned}$$

Hence

$$\lim_{t\to\pm\infty} \|v_1(t)-v_0(t)\|=\|W_\pm h_1-h_0\| \tag{6.66}$$

where

$$W_\pm := \lim_{t\to\pm\infty} W(t)=\lim_{t\to\pm\infty} U_0^*(t)U_1(t)=\lim_{t\to\pm\infty} \exp(itA_0^{1/2})\exp(-itA_1^{1/2}) \tag{6.67}$$

are the WOs associated with A_0 and A_1.

The manipulations leading to (6.66) have always to be justified but are certainly valid when the $U_j(t), j=0,1$ are unitary operators. In practice we endeavour to ensure that this is the case. We notice that if the A_j, $j = 0. 1$ are self-adjoint operators then Stone's theorem [14] ensures that the $U_j(t)$, $j = 0. 1$ are indeed unitary operators.

Thus we see that the limit on the left-hand side of (6.66) will be zero, and so the FP and the PP will be AE, provided that the initial data for the FP and the PP are related according to

$$h_\pm = W_\pm h_1 \tag{6.68}$$

where, for the sake of clarity, h_0 has been replaced by $h_\pm$ to indicate that different initial values might have to be considered for the FP when investigating $t \to +\infty$ and $t \to -\infty$.

Before indicating a means of constructing the wave operators $W_\pm$ we first recall some features of waves on a semi-infinite string.

Example 6.4. The wave motion of a semi-infinite string is governed by an equation of the form

$$(\partial_t^2 - \partial_x^2)u(x, t) = 0, \quad (x, t) \in \Gamma \times \mathbf{R} \tag{6.69}$$

where $\Gamma = (0, \infty)$.

Since Γ is an unbounded region then any solution of (6.69) will, in practice, be required to satisfy certain growth conditions, called **radiation conditions**, as $|x| \to \infty$. To indicate the nature of these conditions we recall that equations like (6.69) have solutions that can be written in the form

$$u(x, t) = f(x - t) + g(x + t) \tag{6.70}$$

where f and g are arbitrary functions characterising a wave of constant profile travelling with unit velocity from left to right and from right to left respectively. The precise form of f and g is settled in terms of the initial and boundary conditions that are imposed on solutions of (6.69). In the particular case when both waves can be assumed to have the same time dependency, $\exp(-i\omega t)$, then we could expect to be able to write (6.70) in the form

$$u(x, t) = e^{-i\omega t}u_+(x) + e^{-i\omega t}u_-(x) \tag{6.71}$$

Direct substitution of (6.71) into (6.69) shows that the two quantities u_+ and u_- must satisfy

$$(d_x^2 + \omega^2)u_\pm(x) = 0 \tag{6.72}$$

Now (6.72) does not imply that the $u_\pm$ are necessarily the same. Indeed,

$$u_+(x) = e^{i\omega x} \quad \text{and} \quad u_-(x) = e^{-i\omega x} \tag{6.73}$$

both satisfy (6.72). Combining (6.71) and (6.73) we obtain

$$u(x, t) = \exp(-i\omega(t - x)) + \exp(-i\omega(t + x)) \tag{6.74}$$

Thus, on recalling (6.70) we see that u_+ characterises a wave moving from left to right and u_- a wave moving from right to left, both having the same time dependency $\exp(-i\omega t)$. Equivalently, we can say that u_+ is an **outgoing wave** since it is moving away from the origin whilst u_- is an **incoming wave** as it is moving towards the origin. This particular feature of wave motion can be neatly encapsulated as follows. (See the Commentary and the References cited there.)

Definition 6.5. Solutions $u_\pm$ of the equation

$$(\Delta + \omega^2)u_\pm(x) = f(x), \quad x \in R^n$$

are said to satisfy the **Sommerfeld radiation conditions** if and only if

$$\left\{\frac{\partial}{\partial r} \mp iw\right\}u_\pm(x) = 0\left(\frac{1}{r^{(n-1)/2}}\right) \quad \text{as } r = |x| \to \infty \tag{6.75}$$

$$u_\pm(x) = 0\left(\frac{1}{r^{(n-1)/2}}\right) \quad \text{as } r - |x| \to \infty \tag{6.76}$$

The estimates in (6.75) and (6.76) are considered to hold uniformly with respect to the direction $x/|x|$.

The estimate (6.75) taken with a minus (plus) sign is called the **Sommerfeld outgoing (incoming) radiation condition**.

With $u_\pm$ defined as in (6.73) it is clear that u_+ is outgoing whilst u_- is incoming. From the practical point of view this is entirely to be expected. Furthermore, it will often be convenient to think of u_- as an incident wave and u_+ as a scattered wave.

Since we are dealing with perturbation processes it is reasonable to assume that w_1, the kernel function in the generalised eigenfunction expansion theorem (6.58) to (6.60), is a perturbation of w_0 the kernel function in the generalised eigenfunction expansion theorem (6.50) to (6.52). Since w_0 characterises a plane wave we shall refer to w_1, a perturbation of w_0, as a **distorted plane wave**.

Definition 6.6. An outgoing (incoming) distorted plane wave $w_+(x, p)$ $(w_-(x, p))$ satisfies.

(i) $(\Delta + \omega^2)w_\pm(x) = 0,\ x, p \in \mathbf{R}^n$

(ii) $w_+(x, p)w_0(x, p)$ satisfies the outgoing radiation condition (resp. $w_-(x, p) - w_0(x, p)$ satisfies the incoming radiation condition).

Consequently, we shall assume here that the kernel $w_1(x, p)$ is either an outgoing or an incoming distorted plane wave and we shall write

$$w_1(x, p) \equiv w_\pm(x, p) = w_0(x, p) + w'_\pm(x, p) \tag{6.77}$$

where w'_+ (w'_-) behaves like an outgoing (incoming) wave.

Of course, when dealing with specific, physical problems the existence and structure of the distorted plane waves must be established.

The existence of the distorted plane waves can be established by means of the **Limiting Absorption Principle** (LAB) [4], [5] which is based on noticing that if A is a self-adjoint, linear operator in a Hilbert space H and if $\lambda = \mu + iv \in \mathbb{C}$, with $v \neq 0$ then the equation

$$(\mathrm{A} - \lambda I)u(x, \lambda) = f(x)$$

has a solution $u(\cdot, \lambda) \in H$ for each $f \in H$ because $\lambda \notin \sigma(A)$. In the LAB method we look for solutions in the form

$$u_{\pm}(x, \mu) = \lim_{v \to 0^{\pm}} u(x, \lambda)$$

The difficulty with this approach is centred on the interpretation of this limit. In general it can only be understood in the sense of convergence in a Hilbert space $H(\varpi)$, where ϖ is an arbitrary subdomain of the region over which functions in H are defined. Physically, the quantity $u(x, \lambda)$, $v \neq 0$ describes a steady state wave in an energy absorbing medium with absorption coefficient proportional to v [17]. We shall deal with this method in more detail when we come to consider the specific problems discussed in later chapters.

If we assume the existence of the $w_{\pm}$ and, moreover, that they form two complete sets of generalised eigenfunctions for A_1 then on substituting (6.77) into (6.58) to (6.60) we obtain

$$\tilde{f}_{\pm}(p) = (F_{\pm}f)(p) = \lim_{r \to \infty} \int_{|x| \leq r} \overline{w_{\pm}(x, p)} f(x) dx \tag{6.78}$$

$$f(x) = (F^*_{\pm}\tilde{f})(x) = \lim_{r \to \infty} \int_{|p| \leq r} w_{\pm}(x, p) \tilde{f}_{\pm}(p) dp \tag{6.79}$$

$$(\Phi(A_1)f)(x) = \lim_{r \to \infty} \int_{|p| \leq r} w_{\pm}(x, p) \Phi(|p|^2) \tilde{f}_{\pm}(p) dp \tag{6.80}$$

provided these limits exist. We refer to F_+ as an **outgoing generalised Fourier transform** and F_- as an **incoming generalised Fourier transform.**

On the basis of these various assumptions we see that the solution $v_1(x, t)$ given in (6.62) has two spectral representations depending on whether w_+ or w_- is used in the expansion theorem (6.78) to (6.80). Specifically, we have

$$v_1(x, t) = \lim_{r \to \infty} \int_{|p| \leq r} w_+(x, p) \exp(-it|p|) \tilde{h}_+(p) dp \tag{6.81}$$

and

$$v_1(x, t) = \lim_{r \to \infty} \int_{|p| \leq r} w_-(x, p) \exp(-it|p|) \tilde{h}_-(p) dp \tag{6.82}$$

where

$$\tilde{h}_{\pm}(p) = \lim_{r\to\infty}\int_{|x|\le r} \overline{w_{\pm}(x, p)} h(x)dx \tag{6.83}$$

Since w_+(resp. w_-) is an outgoing (resp. incoming) distorted plane wave we refer to (6.81) (resp. (6.82)) as the outgoing (resp. incoming) spectral representations of v_1.

We are now in a position to construct a useful form for the wave operators $W_{\pm}$. If we substitute the decomposition (6.77) for w_- into (6.82) then we obtain

$$v_1(x, t) = v_0^-(x, t) + v^-(x, t) \tag{6.84}$$

where

$$v_0^-(x, t) = \lim_{r\to\infty}\int_{|p|\le r} w_0(x, p)\exp(-it|p|)\tilde{h}_-(p)dp \tag{6.85}$$

$$v^-(x, t) = \lim_{r\to\infty}\int_{|p|\le r} w_-'(x, p)\exp(-it|p|)\tilde{h}_-(p)dp \tag{6.86}$$

We now notice that since the kernel function in the integral (6.85) is w_0 then it follows that v_0^- represents a free wave. Therefore we can write

$$v_0^-(x, t) = U_0(t)h_0^-(x) = \exp(-itA_0^{1/2}h_0^-(x) \tag{6.87}$$

where

$$h_0^-(x) = v_0^-(x, 0) \tag{6.88}$$

Hence, bearing in mind (6.85), (6.68) and (6.58) we find

$$h_0^-(x) = v_0^-(x, 0) = (F_0^*\tilde{h}_-)(x) = (F_0^*F_-h_1)(x) \tag{6.89}$$

Now, (6.89) relates the initial data for a FP and the initial data for an associated PP. Therefore, we conclude that as $t \to -\infty$ we might expect that

$$h_0^-(x) = (F_0^*F_-h)(x) = W_-h(x) \tag{6.90}$$

that is, we might expect that

$$W_- = F_0^*F_- \tag{6.91}$$

It turns out that this is indeed the case provided we have local energy decay of the form

$$\lim_{t\to-\infty} v^-(\cdot, t) = 0 \tag{6.92}$$

Using (6.84) we see that (6.92) is equivalent to

$$\lim_{t\to-\infty}\left\|\left[v_1(\cdot,t)-v_0^-(\cdot,t)\right]\right\|=0 \tag{6.93}$$

where $\|\cdot\|$ is the $H(\mathbf{R}^3)$ norm. It now follows that

$$\begin{aligned}\left\|v_1(\cdot,t)-v_0^-(\cdot,t)\right\| &= \left\|\exp(-itA_1^{1/2})h_1-\exp(-itA_0^{1/2})h_0^-\right\| \\ &= \left\|\{\exp(itA_0^{1/2})\exp(-itA_1^{1/2})-F_0^*F_-\}h_1\right\|\end{aligned} \tag{6.94}$$

Equation (6.94) together with (6.93) and the definition of the WO given in (6.67) implies that W_- exists and is given by

$$W_- = F_0^*F_- \tag{6.95}$$

If we substitute the decomposition (6.77) for w_+ into (6.81) then we obtain

$$v_1(x,t) = v_0^+(x,t) + v^+(x,t) \tag{6.96}$$

where

$$v_0^+(x,t)=\lim_{r\to\infty}\int_{|p|\le r} w_0(x,p)\exp(-it|p|)\tilde{h}_+(p)dp \tag{6.97}$$

$$v^+(x,t)=\lim_{r\to\infty}\int_{|p|\le r} w_+'(x,p)\exp(-it|p|)\tilde{h}_+(p)dp \tag{6.98}$$

Arguing as before we see that $v_0^+(x,t)$ represents a free wave and that we can write

$$v_0^+(x,t) = U_0(t)h_0^+(x) = \exp(-itA_0^{1/2})h_0^+(x) \tag{6.99}$$

where

$$h_0^+(x) = v_0^+(x,0) = (F_0^*\tilde{h}_+)(x) = (F_0^*F_+h_1)(x) \tag{6.100}$$

This results implies that we might expect that

$$W_+ = F_0^*F_+ \tag{6.101}$$

We can show that W_+ exists and that (6.100) is indeed the case provided that we have local energy decay of the form

$$\lim_{t\to+\infty} v^+(\cdot,t)=0 \tag{6.102}$$

The proof follows as for the case of W_- and the details are left as an exercise.

Once we have determined the existence and the form of the wave operators $W_\pm$ then a scattering operator, S, which links the initial conditions $h_0^\pm$ can be introduced as follows.

The above results indicate that

$$h_0^\pm = W_\pm h_1 = F_0^* F_\pm h_1 \tag{6.103}$$

This in turn implies

$$F_0 h_0^\pm = \hat{h}_0^\pm = F_\pm h_1$$

Hence

$$\hat{h}_0^+ = F_+ h_1 = F_+ F_-^* \hat{h}_0^- =: S\hat{h}_0^- \tag{6.104}$$

and we see that

$$S := F_+ F_-^* : \hat{h}_0^- \to \hat{h}_0^+ \tag{6.105}$$

This operator and the unitarily equivalent operator

$$F_0^* S F_0 := F_0^* F_+ F_-^* F_0 : h_0^- \to h_0^+ \tag{6.106}$$

are particularly useful when discussing the theoretical and practical details of the asymptotic condition and the associated AE results.

6.5 Asymptotic Conditions

We introduced in Chapter 1 the notion of AE. The aim in this section is to provide a more precise formulation of this asymptotic property.

The requirement that the scattered waves can be characterised, at large positive and large negative times, in terms of free waves which are totally unaffected by any scatterer, is called the **asymptotic condition**. To place this in a mathematical framework we again consider a simple case. Specifically, we consider a FP characterised by an operator A_0 and an associated group $\{U_0(t)\}$ and a PP, describing the wave scattering, which is characterised in terms of an operator A_1 and a group $\{U_1(t)\}$. Here, as introduced earlier

$$U_0(t) = \exp\{-itA_0^{1/2}\},\ U_1(t) = \exp\{-itA_1^{1/2}\} \tag{6.107}$$

Furthermore, we have seen that the FP has an initial state vector h_0 given by (6.37) whilst the PP has an initial state vector h_1 given by (6.57). We shall assume that A_0 and A_1 both act in the same Hilbert space H.

In a typical scattering situation the time evolution of the scattered wave which has initial state $h_1 \in H$ is governed by the group $\{U_1(t)\}$, that is, the state of the scattered wave at some other time t is $U_1(t)\, h_1$. In the absence of any scattering mechanism the time evolution of the free wave having a initial state $h_0 \in H$ is governed by the group $\{U_0(t)\}$. The state of the free wave at some other time t is then $U_0(t)h_0$. Our aim is to see if we can approximate, as t $\to \pm\infty$, the waves arising from the PP by waves arising from the free evolution of suitable FPs. This is most readily done by assuming that for $h_1 \in H$ there exist two initial states $h_\pm$ such that the wave $U_1(t)\, h_1$ converges to $U_0(t)h_\pm$ as $t \to \pm\infty$. Symbolically, we mean that we should be able to satisfy the requirements

$$\lim_{t\to-\infty} \|U_1(t)h_1 - U_0(t)h_-\| = 0, \quad \lim_{t\to+\infty} \|U_1(t)h_1 - U_0(t)h_+\| = 0 \tag{6.108}$$

where $\|\cdot\|$ is the norm on H. We refer to (6.108) as the **asymptotic conditions**. When these requirements are satisfied then it will mean that the scattered wave, characterised by $U_1(t)h_1$, is virtually indistinguishable from the wave $U_0(t)h_-$ in the remote past and from $U_0(t)h_+$ in the distant future. The requirement that vectors such as $h_\pm$ should exist and moreover exist as elements of H is really quite a severe restriction. However, we would remark that an indication of how this requirement can be met has already been given in the previous subsection.

The set of vectors $h_1 \in H$ for which (6.108) can be satisfied is called the **set of scattering states** for A_1, and will be denoted by $M(A_1)$. We shall assume that for each self-adjoint operator A with which we shall be concerned the set of scattering states, $M(A)$, has the following properties.

(i) $M(A)$ is a subspace of H.

(ii) $M(A)$ is invariant under the group $\{U(t)\}$, that is, if $h \in M(A)$ then

$$U(t)h = \exp(-itA^{1/2})h \in M(A) \quad \text{for all} \quad t \in \mathbf{R}$$

We notice that if h is an eigenfunction of $A^{1/2}$, that is, for some $\mu \in H$

$$A^{1/2}h = \mu h$$

then the state $U(t)h = \exp(-itA^{1/2})h = \exp(-it\mu)h$ is simply a multiple of the state h and as such it cannot define a scattering process. It defines a so-called **bound state** of A. For this reason scattering states are expected to be associated with the continuous spectrum of A.

In this description of the asymptotic condition we are associating with each initial state $h_- \in M(A_0)$ another state vector $h_+ \in M(A_0)$, both state vectors being interpreted as initial states at the time $t = 0$. If there is no scattering taking place then $U_1(t) = U_0(t)$ and clearly we have $h_- = h_+$. However, when scattering does occur then the correspondence between h_- and h_+ is effected by means of a scattering operator S. Typical forms for the scattering operator have been indicated in (6.105) and (6.106).

When dealing with specific problems it is sometimes convenient to alter the various requirements mentioned above. This is because h_1 is usually associated with the PP and as such is a given quantity. Our task then is to determine the states $h_\pm$ so that the asymptotic conditions (6.108) are satisfied. This is done by means of the WO, as in (6.68), and using an SO to determine the relation between h_- and h_+, as in (6.105) or (6.106). We can summarise the above discussion in the following manner.

Definition 6.7. A wave evolving according to

$$U_1(t)h_1(x) = \exp(-itA_1^{1/2})h_1(\mathrm{x})$$

is said to satisfy the asymptotic condition for t $\to +\infty$, defined with respect to the family of operators

$$\{U_0(t) = \exp(-itA_0^{1/2})\}$$

if there exists an element $h_+ \in M(A_0) \subseteq H$ such that as $t \to +\infty$ the wave $U_1(t)h_1(x)$ is asymptotically indistinguishable from the wave $U_0(t)h_+(x)$.

A similar definition holds as $t \to -\infty$.

This definition implies that an element $h_\pm \in H$ is such that the free evolution $U_0(t)h_\pm$ defines, as $t \to \pm\infty$, the asymptotes of some evolution $U_1(t)h_1$. However, we need to determine whether or not every $h_1 \in H$, evolving according to $U_1(t)h_1$ has asymptotes, as t $\to \pm\infty$, of the form $U_0(t)h_\pm$. To be able to settle this question we need some preparation which involves properties of the wave operators $W_\pm$.

First we show that the wave operators $W_\pm$ satisfy the so-called **intertwining relation**

$$A_0 W_\pm = W_\pm A_1 \tag{6.109}$$

To see that this is the case we notice that

$$\begin{aligned}
\exp(-i\tau A_0^{1/2})W_\pm &= \exp(-i\tau A_0^{1/2}) \lim_{t\to\pm\infty} \{\exp(itA_0^{1/2})\exp(-itA_1^{1/2})\} \\
&= \lim_{t\to\pm\infty} \left\{\exp(i(t-\tau)A_0^{1/2})\exp(-itA_1^{1/2})\right\} \\
&= \lim_{t\to\pm\infty} \left\{\exp(i(t-\tau)A_0^{1/2})\exp(-i(t-\tau)A_1^{1/2})\right\}\exp(-i\tau A_1^{1/2}) \\
&= W_\pm \exp(-i\tau A_1^{1/2})
\end{aligned}$$

Differentiate with respect to τ and set $\tau = 0$ to obtain (6.109).

We next notice, using the properties of inner products on Hilbert spaces, that for all f, g $\in D(W_\pm^*)$ we have

$$(W_\pm^* f, W_\pm^* g) = \lim_{t\to\pm\infty} (\exp(itA_0^{1/2})\exp(-itA_1^{1/2})f, \exp(itA_0^{1/2})\exp(-itA_1^{1/2})g) = (f, g) \tag{6.110}$$

Hence the wave operators $W_\pm^*$ are isometries. Furthermore, we can use (6.110) to obtain

$$(W_{\pm}^{*}f, W_{\pm}^{*}g) = (f, W_{\pm}W_{\pm}^{*}g) = (f, g)$$

from which it follows that

$$W_{\pm}W_{\pm}^{*} = I \tag{6.111}$$

We remark that from (6.111) we might expect that $W_{\pm}^{*}$ should behave like $W_{\pm}^{*}$. We obtain conditions below which can ensure this.

We notice that for f, $g \in H$ related according to

$$g := W_{\pm}^{*}f \in R(W_{\pm}^{*}) = \text{range of } W_{\pm}^{*}$$

we have

$$W_{+}^{*}W_{+}g = W_{+}^{*}(W_{+}W_{+}^{*}f) = W_{+}^{*}f = g \tag{6.112}$$

If, however, an element $h \in H$ is orthogonal to $R(W_{+}^{*})$ then

$$0 = (h, W_{+}^{*}f) = (W_{+}h, f) \quad \text{for all} \quad f \in H \tag{6.113}$$

Hence it follows that

$$W_{+}h = 0, \quad h \in R(W_{\pm}^{*})^{\perp} \tag{6.114}$$

and that

$$W_{+}^{*}W_{+}h = W_{+}^{*}(W_{+}h) = 0$$

Hence we have shown that $W_{+}^{*}W_{+}$ is a projection onto $R(W_{+}^{*})$.

Similarly, $W_{-}^{*}W_{-}$ is a projection onto $R(W_{-}^{*})$.

With this preparation we return to the question of the availability of the initial data, $h_{\pm}$, for the FP which will yield the required AE for the PP.

We emphasise that, as always in this monograph, the FP will be concerned with an incident wave (signal) in the absence of any perturbation whilst the PP will be concerned with scattered waves which are a consequence of some perturbation of the incident wave.

We have seen that when developing a scattering theory we try to relate the evolution of a given PP and the evolution of a rather simpler, associated FP. If the evolution of the PP is governed by the group $\{U_1(t)\}$ and that of the FP by the group $\{U_0(t)\}$ and if the PP has given initial data h_1 then the solutions of the PP and the FP will be AE as $t \to -\infty$ provided the FP has initial data, h_{-}, given by

$$h_{-} = W_{-}h_1 \tag{6.115}$$

where $W_{-} = \lim_{t \to -\infty} U_0^{*}(t)U_1(t)$ (see (6.67), (6.68)).

Once the initial data, h_-, is determined according to (6.115) it then remains to determine whether there exist some, possibly different, initial data, h_+, for the FP that will ensure that we *also* have AE of the PP and the FP as $t \to +\infty$. With this in mind we first notice, from (6.115), that to ensure AE as $t \to -\infty$ the initial data for the PP must satisfy

$$h_1 = W_-^* h_- \in R(W_-^*) \tag{6.116}$$

However, with AE as $t \to +\infty$ in mind, we see that we can also introduce the influence of the wave operator W_+ by expressing h_1 in the form

$$h_1 = h_+ + h^\perp, \quad h_+ \in R(W_+^*), \quad h^\perp \in R(W_+^*)^\perp \tag{6.117}$$

Since $W_+^* W_+$ is a projection onto $R(W_+^*)$ the first component of (6.117) indicates, remembering (6.116), that the following must also hold

$$R(W_+^*) \ni h_+ = W_+^* W_+ h_1 = W_+^* W_+ W_-^* h_- =: W_+^* S h_- \tag{6.118}$$

where

$$S := W_+ W_-^* : h_- \to h_+ \tag{6.119}$$

is the scattering operator which, when known, enables us to determine the required initial data from the previously obtained data h_-.

As $t \to +\infty$ the evolution of the initial data h_1 will still be governed by the group $\{U_1(t)\}$ and from (6.117) we have

$$U_1(t)h_1 = U_1(t)h_+ + U_1(t)h^- \tag{6.120}$$

Consequently,

$$W_+U_1(t)h_1 = W_+U_1(t)h_+ + W_+U_1(t)h^\perp$$

Now, using the intertwining relation (6.109) and (6.114) we find

$$W_+U_1(t)h^\perp = U_0(t)W_+h^\perp = 0$$

Consequently, we see that as $t \to +\infty$ the component $U_1(t)h^\perp$ of the state of the system remains orthogonal to the scattering subspace generated by h_+.

Finally, in this section we demonstrate some useful connections between the ranges of the WO and the properties of the SO.

Theorem 6.8. (i) $R(W_-^*) \subseteq R(W_+^*)$ *if and only if S is isometric*

(ii) $R(W_+^*) \subseteq R(W_-^*)$ *if and only if S^* is isometric*

(iii) $R(W_+^*) = R(W_-^*)$ *if and only if S is unitary.*

Proof. (i) Assume $R(W_-^*) \subseteq R(W_+^*)$. For any element $f \in H$ we then have $W_-^* f \in R(W_+^*)$. Since $W_+^* W_+$ is a projection onto $R(W_+^*)$ this implies

$$W_+^* W_+ W_-^* f = W_-^* f \tag{6.121}$$

Furthermore, since we always have

$$\|W_\pm h\| = \lim_{t \to \pm\infty} \left\| \exp\left(itA_0^{1/2}\right)\exp\left(-itA_1^{1/2}\right)h \right\| = \|h\| \tag{6.122}$$

then using (6.119), (6.122) and properties of projection operators we obtain

$$\|Sf\| = \|W_+^* Sf\| = \|W_+^* W_+ W_-^* f\| = \|W_-^* f\| = \|f\| \tag{6.123}$$

Hence S is isometric.

Conversely, assume that S is isometric. Then (6.123) indicates that the projection of $W_-^* f$ onto $R(W_+^*)$ has a norm which is identical with $\|W_-^* f\|$. Hence $W_-^* f \in R(W_+)$. Hence $R(W_-^*) \subseteq R(W_+^*)$.

(ii) This follows by noticing that $S^* = W_- W_+^*$ and using the same argument as for (i) with plus and minus interchanged.

(iii) This follows by noticing that S is unitary if and only if $SS^* = S^*S = I$, that is, if and only if both S and S^* are isometric. Consequently (ii) will follow from (i) and (ii). ■

A physical interpretation of this theorem can be obtained, for instance, by considering the condition $R(W_-^*) \subseteq R(W_+^*)$. We have seen that if the initial data for the PP and the FP are related as $t \to -\infty$ according to

$$h_1 = W_-^* h_- \tag{6.124}$$

then the PP and the FP are AE as $t \to -\infty$. In the above theorem (6.121) implies that we also have

$$h^1 = W_+^* S h_- \tag{6.125}$$

For AE of the PP and the FP as $t \to +\infty$ when h_- is already fixed to ensure AE as $t \to -\infty$ we consider

$$\begin{aligned}\lim_{t \to +\infty} \|U_1(t)h_1 - U_0(t)h_+\| &= \|W_+ h_1 - h_+\| \\ &= \|W_+ W_-^* h_- - h_+\| \\ &= \|Sh_- - h_+\|\end{aligned}$$

The right-hand side vanishes by virtue of (6.125) and (6.68) which implies that we have the required AE as $t \to +\infty$. Therefore, we see that the condition $R(W_-^*) \subseteq R(W_+^*)$ implies that if solutions of the PP can be shown to be asymptotically free as $t \to -\infty$ then they become asymptotically free again as $t \to +\infty$. The scattering operator S provides the transformation between the initial state for the asymptotically free state as $t \to -\infty$ and the initial state for the asymptotically free state as $t \to +\infty$. Similar interpretations for (ii) and (iii) can also be given.

In summary, we have seen that there are conditions which ensure that a given initial state h_1 evolving according to $U_1(t)h_1$ at some general time t will approach $U_0(t)h_-$ as $t \to -\infty$ and $U_0(t)Sh_- = U_0(t)h_+$ as $t \to +\infty$.

In many cases of practical interest the wave operators can be shown to have the following property.

Definition 6.9. The wave operators $W_\pm$ defined as in (6.67) are said to be **asymptotically complete** if $R(W_+^*) = R(W_-^*)$.

Asymptotic completeness of the wave operators is thus equivalent to the unitarity of the scattering operator. However, as we shall see in later chapters, to prove, for a given problem, that the associated SO is indeed unitary is not always a simple matter.

6.6 A Remark about Spectral Families

A comparison of (6.44) and (6.47), bearing in mind (6.48), would seem to suggest that

$$dE_0(\lambda)f(x) = w_0(x,p)\hat{f}(p)dp \tag{6.126}$$

To make this more precise we first recall that $E_0(\mu)$ satisfies

$$E_0(\mu) = \begin{cases} I, & \mu \geqslant 0 \\ 0, & \mu < 0 \end{cases}$$

Consequently, $E_0(\mu)$ has the property of the Heaviside unit function $H(\tau)$. Thus, if in (6.47) we take $\Phi(\lambda) = H(\mu - \lambda)$ then we obtain

$$E_0(\mu)f(x) = \lim_{R\to\infty} \int_{|p|\leq R} w_0(x,p)H(\mu - |p|^2)f(p)dp, \quad u \geqslant 0$$

from which it follows that

$$E_0(\mu)f(x) = \begin{cases} \int_{|p|\leq\sqrt{\mu}} w_0(x,p)f(p)dp, & \mu \geqslant 0 \\ 0, & \mu < 0 \end{cases} \tag{6.127}$$

Differentiating (6.127) we recover (6.126).

Even for the FP the practical determination of the spectral family $\{E_0(\lambda)\}$ is a difficult matter. It would seem best left to abstract analytical discussions where it can be of considerable use [7], [12].

The results (6.45) to (6.47) are often referred to as a **Fourier inversion theorem**. We have already mentioned that the results (6.45) to (6.47) can be obtained quite independently of any scattering considerations by means of the Plancherel theory of Fourier transforms. We shall endeavour, for both FP and PP, to use the Fourier inversion theorem approach rather than the spectral family approach.

6.7 Some Comparisons of the Two Approaches

In the last few sections we have worked directly with wave equations and their solutions. However, we could have worked, throughout, in terms of the equivalent first order systems which were introduced earlier to settle the wellposedness of the problems.

For convenience we gather together here the salient results from the treatment of wave equations and their associated first order systems. We then indicate how they are related.

When working with IVPs of the form

$$\{\partial_t^2 + A_j\}u_j(x, t) = 0, \; (x, t) \in \mathbf{R}^n \times \mathbf{R}, \quad j = 0.1 \tag{6.1}$$

$$u_j(x, 0) = \varphi_j(x), \; u_{jt}(x, 0) = \psi_j(x), \quad j = 0.1 \tag{6.2}$$

we noticed that their solutions could be written in the form (see (6.37) and (6.57))

$$u_j(t) = (\cos tA_j^{1/2})\varphi_j + A_j^{-1/2}(\sin tA_j^{1/2})\psi_j, \quad j = 0, 1$$

We then defined

$$h_j(x) = \varphi_j(x) + iA_j^{-1/2}\psi_j(x), \quad j = 0, 1$$

and combined these results to obtain(see (6.37) and (6.57))

$$u_j(t) \equiv u_j(\cdot, t) = \mathrm{Re}(v_j(\cdot, t)), \quad j = 0, 1 \tag{6.38}$$

where

$$v_j(t) \equiv v_j(\cdot, t) = \exp\{-itA_j^{1/2}\}h_j =: U_j(\mathrm{t})h_j, \quad j = 0, 1 \tag{6.39}$$

The quantity v_j is referred to as the complex-valued solution of (6.1).

We remark that if the initial time is $t = s$ rather than zero then in the above t has to be replaced by $(t - s)$.

We then went on to discuss the AE of the solutions $v_j(x, t)$ and, as a consequence, introduced the wave operators $W_{\pm}$ defined by

$$W_{\pm} := \lim_{t\to\pm\infty} W(t) = \lim_{t\to\pm\infty} U_0^*(t)U_1(t) = \lim_{t\to\pm\infty} \exp(itA_0^{1/2})\exp(-itA_1^{1/2}) \quad (6.67)$$

The wave operators were then determined in the form (see (6.91) and (6.101))

$$W_{\pm} = F_0^* F_{\pm}$$

where F_0 is the Fourier transform defined in (6.45) and $F_{\pm}$ are the incoming and outgoing generalised Fourier transforms defined in (6.78) bearing in mind (6.77) and (6.58).

The IVPs (6.1) and (6.2) can be written as first order systems of the following form

$$\begin{bmatrix} u_j \\ u_{jt} \end{bmatrix}_t (x,t) + \begin{bmatrix} 0 & -I \\ A_j & 0 \end{bmatrix} \begin{bmatrix} u_j \\ u_{jt} \end{bmatrix} (x,t) = \begin{bmatrix} 0 \\ 0 \end{bmatrix}, \quad j = 0, 1 \quad (6.15)$$

$$\begin{bmatrix} u_j \\ u_{jt} \end{bmatrix} (x,0) = \begin{bmatrix} \varphi_j \\ \Psi_j \end{bmatrix} (x), \quad j = 0, 1 \quad (6.16)$$

This array can be conveniently written as an IVP for an ordinary differential equation in H_{E} of the form

$$\{d_t - i\mathbf{G}_j\}\mathbf{u}_j(t) = 0, \quad \mathbf{u}_j(0) = \mathbf{u}_j^0 \quad (6.21)$$

where, for $j = 0, 1$.

$$\mathbf{G}_j : H_E(\mathbf{R}^n) \supseteq D(\mathbf{G}_j) \to H_E(\mathbf{R}^n)$$

$$G_j \xi = i \begin{bmatrix} 0 & -1 \\ A_j & 0 \end{bmatrix} \begin{bmatrix} \xi_1 \\ \xi_2 \end{bmatrix}, \quad \xi = \langle \xi_1, \xi_2 \rangle \in D(G_j)$$

$$D(\mathbf{G}_j) = \{\xi = \langle \xi_1, \xi_2 \rangle \in H_E(\mathbf{R}^n) : A_j \xi_1 \in L_2(\mathbf{R}^n), \ \xi_2 \in H_D(\mathbf{R}^n)\}$$

and where we understand that

$$\mathbf{u}_j \equiv \mathbf{u}_j(\cdot, \cdot) : \mathrm{t} \to \mathbf{u}_j(\cdot, t) =: \mathbf{u}_j(t) \in H_E(\mathbf{R}^n), \quad j = 0, 1$$

The IVPs (6.21) have solutions of the form

$$\mathbf{u}_j(t) = \exp\{it\mathbf{G}_j\}\mathbf{u}_j^0 = \mathbf{U}_j(t)\mathrm{u}_j^0 \quad (6.28)$$

where

$$\mathbf{U}_j(t) = \exp(it\mathbf{G}_j) = \cos(tA_j^{1/2}) \begin{bmatrix} I & 0 \\ 0 & I \end{bmatrix} - A_j^{-1/2} \sin(tA_j^{1/2}) \begin{bmatrix} 0 & -I \\ A_j & 0 \end{bmatrix} \quad (6.24)$$

$$= \begin{bmatrix} \cos(tA_j^{1/2}) & A_j^{-1/2} \sin(tA_j^{1/2}) \\ -A_j^{1/2} \sin(t A_j^{1/2}) & \cos(tA_j^{1/2}) \end{bmatrix} \quad (6.25)$$

Acknowledging (6.37) and (6.57) we obtain, for $j = 0, 1$

$$\mathbf{u}_j(x,t) = \mathbf{U}_j(t)\mathbf{u}_j^0(x) = \mathrm{Re}\begin{bmatrix} (\exp\{-itA_j^{1/2}\})h_j(x) \\ (-iA_j^{1/2})(\exp\{-itA_j^{1/2}\})h_j(x) \end{bmatrix} = \mathrm{Re}\{\mathbf{v}_j(x,t)\}$$

where the complex-valued solutions of (6.21) can be written in the form

$$\mathbf{v}_j(x,t) := U_j(t)\begin{bmatrix} I \\ -iA_j^{1/2} \end{bmatrix} h_j(x) =: U_j(t)\mathbf{g}_j(x), \quad j=0,1$$

with

$$\mathbf{g}_j(x) := \begin{bmatrix} I \\ -iA_j^{1/2} \end{bmatrix} h_j(x), \quad j=0,1$$

and the $U_j(t)$ are defined as in (6.39) and (6.56).

In a similar manner to that outlined above when discussing the AE of solutions $v_j(x, t)$ we investigate the AE of the solutions $\mathbf{v}_j(x, t)$ by requiring

$$\begin{aligned} 0 &= \lim_{t\to\pm\infty} \|\mathbf{v}_1(\cdot,t) - \mathbf{v}_0(\cdot,t)\|_{H_E} \\ &= \lim_{t\to\pm\infty} \|U_1(t)\mathbf{g}_1(\cdot) - U_0(t)\mathbf{g}_0(\cdot)\|_{H_E} \\ &= \|W_\pm \mathbf{g}_1(\cdot) - \mathbf{g}_0(\cdot)\|_{H_E} \end{aligned}$$

where $\|\cdot\|_{H_E}$ denotes the norm in the energy space $H_E(\mathbf{R}^n)$, (6.11), and, as in (6.67),

$$W_\pm = \lim_{t\to\pm\infty} U_0^*(t)U_1(t)$$

Thus we see that we will have the required AE if

$$\mathbf{g}_0(x) = W_\pm \mathbf{g}_1(x)$$

that is, if

$$\begin{bmatrix} I \\ -iA_0^{1/2} \end{bmatrix} h_0(x) = W_\pm \begin{bmatrix} I \\ -iA_1^{1/2} \end{bmatrix} h_1(x)$$

We see that the first component of this result yields, as might have been expected, the same result as that obtained in Subsection 6.2.

6.8 Summary

In the above sections we have outlined in a reasonably precise way a strategy, already hinted at in Chapter 1, for the analysis of wave scattering problems in the time domain. We now see that this strategy, written more compactly, consists of the following fundamental problems.

- Settle the existence and uniqueness of solutions to (6.1), (6.2) and determine their propagation properties. We have seen that this can be achieved by using Stone's theorem to show that the solution forms (6.28) are valid.
- Establish the existence and uniqueness of the wave operators $W_{\pm}$ defined in (6.67). We have seen, in the above sections, that this can be achieved using generalised eigenfunction (generalised Fourier transform) techniques in conjunction with certain energy decay requirements.
- Provide a spectral analysis of the associated spatial operators A_j, $j = 0, 1$ in order to be able to generate appropriate generalised eigenfunction expansion theorems.
- Prove the limiting absorption principle for the operators A_j, $j = 0, 1$ and as a consequence settle the existence and uniqueness of appropriate distorted plane waves.
- Investigate the completeness of the wave operators. This means determining whether or not all solutions of the PP are asymptotically free as $t \to \pm\infty$. This is closely related to establishing the existence of the quantities $h_{\pm}$ introduced in (6.68).

These problems have been discussed, with various degrees of generality, see for example [10], [12], [14], [2], [13]. We will find that by working through the above programme we will be able to develop promising methods for the practical construction of solutions, wave operators and the scattering operator.

References and Further Reading

[1] R.A. Adams: *Sobolev Spaces*, Academic Press, New York, 1975.

[2] W.O. Amrein, J.M. Jauch and K.B. Sinha: *Scattering Theory in Quantum Mechanics*. Lecture Notes and Supplements in Physics, Benjamin, Reading, 1977.

[3] G.R. Baldock and T. Bridgeman: *Mathematical Theory of Wave Motion*, Ellis Horwood, Chichester, 1981.

[4] D.M. Eidus: The principle of limiting absorption, *Math. Sb.*, **57**(99), 1962 and *AMS Transl.*, **47**(2), 1965, 157–191.

[5] D.M. Eidus: The principle of limiting amplitude, *Uspekhi Mat. Nauk.* **24**(3), 1969, 91–156 and *Russ. Math. Surv.* **24**(3), 1969, 97–167.

[6] J.A. Goldstein: *Semigroups of Linear Operators and Applications*, Oxford University Press, Oxford, 1986.

[7] T. Kato: *Perturbation Theory for Linear Operators*, Springer, New York, 1966.

[8] T. Ikebe: Eigenfunction expansions associated with the Schrodinger Operators and their application to scattering theory, *Arch. Rat. Mech. Anal.* **5**, 1960, 2–33.

[9] P.D. Lax and R.S. Phillips: *Scattering Theory*, Academic Press, New York, 1967.

[10] R. Leis: *Initial Boundary Value Problems of Mathematical Physics*, John Wiley & Sons, Chichester, 1986.

[11] A. Pazy: *Semigroups of Linear Operators and Applications to Partial Differential Equations*, Springer, New York, 1983.

[12] D.B. Pearson: *Quantum Scattering and Spectral Theory*, Academic Press, London, 1988.
[13] M. Reed and B. Simon: *Methods of Mathematical Physics, Vols 1–4*, Academic Press, New York, 1972–1979.
[14] G.F. Roach: *An Introduction to Linear and Nonlinear Scattering Theory*, Pitman Monographs and Surveys in Pure and Applied Mathematics, Vol. 78, Longman, Essex, 1995.
[15] E.J.P. Schmidt: On scattering by time-dependent potentials, *Indiana Univ. Math. Jour.* **24**(10), 1975, 925–934.
[16] H. Tanabe: *Evolution Equations*, Pitman Monographs and Studies in Mathematics, Vol. 6, Pitman, London, 1979.
[17] C.H. Wilcox: *Scattering Theory for the d'Alembert Equation in Exterior Domains*, Lecture Notes in Mathematics, No. 442, Springer, Berlin, 1975.
[18] C.H. Wilcox: Scattering states and wave operators in the abstract theory of scattering, *Jour. Functional Anal.* **12**, 1973, 257–274.

7

An Approach to Echo Analysis

7.1 Introduction

As we mentioned earlier we interested in the manner in which a signal emitted by a transmitter evolves through a medium and in the form that it assumes at a receiver. In this chapter we illustrate how this information can be obtained. In the approach adopted here a given wave problem, defined in $\mathbf{R}^n \times \mathbf{R}$, is reduced to a first order system which is defined in a suitably chosen abstract space. From a practical point of view there are two main advantages in adopting this approach. First, by carefully selecting the abstract space setting energy aspects can be very simply accommodated. A first order system structure can allow a ready application of results from the theory of semigroups given in Chapter 5. In particular, we have seen that conditions can be given which ensure that the first order system, and hence the given wave problem, is well-posed. We shall assume in this chapter that these conditions are satisfied.

7.2 A Typical Mathematical Model

Let

$$Q \subset \{(x, t) \in \mathbf{R}^n \times \mathbf{R}\}$$

$$\Omega = \{x \in \mathbf{R}^n : (x, t) \in Q\}, \quad B = \{x \in \mathbf{R}^n : (x, t) \notin Q\}$$

The region Q is assumed to be open in $\mathbf{R}^n \times \mathbf{R}$ and Ω denotes the exterior of a scattering target B.

The scattering problems with which we shall be concerned are centred on the following IBVPs.

For $j = 0, 1$ determine a quantity $u_j(x, t)$ satisfying the IBVP

$$\{\partial_t^2 + L_j(x)\}u_j(x, t) = f_j(x, t), \quad (x, t) \in Q \tag{7.1}$$

$$u_j(x, s) = \varphi_j(x, s), \quad u_{jt}(x, s) = \psi_j(x, s), \quad x \in \Omega_j, \quad s \in \mathbf{R} \tag{7.2}$$

$$u_j(x, t) \in (bc)_j, \quad (x, t) \in \partial\Omega_j \times \mathbf{R} \tag{7.3}$$

where

$L_j(x)$ is a differential expression characterising the wave field

f_j, $\partial_j(\cdot, s)$, $\psi_j(\cdot, s)$ are given data functions

$s \in \mathbf{R}$ is a fixed initial time

$\partial\Omega_j$ is the boundary of the target B_j

$(bc)_j$ indicates boundary conditions to be satisfied by $u_j(\cdot, \cdot)$.

We remark that in (7.1) the inhomogeneous term f_j characterises the transmitter and the signals which it emits.

We shall assume that the case $j = 0$ denotes an FP whilst the case $j = 1$ denotes a PP.

In this chapter we will reduce the generality of the problem (7.1) to (7.3) by confining attention to **acoustic Dirichlet problems**. In this case we take, for $j = 0, 1$

$$L_j(x) = -\Delta, \quad \text{for all } x \tag{7.4}$$

$$(bc)_j = \{u_j(\cdot, \cdot): u_j(x, t) = 0, \quad (x, t) \in \partial\Omega_j \times \mathbf{R}\} \tag{7.5}$$

$$u_j(x, s) = 0, \quad u_{jt}(x, s) = 0 \tag{7.6}$$

Our aim here is to discuss and determine solutions of the acoustic Dirichlet problems defined above and, in particular, investigate the nature of the solution at large distances from both the target and the transmitter. We have already indicated that a very convenient method of tackling such problems is to represent them in some suitable energy space. The appropriate settings and associated results are distinguished by employing subscripts.

As before, we use the notation

$$\mathbf{g} = \begin{bmatrix} g_1 \\ g_2 \end{bmatrix} = \langle g_1, g_2 \rangle$$

and introduce the energy norm

$$\|\mathbf{g}\|_E^2 = \|\langle g_1, g_2 \rangle\|_0^2 = \frac{1}{2}\int_{R^n} \left\{|\nabla g_1(x)|^2 + |g_2(x)|^2\right\} dx \tag{7.7}$$

where we assume $g_1, g_2 \in C_0^\infty(\mathbf{R}^n)$.

Associated with the norm in (7.7) is the inner product

$$(\mathbf{f}, \mathbf{g})_E = (\nabla f_1, \nabla g_1) + (f_2, g_2) \tag{7.8}$$

where on the right-hand side of (7.8) the notation $(\cdot, \cdot)$ denotes the usual $L_2(\mathbf{R}^n)$ inner product.

We shall write $H_E = H_E(\mathbf{R}^n)$ to denote the completion of $C_0^\infty(\mathbf{R}^n) \times C_0^\infty(\mathbf{R}^n)$ with respect to the energy norm (7.7) and introduce, for $j = 0, 1$

$$H_j = \{\mathbf{g} \in H_E : \mathbf{g} = 0 \text{ on } B_j\} \tag{7.9}$$

$$H_j^{\text{loc}} = \{\mathbf{g} = \langle g_1, g_2 \rangle : \zeta \mathbf{g} \in H_j \ \forall \zeta \in C_0^\infty(\mathbf{R}^n)\} \tag{7.10}$$

In most practical cases the FP models a free space problem. In this case solutions to the FP will indicate the evolution of a signal through the medium in the absence of any scatterers. We shall assume that this is the case here. We notice that the first component of $\mathbf{g} \in H_j^{\text{loc}}(t)$ must vanish on $\partial\Omega_j$ and on B_j.

For $j = 0, 1$ the function $u_j = u_j(\cdot, \cdot)$ is a solution of **locally finite energy** of the IBVP (7.1) to (7.3) if

(i) $\mathbf{u}_j = \langle u_{1j}, u_{2jt} \rangle \in C(\mathbf{R}, H_j^{\text{loc}})$

(ii) $\{\partial_t^2 + L_j(x)\}u_j(x, t) = f_j(x, t), \ (x, t) \in Q$

(in the sense of distributions).

We shall say that u_j defines a **solution of finite energy** if $\mathbf{u}_j \in C(\mathbf{R}, H_E)$ and $\mathbf{u}_j(t) \in H_j$ for each t whilst u_j defines a **free solution of finite energy** if $\mathbf{u}_j \in C(\mathbf{R}, H_E)$.

If $\mathbf{u}_j \in H_E$ then u_j has finite energy and we write

$$\|\mathbf{u}_j(t)\|^2 = \|u_j(t)\|^2 = \frac{1}{2}\int_{\mathbf{R}^n} \{|\nabla u_j(x,t)|^2 + |u_{jt}(x,t)|^2\} dx \tag{7.11}$$

for the total energy of u_j at time t.

The wave energy in a sphere is obtained from (7.11) by restricting the range of integration appropriately.

With this preparation we find that the IBVPs (7.1) to (7.6) lead to IVPs of the following form. For $j = 0, 1$

$$\{\partial_t + \mathbf{N}_j\}\mathbf{u}_j(x, t) = \mathbf{F}_j(x, t), \quad (x, t) \in Q \tag{7.12}$$

$$\mathbf{u}_j(x, s) = \langle \varphi_j(\cdot, s), \psi_j(\cdot, s) \rangle(x), \quad x \in \Omega_j \tag{7.13}$$

where

$$\mathbf{u}_j(x, t) = \langle u_j, u_{jt} \rangle(x, t), \ \mathbf{F}_j(x, t) = \langle 0, f_j \rangle(x, t)$$

$$\mathbf{N}_j := \begin{bmatrix} 0 & -I \\ A_j & 0 \end{bmatrix}$$

and on denoting by $L_2^D(\Omega_j)$ the completion of $C_0^\infty(\Omega_j)$ in $L_2(\Omega_j)$ we have introduced

$$A_j: L_2^D(\Omega_j) \to L_2^D(\Omega_j)$$

$$A_j u(\cdot, t) - L_j(\cdot) u_j(\cdot, t), \quad u_j(\cdot, t) \in D(A_j)$$

$$D(A_j(t)) = \{u \in L_2^D(\Omega_j): L_j(\cdot, \mathrm{t}) u(\cdot, t) \in L_2^D(\Omega_j)\}$$

Throughout we shall assume that the receiver and the transmitter are in the far field of the B_j and, furthermore, that supp $f_j \subset \{|(x, t)|, t_0 \leq t \leq T, |x - x_0| \leq \delta_0\}$ where x_0 denotes the position of the transmitter and t_0, T, δ_0 are constants.

If we introduce

$$\mathbf{G}_j: H_j \to H_j$$

$$\mathbf{G}_j \mathbf{u}_j(t) = i\mathbf{N}_j \mathbf{u}(t), \quad \mathbf{u}_j(t) \in D(\mathrm{G}_j)$$

$$D(\mathbf{G}_j) = \{\mathbf{u}_j(t) \in H_j: \mathbf{N}_j \mathbf{u}_j(t) \in H_j\}$$

then the IVP (7.12) can be realised as a first order system in $H_j(t)$ in the form

$$\{d_t - i\mathbf{G}_j\} \mathbf{u}_j(t) = \mathbf{F}_j(t), \quad t \in \mathbf{R} \tag{7.14}$$

$$\mathbf{u}_j(s) = \mathbf{u}_{js}$$

In Chapter 5 and Chapter 6 we indicated conditions which ensured that the IVPs (7.14) were well-posed and furthermore had solutions which could be written in the form

$$\mathbf{u}_j(t) = \mathbf{U}_j(t-s)\mathbf{u}_s + \int_s^t \mathbf{U}_j(t-\tau)\mathbf{F}_j(\tau)d\tau, \quad j = 0, 1 \tag{7.15}$$

where $\mathbf{U}_j(t - s)$ is the propagator for (7.14) which we have determined in the form

$$\mathbf{U}_j(t - s) = \exp\{i(t - s)\mathbf{G}_j\}, \quad j = 0, 1 \tag{7.16}$$

It is a straightforward matter to show that the propagators $\mathbf{U}_j(t - s), j = 0, 1$ have the properties

$$\mathbf{U}_j(t - r)\mathbf{U}_j(r - s) = \mathbf{U}_j(t - s)$$

$$\mathbf{U}_j(0) = \mathbf{I}$$

$$\partial_t \mathbf{U}_j(t - s) = i\mathbf{G}_j \mathbf{U}_j(t - s)$$

$$\partial_s \mathbf{U}_j(t - s) = -i\mathbf{U}_j(t - s)\mathbf{G}_j$$

The relation (7.15) is aversion of the familiar variation of parameter formula and the integral involved is an associated Duhamel type integral.

7.3 Scattering Aspects and Echo Analysis

As we have seen in the previous chapter, scattering theory is concerned with the (asymptotic) comparison of two systems. This is the type of theory we want to have available in practice since experimental measurements are usually made in the far field, that is, far distant from the receiver and the transmitter. In the present case these systems are as summed to be characterized by the operators $i\mathbf{G}_j$, $j = 0, 1$ respectively.

We shall assume

(i) $i\mathbf{G}(t)$, $j = 0, 1$ are self-adjoint operators defined on suitable Hilbert space(s)

(ii) The operators $i\mathbf{G}_j$, $j = 0, 1$ satisfy conditions which ensure that the IVPs

$$\{d_t - i\mathbf{G}_j\}\mathbf{u}_j(t) = \mathbf{F}_j(t), \quad \mathbf{u}_j(s) = \mathbf{u}_{sj}, \quad j = 0, 1 \tag{7.17}$$

are well-posed

(iii) $\mathbf{U}_j(t - s)$, $j = 0, 1$ denote the associated propagators (evolution operators).

Of course, when dealing with specific problems it has to be *prove*d that these assumptions are valid and available.

Following the development in Chapter 1 and Subsection 6.4.1 we now introduce (see also [8])

Wave Operators(WO):

$$\begin{aligned}\mathbf{W}_{\pm s}(\mathbf{G}_0, \mathbf{G}_1) &= s - \lim_{t \to \pm\infty} \mathbf{U}_0^*(t-s)\mathbf{U}_1(t-s) \\ &= s - \lim_{t \to \pm\infty} \mathbf{U}_0(s-t)\mathbf{U}_1(t-s)\end{aligned} \tag{7.18}$$

Scattering Operator (SO):

$$\mathbf{S}_s(\mathbf{G}_0, \mathbf{G}_1) = \mathbf{W}_{+s}(\mathbf{G}_0, \mathbf{G}_1)\mathbf{W}^*_{-s}(\mathbf{G}_0, \mathbf{G}_1) \tag{7.19}$$

We would point out that (ii) and Theorem 5.21 (Stone's theorem) ensure that (7.18) is meaningful.

In developing here an echo analysis for an IBVP of the form (7.1) to (7.5) we shall assume, for the purposes of illustration, that $n = 3$, that the medium is initially at rest and concern ourselves with the IVPs

$$\{d_t - i\mathbf{G}_j\}\mathbf{u}_j(t) = \mathbf{F}_j(t), \quad \mathbf{u}_j(s) = 0, \quad j = 0, 1 \tag{7.20}$$

The Free (unperturbed) Problem (FP) obtains when $j = 0$ and $\Omega = \mathbf{R}^3$.

The Perturbed Problem (PP) obtains when $j = 1$ and $\Omega \subset \mathbf{R}^3$.

Using the properties of propagators listed after (7.16) together with Theorem 5.50 and the fact that $\mathbf{U}_j(t - s): H_j \to H_j$ then the variation of parameters formula indicates that solutions of (7.20) can be written, for $j = 0, 1$, in the form

$$\mathbf{u}_j(t) = \mathbf{U}_j(t-s)\int_s^t \mathbf{U}_j(s-\tau)\mathbf{F}_j(\tau)d\tau =: \mathbf{U}_j(t-s)\mathbf{h}_j(s) \tag{7.21}$$

Scattering phenomena involve three fundamental items: the incident field, $\mathbf{u}_0$, the total field, $\mathbf{u}_1$, and the scattered wave field, $\mathbf{u}^s$. In the present case we have

Free wave field = Incident wave field	$\mathbf{u}_0(t) = \mathbf{U}_0(t - s)\mathbf{h}_0(s)$
Perturbed wave field = Total wave field	$\mathbf{u}_1(t) = \mathbf{U}_1(t - s)\mathbf{h}_1(s)$
Scattered wave field	$\mathbf{u}_s(t) = \mathbf{u}_1(t) - \mathbf{u}_0(t)$

The definition of the WOs and SO enables us to write

$$\begin{aligned}\mathbf{u}_1(t) &= \mathbf{U}_1(t - s)\mathbf{h}1(s)\\ &= \mathbf{U}_0(t - s)\mathbf{U}_0(s - t)\mathbf{U}_1(t - s)\mathbf{h}_1(s)\\ &= \mathbf{U}_0(t - s)\mathbf{W}_{+s}(\mathbf{G}_0, \mathbf{G}_1)\mathbf{h}_1(s) + \sigma_t(1) \text{ as } t \to \infty\end{aligned}$$

where $\sigma_t(1)$ is an $L_2(\mathbf{R}^n)$ function of t with $\sigma_t(1) \to 0$ as $t \to \infty$.

Similarly we can obtain

$$\mathbf{u}_s(t) = \mathbf{U}_0(t - s)\{\mathbf{W}_{+s}(\mathbf{G}_0, \mathbf{G}_1)\mathbf{h}_1(s) - \mathbf{h}_0(s)\} + \sigma_t(1) \tag{7.22}$$

For $t \ll 0$ there is no scattered field. Hence $\mathbf{u}_1(t) = \mathbf{u}_0(t)$ that is, $\mathbf{U}_1(t - s)\mathbf{h}_1(s) = \mathbf{U}_0(t - s)\mathbf{h}_0(s)$. Using the properties (7.16), (7.17) and (7.18) we can obtain the relation

$$\mathbf{h}_0(s) = \mathbf{W}_{-s}(\mathbf{G}_0,\mathbf{G}_1)\mathbf{h}_1(s) + \sigma_{x_0}(1) \quad \text{as} \quad |x_0| \to \infty$$

where $\sigma_{x_0}(1)$ is an $L_2(\mathbf{R}^n)$ function of x with $\sigma_{x_0}(1) \to 0$ as $|x_0 \to \infty$. Consequently, operating on both sides of this result with $\mathbf{S}_s$ and recalling (7.19) we obtain

$$\begin{aligned}\mathbf{S}_s(\mathbf{G}_0, \mathbf{G}_1)\mathbf{h}_0(s) &= \mathbf{S}_s(\mathbf{G}_0, \mathbf{G}_1)\mathbf{W}_{-s}(\mathbf{G}_0, \mathbf{G}_1)\mathbf{h}_1(s) + \sigma_{x_0}(1)\\ &= \mathbf{W}_{+s}(\mathbf{G}_0, \mathbf{G}_1)\mathbf{h}_1(s) + \sigma_{x_0}(1)\end{aligned} \tag{7.23}$$

If we now substitute (7.23) in (7.22) then we obtain

$$\mathbf{u}_s(t) = \mathbf{U}_0(t, s)\{\mathbf{S}_s(\mathbf{G}_0, \mathbf{G}_1) - \mathbf{I}\}\mathbf{h}_0(s) + \sigma_t(1) + \sigma_{x_0}(1) \tag{7.24}$$

Thus we see that the scattered (echo) field is determined, in the far field, by the SO and the FP data.

7.4 Construction of the Echo Field

In the previous section, (7.24) provides a representation of the echo field in an abstract setting. To obtain a physical representation of this field, that is a representation in $\mathbf{R}^n \times \mathbf{R}$, we follow the procedure outlined in Chapter 2. We have seen that a problem such as (7.1)–(7.3) defined in $\mathbf{R}^n \times \mathbf{R}$ can be represented as the IVP (7.14) in the abstract (energy) space H_j (see (7.9)). The problem (7.14) has a solution which can be expressed uniquely in the form (7.15) where the propagators $U(t-s)$, $j = 0, 1$ are defined in (7.16). To give an indication of how these echo fields can be obtained we consider, in $\mathbf{R}^3$, the acoustic Dirichlet problem (7.1)–(7.6) in the specific case of the scattering of a single pulse of duration T emitted at time t_0 by a transmitter localised near a point x_0. Hence, the source functions f_j, $j = 0, 1$ which characterise the transmitter will be assumed to have the space-time support

$$\operatorname{supp} f_j \subset \{|(x,t)| : t_0 \leq t \leq t_0 + T \quad \text{and} \quad |x_0 - x| \leq \delta_0\}$$

where t_0 and δ_0 are constants. We shall also assume, in keeping with the formulation of (7.1)–(7.6)

(i) The scatterer $B(t)$ is contained in a closed, bounded set in $\mathbf{R}^3$ with complement $\Omega_1(t) = \mathbf{R}^3 - B(t)$. Hence

$$B(t) \subset \{x : |x| \leq \delta\} \quad \text{where } \delta \text{ is fixed for all } t.$$

(ii) The origin of coordinates lies in $B(t)$.
(iii) $\partial\Omega_1(t)$, the boundary of $\Omega_1(t)$, is for all t, a smooth surface.
(iv) The scatterer and transmitter are disjoint which implies

$$\delta + \delta_0 < |x_0|$$

(v) The transmitter stops transmitting before the signal reaches the scatterer which implies

$$T < |x_0| - \delta - \delta_0$$

With these assumptions in mind we are led, as above, to the problems (7.20) and we see that (7.21) now assumes the form

$$\mathbf{u}_j(t) = \mathbf{U}_j(t,s)\int_{t_0}^{t_0+T} \mathbf{U}_j(s,\tau)\mathbf{F}_j(\tau) =: \mathbf{U}_j(t,s)\mathbf{h}_j(s) \tag{7.25}$$

which leads to the equivalent representation

$$\mathbf{u}_j(t) = \int_{t_0}^{t_0+T} \mathbf{U}_j(t,\tau)\mathbf{F}_j(\tau)d\tau \tag{7.26}$$

Consequently, bearing in mind the representation (2.82) and the definitions of the terms in (7.13) and (7.14) the required physical wave fields, u_j, $j = 0, 1$, are obtained as the first components of the abstract quantities $\mathbf{u}_j$, $j = 0, 1$, in the form

$$u_j(t) = \mathrm{Re}\{v_j(t,s)\} = \mathrm{Re}\{[\exp(-itA_j^{1/2})]h_j\}, \quad j = 0, 1$$
$$h_j = \int_{t_0}^{t_0+T} A_j^{1/2}\{\exp(i\tau A_j^{1/2}\}f_j(\tau)d\tau \tag{7.27}$$

For this scattering problem the FP is taken to be the special case when there are no scatterers present, that is, when $\Omega = \mathbf{R}^3$. Thus, $A_0 : L_2(\mathbf{R}^3) \to L_2(\mathbf{R}^3)$ defined by

$$A_0 u_0 = -\Delta u_0 \quad \text{for all} \quad u_0 \in D(A_0) \tag{7.28}$$

$$D(A_0) = \{u \in L_2(\mathbf{R}^3) : -\Delta u \in L_2(\mathbf{R}^3)\}$$

is self-adjoint in $L_2(\mathbf{R}^3)$ [9], [4].

Our main aim is to calculate the scattered (echo) field $u_s(x, t)$ produced by the signal (incident) field $u_0(x, t)$ where

$$u_s(x, t) = u_1(x, t) - u_0(x, t) \tag{7.29}$$

Now u_1 is defined on a region $\Omega_1 \subset \mathbf{R}^3$ and u_0 on $\Omega = \mathbf{R}^3$. To be able to compare these two quantities we introduce the operator

$$J : L_2(\Omega_1) \to L_2(\mathbf{R}^3)$$
$$Jg(x) = \begin{cases} j(x)g(x) & \text{for } x \in \Omega_1 \\ 0 & \text{for } x \in \mathbf{R}^3 - \Omega_1 \end{cases} \tag{7.30}$$

where $j \in C^\infty(\mathbf{R}^3)$ is such that $0 \le j(x) \le 1$ with $j(x) = 1$ for $|x| \geqslant \delta$ and $j(x) = 0$ in a neighbourhood of B_1. It will be convenient to extend the definition of u_s in (7.29) to the complex plane by defining

$$u_s(x, t) = \mathrm{Re}\{v_s(x, t)\} \tag{7.31}$$

where

$$v_s(x, t) = Jv_1(x, t) - v_0(x, t)$$

with v_j, $j = 0, 1$ defined as in (7.27). Clearly, the far field form of the echo u_s can be obtained from that of v_s.

The calculation of the far field form of u_s and v_s can be based on the theory of wave operators introduced in Chapter 6 and [4] and developed fully in [9].

For the present scattering problem the wave operators are defined by (see Subsection 6.4.1)

$$W_{\pm} = s - \lim_{t \to \pm\infty} \{\exp(itA_0^{1/2})\} J \{\exp(-itA_1^{1/2})\} \tag{7.32}$$

It is proved in [9] that these limits exist and that they define unitary operators $W_{\pm}: L_2(\Omega_1) \to L_2(\mathbf{R}^3)$.

It now follows that for each $h_1 \in L_2(\Omega_1)$

$$\begin{aligned} Jv_1(t) &= J\{\exp(-itA_1^{1/2})\} h_1(x) \\ &= \{\exp(-itA_0^{1/2})\} W_+ h_1(x) + \sigma_t(1) \quad \text{as} \quad t \to +\infty \end{aligned} \tag{7.33}$$

where $\sigma_t(1)$ is an $L_2(\mathbf{R}^3)$ valued function of t which tends to zero in $L_2(\mathbf{R}^3)$ as $t \to +\infty$.

The equations (7.27), (7.31) and (7.33) combine to give

$$v_s(x, t) = \{\exp(-itA_0^{1/2})\} \{W_+ h_1(x) - h_0(x)\} + \sigma_t(1) \quad \text{as} \quad t \to +\infty \tag{7.34}$$

With the assumptions (i) to (v) in mind we see that $v_s(x, t) = 0$ for $t_0 + T \leq t \leq t_0 + |x| - \delta - \delta_0$ and $x \in \mathbf{R}^3$. Furthermore, if we choose

$$t_0 = -|x| + \delta + \delta_0$$

then the arrival time of the signal at the scatterer B_1 will be non-negative. With this understanding we see that

$$J\{\exp(-itA_1^{1/2})\} h_1(x) = \{\exp(-itA_0^{1/2})\} h_0(x) \quad \text{for} \quad t_1 \leq t \leq 0 \tag{7.35}$$

where

$$t_1 = t_0 + T = -|x| + \delta + \delta_0 + T \tag{7.36}$$

Setting $t = 0$ in (7.36) yields

$$Jh_1 = h_0$$

whilst for $t = t_1$

$$\{\exp(it_1 A_0^{1/2})\} J \{\exp(-it_1 A_1^{1/2})\} h_1(x) = h_0(x) \tag{7.37}$$

The scatterer B_1 will be in the far field of the transmitter if either $|x_0| >> 1$ or, by (7.36), $t_1 << -1$. Combining this observation with (7.37) and (7.32) indicates that

$$h_0(x) = W_- h_1(x) + \sigma_{x_0}(1) \quad \text{as} \quad |x_0| \to \infty \tag{7.38}$$

where $\sigma_{x_0}(1)$ is an $L_2(\mathbf{R}^3)$ valued function of x_0 such that $\sigma_{x_0}(1)$ tends to zero in $L_2(\mathbf{R}^3)$ when $|x_0| \to \infty$.

We now introduce, as in Subsection 6.4.1, the scattering operator S defined by

$$S = W_+ W_-^* \tag{7.39}$$

where W_-^* denotes the adjoint of W_-.

Multiplying (7.38) by S gives

$$W_+ h_1(x) = S h_0(x) + \sigma_{x_0}(1) \quad \text{as} \quad |x_0| \to \infty \tag{7.40}$$

This follows since by the unitarity of W_- we have $W_- W_-^* = I$ [9],[4].

We now combine (7.34) and (7.40) to obtain

$$v_s(x, t) = \{\exp(-itA_0^{1/2})\}\{(S - I)h_0(x)\} + \sigma_t(1) + \sigma_{x_0}(1) \tag{7.41}$$

The result (7.41) shows that in the far field an approximation is given in terms of the scattering operator S and the FP data $h_0(x)$. (Notice the symbolic similarity with (7.24).)

The construction of the scattering operator for B_1 can be achieved in terms of an associated generalised eigenfunction expansion. The expansions were introduced in Chapter 6. For convenience we recall some of the salient features here. For details see [9],[4] and Chapter 6.

The operator A_0 is a self-adjoint operator on $L_2(\mathbf{R}^3)$ and has a purely continuous spectrum.

The plane waves

$$w_0(x, p) := (2\pi)^{1/2}\{\exp((ix \cdot p)\}, \quad x, p \in \mathbf{R}^3 \tag{7.42}$$

form a complete family of generalised eigenfunctions for A_0.

For scattering by bounded objects the generalised eigenfunctions are distorted plane waves [9],[6],[4]

$$w_\pm(x, p) = w_0(x, p) + \mathrm{w}_\pm^s(x, p), \quad x \in \Omega_1 \text{ and } p \in \mathbf{R}^3 \tag{7.43}$$

We have seen in Chapter 6 that these distorted plane waves satisfy

$$(\Delta + |p|) w_\pm(x, p) = 0, \quad x \in \Omega_1 \tag{7.44}$$

$$\left\{\frac{\partial w_\pm^s}{\partial |x|} \mp i|p| w_\mp^s\right\}(x, p) = O\left\{\frac{1}{|x|^2}\right\}, \quad |x| \to \infty \tag{7.45}$$

The existence and uniqueness properties of these distorted plane waves can be found in [9],[6] and Chapter 6. Physically, $w_\pm^s(x, p)$, represents the steady state scattered (echo) field when the plane wave (7.42) is scattered by B_1. The far field form of $w_{\pm}^{s}(x, p)$ can be shown to be [9]

$$w_{\pm}^{s}(x,p)=\frac{\exp(\pm|p||x|)}{4\pi|x|}T_{\pm}(|p|\theta,p)+O\left\{\frac{1}{|x|^{2}}\right\} \quad \text{as} \quad |x|\to\infty \tag{7.46}$$

where $\theta = x/|x|$ and $T(p, p')$, the **scattering amplitude** or **differential cross-section** of B_1, is defined for all $p, p' \in \mathbf{R}^3$ such that $|p| = |p'|$.

Following the development in Chapter 6 the plane and distorted plane waves w_0 and $w_{\pm}$ generate the generalised eigenfunction expansions (6.50) to (6.52) and (6.81) to (6.83) respectively. As a consequence, we saw in Subsection 6.4.1 that the wave operators, $W_{\pm}$, had the representation

$$W_{+} = F_0^{*}F_{-} \quad \text{and} \quad W_{-} = F_0^{*}F_{+} \tag{7.47}$$

Combining (7.39) and (7.47) we obtain

$$S = W_{+}W_{-}^{*} = F_0^{*}\hat{S}F_0 \tag{7.48}$$

where

$$\hat{S} = F_{-}F_{+}^{*} \tag{7.49}$$

is called the S-matrix for the scatterer B_1.

In order to be able to make use of (7.41) we need an interpretation of the term $(S - I)h_0$. This can be achieved by first noticing that

$$(S - I)h_0 = (F_0^{*}\hat{S}F_0 - I)h_0 = F_0^{*}(\hat{S} - I)\hat{h}_0 \tag{7.50}$$

For acoustic scattering problems in $\mathbf{R}^3$ it has been shown that

$$(\hat{S}-I)\hat{h}_0=\frac{i|p|}{2(2\pi)^{1/2}}\int_{S^2}T_{+}(p,|p|\theta)\hat{h}_0(|p|\theta)d\theta \tag{7.51}$$

The integration in (7.51) is over points θ of the unit sphere S^2 in $\mathbf{R}^3$. The first proof of the integral representation was given by Shenk [6].

Finally, in this section we investigate the nature of the signal and of the echoin the far field.

The complex wave function $v_0(x, t)$ defined by (7.27) has the Fourier representation

$$v_0(x,t)=(2\pi)^{-1/2}\int_{\mathbf{R}^3}\{\exp(i(x,p-t|p|))\}\hat{h}_0(p)dp \tag{7.52}$$

where

$$\hat{h}_0(p)=(2\pi)^{1/2}i|p|^{-1}\hat{f}_0(-|p|,p) \tag{7.53}$$

and

$$\hat{f}_0(\omega, p) = (2\pi)^{-2} \int_{\mathbf{R}^4} \{\exp(-i(x.p + \omega t))\} f_0(x, t) dx dt \tag{7.54}$$

denotes the four-dimensional Fourier transform of f_0.

The notion of an asymptotic wave function was introduced in Chapters 4 and 6. For this particular problem the asymptotic wave function, u_0^∞, associated with the signal wave field $u_0(x, t) = \text{Re}\{v_0(x, t)\}$ is defined to be (see [9, Chapter 2])

$$u_0^\infty(x, t) = \frac{s(|x| - t, \theta)}{|x|}, \quad x = |x|\theta \tag{7.55}$$

where $s \in L_2(\mathbf{R} \times S^2)$ is defined by

$$\begin{aligned} s(\tau, \theta) &= \text{Re}\left\{(2\pi)^{-1/2} \int_0^\infty \{\exp(i\tau\omega)\}\{-i\omega\} \hat{h}_0(\omega\theta) dw\right\} \\ &= \text{Re}\left\{\int_0^\infty \{\exp(i\tau\omega)\} \hat{f}_0(-\omega, \omega\theta) dw\}\right\} \end{aligned} \tag{7.56}$$

It is proved in [9] that u_0^∞ describes the asymptotic behaviour of u_0 in $L_2(\mathbf{R}^3)$ as $t \to \infty$ in the sense that

$$u_0(\cdot, t) = u_0^\infty(\cdot, t) + \sigma_t(1) \quad \text{as} \quad t \to \infty \tag{7.57}$$

From (7.41) we see that the echo wave field can be represented in the form

$$u_s(x, t) = \text{Re}\{[\exp(itA_0^{1/2})](S - I)h_0(x, t)\} + \sigma_t(1) + \sigma_{x_0}(1) \tag{7.58}$$

The first term on the right-hand side is the same as that for the signal u_0, as defined in (7.27), but with h_0 replaced by $(S - I)h_0$. Consequently, with this in mind, it follows from our treatment of u_0 that we can write

$$u_s(x, t) = u_s^\infty(x, t) + \sigma_t(1) + \sigma_{x_0}(1) \quad \text{as} \quad t \to \infty \tag{7.59}$$

where

$$u_s^\infty(x, t) := \frac{e(|x| - t, \theta)}{|x|}, \quad x = |x|\theta \tag{7.60}$$

and

$$e(\tau, \theta) = \text{Re}\left\{(2\pi)^{-1/2} \int_0^\infty [\exp(i\tau\omega)][(S - I)h_0]^\wedge(\omega, \theta) d\omega\right\} \tag{7.61}$$

where $[\ldots]^\wedge$ denotes the Fourier transform of $[\ldots]$.

From (7.50) and (7.51) it follows that

$$e(\tau : \theta) = \frac{1}{4\pi} \text{Re}\left\{\int_0^\infty [\exp(i\tau\omega)]\omega^2 \int_{S^2} T_+(\omega\theta, \omega\theta')\hat{h}_0(\omega\theta') d\theta' dw\right\} \tag{7.62}$$

and, by using (7.53),

$$e(\tau:\theta)=\frac{1}{2(2\pi)^{1/2}}\mathrm{Re}\left\{i\int_0^{\infty}[\exp(i\tau\omega)]\omega\int_{S^2}T_+(\omega\theta,\omega\theta')\hat{f}_0(-\omega,\omega\theta')d\theta'dw\right\} \tag{7.63}$$

Thus (7.56) and (7.63) provide a representation of the asymptotic signal wave form and the asymptotic echo wave form respectively.

We are now at the stage where we have, in principle at least, a structure for determining the echo field in a scattering problem. However, for it to be a workable proposition in practice we see that we have to be able to determine the WOs and SO associated with the scattering problem of interest. In earlier chapters we have shown how this can be done using generalised eigenfunction expansion theorems. Consequently, we shall, in the remaining chapters of this monograph, tend to concentrate on this aspect for some specific problems which are of practical interest in the physical sciences.

References and Further Reading

[1] H. Amann: *Ordinary Differential Equations, An Introduction to Nonlinear Analysis*, W. de Gruyter, Berlin, 1990.

[2] T. Kato: *Perturbation Theory for Linear Operators*, Springer, New York, 1966.

[3] G.F. Roach: *Greens Functions* (2nd Edn). Cambridge Univ. Press, London, 1970/1982.

[4] G.F. Roach: An *Introduction to Linear and Nonlinear Scattering Theory*, Pitman Monographs and Surveys in Pure and Applied Mathematics **78**, Longman, Essex, 1995.

[5] E.J.P. Schmidt: On scatteringby time dependent potentials, *Indiana Univ. Math. Jour.* **24**(10), 1975, 925–934.

[6] N.A. Shenk: Eigenfunction expansions and scattering theory for the wave equation in an exterior domain, *Arch. Rational Mech. Anal.* **21**, 1966, 120–150.

[7] P.E. Sobolevski: Equationsof parabolic typeina Banach space. *Amer. Math. Soc. Transl.* **49**, 1996, 1–62.

[8] H. Tanabe: *Evolution Equations*, Pitman Monographs and Studies in Mathematics **6**, Pitman, London, 1979.

[9] C.H. Wilcox: *Scattering Theory for the d'Alembert Equation in Exterior Domains*, Lecture Notes in Mathematics, No. 442, Springer, Berlin, 1975.

8

Scattering Processes in Stratified Media

8.1 Introduction

The propagation of acoustic waves in an inhomogeneous medium occupying a domain $\Omega \subset \mathbf{R}^3$ is governed by two functions of $x := (x_1, x_2, x_3) \in \mathbf{R}^3$ namely $\rho(x)$, the density of the medium, and $c(x)$, the local wave speed in the medium.

We have seen that acoustic waves propagating in an inhomogeneous medium can be characterised in terms of an acoustic potential, $u(x, t)$, $(x, t) \in \mathbf{R}^3 \times \mathbf{R}$ which satisfies the partial differential equation [4]

$$\{\partial_t^2 + L(x)\}u(x, t) = F(x, t) \tag{8.1}$$

where

$$L(x) := -c^2(x)\rho(x)\nabla \cdot \left(\frac{1}{\rho(x)}\nabla\right)$$

$F(x, t) :=$ source density function which characterises a signal transmitted into the medium

The equation (8.1) must be supplemented by certain initial conditions which describe the state of the medium before an external signal is incident upon it and by certain conditions at the boundary $\partial\Omega$ of the domain $\Omega \subset \mathbf{R}^3$. With respect to the latter there are two particularly important boundary conditions:

Free Boundary Condition:

$$u(x, t) = 0, \quad x \in \partial\Omega \tag{8.2}$$

Rigid Boundary Condition:

$$\frac{\partial u}{\partial n}(x, t) = 0, \quad x \in \partial\Omega \tag{8.3}$$

where $n := (n_1, n_2, n_3)$ denotes the unit exterior normal to $\partial\Omega$ at points $x \in \partial\Omega$ whenever the normal exists at $x \in \partial\Omega$.

Throughout this chapter we shall assume that the density $\rho(x)$ and wave speed $c(x)$ are bounded in the sense

$$0 < \rho_m \leq \rho(x) \leq \rho_M < \infty \quad \text{and} \quad 0 < c_m \leq c(x) \leq c_M < \infty \tag{8.4}$$

where ρ_m, ρ_M, c_m and c_M are suitable constants. We would remark that we shall also assume in this chapter that ρ and c are continuous. This assumption is sufficient for our present purposes. However, it can be relaxed provided that the meaning of the wave equation (8.1) under such circumstances is clarified.

As in Section 7.4, the signal source, characterised by the term F in (8.1), will be assumed to be localised in space and time. It will be sufficient here to assume that F has the property

$$\text{supp}F \subset \{(x, t) : x_1^2 + x_2^2 + (x_3 - x_3^0)^2 \leq \delta_0,\ T \leq t \leq 0\} \tag{8.5}$$

This being the case then $u(x, t)$ can be characterised as that solution of (8.1) which satisfies

$$u(x, t) \equiv 0 \quad \text{for all} \quad t < T \tag{8.6}$$

Furthermore, (8.5) implies that if initial conditions

$$u(x, 0) = \varphi(x), \quad u_t(x, 0) = \psi(x) \tag{8.7}$$

are imposed on solutions of (8.1) then

$$\text{supp } \varphi \cup \text{supp } \psi \subset \{x \in \mathbf{R}^3 : x_1^2 + x_2^2 + (x_3 - x_3^0)^2 \leq \delta\} \tag{8.8}$$

where $\delta = \delta_0 + c_M\,|T|$. It will then follow that $u(x, t)$ can be characterised for $t \geqslant 0$ as the solution of the homogeneous IVP

$$\{\partial_t^2 + L(x)\}u(x, t) = 0, \quad x \in \Omega, \quad t > 0 \tag{8.9}$$

$$u(x, 0) = \varphi(x), \quad u_t(x, 0) = \psi(x), \quad x \in \Omega \tag{8.10}$$

where it is understood that the initial conditions (8.10) satisfy (8.8). This reduction of an IVP for (8.1) to the homogeneous IVP (8.9), (8.10) is another use of the Duhamel Principle mentioned in Chapters 2 and 7. For the time being we shall confine attention to the homogeneous equation (8.9).

When investigating wave processes in nonhomogeneous media the intention will be to parallel, as far as possible, the analysis used when studying wave processes in homogeneous media.

For homogeneous media problems we saw in Chapters 1, 6 and 7 that the analysis of the given physical problem was centred on representing the physical problem as an abstract problem in some suitably chosen Hilbert space. This being done, the independently established Plancherel theory of Fourier transforms then enabled generalised eigenfunction expansion theorems to be established which provided the means for interpreting the abstract solution forms as physically meaningful quantities. However, when we come to deal with nonhomogeneous media problems then we have to prove that appropriate generalised eigenfunction expansion theorems are available for our use.

In this chapter we confine attention to one of the simpler nonhomogeneous media, namely, stratified media. For simplicity and ease of presentation the construction of solutions will be illustrated for a particular type of nonhomogeneous medium which is typical of a large class of stratified media. However, even in this case the analysis can become quite technical and associated proofs are frequently very lengthy. Consequently, we shall often simply state results. Full details can always be found in the references cited in the text or in the Commentary.

8.2 Hilbert Space Formulation

The evolution of acoustic waves in a medium filling a domain $\Omega \subset \mathbf{R}^3$ is described by the solutions of an IBVP of the form

$$\{\partial_t^2 + L(x)\}u(x, t) = 0, \quad x, t \in \Omega \times \mathbf{R}, \quad t > 0 \tag{8.11}$$

$$u(x, 0) = \varphi(x), \quad u_t(x, 0) = \psi(x), \quad x \in \Omega \tag{8.12}$$

$$Bu(x, t) = 0, \quad (x, t) \in \partial\Omega \times \mathbf{R} \tag{8.13}$$

where (8.13) characterises one or other of the boundary conditions (8.2) or (8.3).

When analysing the IBVP (8.11) to (8.13) we shall make use of the divergence theorem which states that for a sufficiently smooth vector $\mathbf{w}$ defined on $\bar{\Omega} = \Omega \cup \partial\Omega$ the following relation holds [2]

$$\int_\Omega \nabla \cdot \mathrm{w}\, dx = \int_{\partial\Omega} \mathrm{w} \cdot \mathbf{n}\, ds \tag{8.14}$$

where

d_x = volume element in Ω
d_s = surface element in $\partial\Omega$
$\mathbf{n} = \mathbf{n}(x)$ = exterior unit named to $\partial\Omega$ at the point $x \in \partial\Omega$

When $\mathbf{w}(x) = f(x)\nabla g(x)$ where f and g are sufficiently differentiable functions of x then

$$\nabla \cdot (f(x)\nabla g(x)) = (\nabla f \cdot \nabla g)(x) + (f\nabla g)(x) \tag{8.15}$$

If we now set in (8.15)

$$f(x) = v(x) \quad \text{and} \quad \nabla g(x) = (\rho^{-1}\nabla u)(x)$$

then it follows that

$$\begin{aligned}\int_{\Omega} \nabla \cdot (v(x)\overline{\rho^{-1}(x)\nabla u(x))}\, dx &= \int_{\Omega} (\nabla_v(x) \cdot \overline{\rho^{-1}(x)\nabla u(x)} \\ &\quad - v(x)\overline{Lu(x)}c^{-2}(x)\rho^{-1}(x))dx \\ &= \int_{\partial\Omega} v(x)\overline{\frac{\partial u}{\partial n}}(x)\rho^{-1}(x)ds \end{aligned} \tag{8.16}$$

This is a particularly convenient form of (8.14) when discussing IBVPs of the form (8.11) to (8.13). If we now repeat the calculation leading to (8.16) but with v and u interchanged then it is a straightforward matter to establish the relation

$$\int_{\Omega} \{v\overline{Lu} - u\overline{Lv}\}(x)c^{-2}(x)\rho^{-1}(x)dx = \int_{\partial\Omega} \left\{\overline{u}\frac{\partial v}{\partial n} - \overline{v}\frac{\partial u}{\partial n}\right\}(x)\rho^{-1}(x)ds \tag{8.17}$$

We remark that (8.16) and (8.17) are particular forms of Greens identities [2].

The results (8.16) and (8.17) are particularly useful as they provide a good indication of an appropriate Hilbert space structure to use when analysing IBVPs of the form (8.11) to (8.13). Consequently, we shall introduce here the weighted Hilbert space

$$H(w) := L_2(\Omega, c^{-2}(x)\rho^{-1}(x)dx) \tag{8.18}$$

consisting of functions defined on Ω which are square integrable with respect to the weighted measure

$$w(x)dx := c^{-2}(x)\rho^{-1}(x)dx \tag{8.19}$$

and which is endowed with an inner product $(\cdot, \cdot)_{H(w)}$ defined by

$$(u, v)_{H(w)} := \int_{\Omega} \overline{u(x)}v(x)c^{-2}(x)\rho^{-1}(x)dx \tag{8.20}$$

We now introduce an operator A as follows

$$\begin{gathered} A : H(w) \to H(w) \\ Au = Lu \quad \text{for all} \quad u \in D(A) \\ D(A) = \{u \in H(w) : Lu \in H(w),\, Bu(x, \cdot) = 0,\, x \in \partial\Omega\} \end{gathered} \tag{8.21}$$

Hence, bearing in mind (8.17) and (8.20), we can conclude that

$$(Au, v)_{H(w)} = (u, Av)_{H(w)} \tag{8.22}$$

and it follows that A is symmetric, that is, formally self-adjoint, on $H(w)$.

Furthermore, we see that (8.16) implies

$$(Au, u)_{H(w)} = \int_{\Omega} (\nabla u \cdot \overline{\nabla u})(x)\rho^{-1}(x)dx \geqslant 0 \tag{8.23}$$

Hence A is a formally self-adjoint, positive operator on $D(A) \subset H(w)$.

With these several results to hand we are now well placed to give IBVPs of the form (8.11) to (8.13) a Hilbert space realisation which will easily allow the powerful analytical notions and techniques introduced in earlier chapters to be used.

First we see that a realisation in $H(w)$ of the given IBVP (8.11) to (8.13) can be the abstract IVP

$$\{\partial_t^2 + A\}u(x, t) = 0 \tag{8.24}$$

$$u(x, 0) = \varphi(x), \quad u_t(x, 0) = \psi(x), \quad x \in \Omega \tag{8.25}$$

where it is understood that the imposed boundary conditions are accommodated in the definition of $D(A)$.

Following the analysis of waves in homogeneous media given earlier we replace the abstract IVP (8.24), (8.25) involving a partial differential equation by an equivalent IVP for an ordinary differential equation, namely

$$\{d_t^2 + A\}u(t) = 0, \quad u(0) = \varphi, \quad u_t(0) = \psi \tag{8.26}$$

where it is understood

$$u \equiv u(\cdot, \cdot) : \mathrm{t} \to u(\cdot, t) =: u(t) \tag{8.27}$$

If we assume that (8.26) is well-posed then (8.26) has a solution which can be written uniquely in the form

$$u(t) = \{\cos tA^{1/2}\}\varphi + A^{-1/2}\{\sin tA^{1/2}\}\psi \tag{8.28}$$

To settle the wellposedness of (8.26) we again parallel the analysis in homogeneous media and reduce it to the first order system

$$\Psi_t(t) - i\mathbf{M}\Psi(t) = \mathbf{0}, \quad \Psi(0) = \Psi_0 \tag{8.29}$$

where

$$\mathbf{\Psi}(t) = \begin{bmatrix} u \\ u_t \end{bmatrix}(t), \quad \mathbf{\Psi}(0) = \mathbf{\Psi}_0 = \begin{bmatrix} \varphi \\ \psi \end{bmatrix} \tag{8.30}$$

$$\mathrm{M} = i\begin{bmatrix} 0 & -I \\ A & 0 \end{bmatrix} \tag{8.31}$$

and it is understood that

$$\begin{gathered} \Psi = \langle \psi_1, \psi_2 \rangle \in H(w) \times H(w), \\ \mathbf{M} : H(w) \times H(w) \supseteq D(\mathbf{M}) \rightarrow H(w) \times H(w) \end{gathered} \tag{8.32}$$

A precise definition of $D(\mathbf{M})$, the domain of $\mathbf{M}$ will be given below. First, since we are ultimately interested in solutions of (8.11) to (8.13) that can be regarded as "solutions with finite energy" we introduce the integral

$$E(u, \Omega, t) := \int_{\Omega} \{|\nabla u(x,t)|^2 + |u_t(x,t)|^2\} w(x) dx \tag{8.33}$$

and interpret it as the energy of acoustic waves in a nonhomogeneous region Ω at time t. This suggests the introduction of a weighted energy space, $H_E(w)$, defined by

$$H_E(w) := H_D(w) \times H(w) \tag{8.34}$$

where

$H_D(w) :=$ closure of $C^{\infty}(\Omega)$ with respect to the norm

$$\|f\|^2_{H_{D(w)}} := \int_{\Omega} \{|\nabla f(x)|^2\} w(x) dx \tag{8.35}$$

The norm on $H_E(w)$ is defined by

$$\|\mathbf{f}\|^2_{H_{E(w)}} := \int_{\Omega} \{|\nabla f_1(x)|^2 + |f_2(x)|^2\} w(x) dx, \quad \mathbf{f} = \langle f_1, f_2 \rangle \in H_{E(w)} \tag{8.36}$$

Associated with this norm is the inner product

$$(\mathbf{f}, \mathbf{g}) H_{E(w)} : = (\nabla f_1, \nabla g_1)_{H(w)} + (f_2, g_2)_{H(w)} \tag{8.37}$$

where $\mathbf{f} = \langle f_1, f_2 \rangle$ and $\mathbf{g} = \langle g_1, g_2 \rangle$.

In terms of the above notation we shall understand

$$\mathbf{M} : H(w) \times H(w) \supseteq D(\mathbf{M}) \rightarrow H(w) \times H(w) \tag{8.38}$$

where for $\mathbf{f} = \langle f_1, f_2 \rangle \in H(w) \times H(w)$

$$\mathbf{M}\mathbf{f} = i\begin{bmatrix} 0 & -I \\ A & 0 \end{bmatrix}\begin{bmatrix} f_1 \\ f_2 \end{bmatrix} = i\begin{bmatrix} -f_2 \\ Af_1 \end{bmatrix} =: i\langle -f_2, Af_1 \rangle \tag{8.39}$$

However, when the analysis is being conducted in the energy space $H_E(w) \subset H(w) \times H(w)$ we shall understand $\mathbf{M}$ to be such that

$$\mathbf{M} : H_E(w) \supset D(\mathbf{M}) \to H_E(w) \tag{8.40}$$

and say that $\mathbf{f} = \langle f_1, f_2 \rangle \in D(\mathbf{M})$ provided $\mathbf{Mf} \in D(\mathbf{M})$, that is provided $\langle -f_2, Af_1 \rangle \in H_E(w)$. We therefore define $D(\mathbf{M})$ to be

$$D(\mathbf{M}) := \{\mathbf{f} = \langle f_1, f_2 \rangle \in H_E(w) : f_2 \in H_D(w),\ Af_1 \in H(w)\} \tag{8.41}$$

It now follows that $\mathbf{M}$ defined in this way is a positive, formally self-adjoint operator on $D(\mathbf{M})$. To see this, notice first that for $\mathbf{f}, \mathbf{g} \in H_E(w)$

$$\begin{aligned}(\mathbf{Mf}, \mathbf{g})_{H_{E(w)}} &= \left(i\begin{bmatrix}-f_2\\ Af_1\end{bmatrix}, \begin{bmatrix}g_1\\ g_2\end{bmatrix}\right)\\ &= (-i\nabla f_2, \nabla g_1)_{H(w)} + (iAf_1, g_2)_{H(w)}\end{aligned} \tag{8.42}$$

and

$$\begin{aligned}(\mathbf{f}, \mathbf{Mg})_{H_{E(w)}} &= \left(\begin{bmatrix}f_1\\ f_2\end{bmatrix}, i\begin{bmatrix}-g_2\\ Ag_1\end{bmatrix}\right)_{H_E(w)}\\ &= (\nabla f_1, -i\nabla g_2)_{H(w)} + (f_2, iAg_1)_{H(w)}\end{aligned} \tag{8.43}$$

If we now apply (8.16) to the terms containing A on the right-hand side of (8.42) and (8.43) and acknowledge that by virtue of the boundary conditions accommodated in the definition of $D(A)$ all integrated terms, that is terms involving $\partial\Omega$, will vanish, then the required symmetry and positivity of $\mathbf{M}$ on $D(\mathbf{M})$ will follow.

For many problems of practical interest it turns out that the associated operator $\mathbf{M}$ on an energy space $H_E(w)$ is, in fact, a self-adjoint operator on its domain $D(\mathbf{M}) \subset H_E(w)$ [3]. This being the case it is then a straightforward matter, using the results in Section 5.3, to settle the wellposedness of (8.29) and hence of (8.11) to (8.13). Furthermore, whenever solutions exist in $H_E(w)$ then, by definition, they will have the required finite energy property. To complete the analysis of problems of the form (8.11) to (8.13) it then only remains to establish the availability of suitable generalised eigenfunction expansion theorems in order to be able to provide a practical interpretation of terms such as $(\cos tA^{1/2})\varphi$ which appear in the abstract solution form (8.28).

Our approach will be modelled on the analysis of wave processes in homogeneous media which we discussed in Chapters 1, 6 and 7. There we studied the IBVP

$$\begin{gathered}\{\partial_t^2 - \Delta\}u(x, t) = 0, \quad (x, t) \in \Omega \times \mathbf{R}\\ u(x, 0) = \varphi(x), \quad u_t(x, 0) = \psi(x), \quad x \in \Omega\\ u(x, t) \in (bc), \quad (x, t) \in \partial\Omega \times \mathbf{R}\end{gathered}$$

For the free, that is, unperturbed problem we took $\Omega = \mathbf{R}^3$ and introduced the operator

$$A_0 : L_2(\mathbf{R}^3) \to L_2(\mathbf{R}^3) = : H(\mathbf{R}^3)$$
$$A_0 u_0 = -\Delta u_0, \quad u_0 \in D(A_0)$$
$$D(A_0) = \{u \in H(\mathbf{R}^3) : \Delta u \in H(\mathbf{R}^3)\}$$

The IVP for the free problem was then represented in $H(\mathbf{R}^3)$ in the form

$$\{d_t^2 + A_0\}u_0(t) = 0, \quad t \in \mathbf{R}$$
$$u_0(0) = \varphi, \quad u_{0t}(0) = \psi$$

This abstract problem was shown to have a solution in the form

$$u_0(t) = (\cos(tA_0^{1/2}))\varphi + A_0^{-1/2}(\sin(tA_0^{1/2}))\psi$$

A physical interpretation of this abstract solution form was then obtained by using the Plancherel theory of Fourier transforms which yields the following, independently obtained, results [5], [7], [3].

$$F(f)(p) := \hat{f}(p) = \lim_{R\to\infty} \int_{|x|\le R} \overline{w_0(x,p)} f(x)dx$$
$$f(x) = F^*(\hat{f})(x) = \lim_{R\to\infty} \int_{|p|\le R} w_0(x,p)\hat{f}(p)dp$$
$$\Phi(A_0)f(x) = \lim_{R\to\infty} \int_{|p\le R|} w_0(x,p)\Phi|p|^2)\hat{f}(p)dp$$

where $F : H(\mathbf{R}^3) \to H(\mathbf{R}^3)$ denotes the Fourier transform, $F^* = F^{-1}$ and Φ is a suitably "nice" function. Furthermore,

$$w_0(x,p) = \frac{1}{(2\pi)^{3/2}} \exp(ix \cdot p), \quad x, p \in \mathbf{R}^3$$

is the usual Fourier kernel. We notice that w_0 satisfies the Helmholtz equation

$$(\Delta + |p|^2)w_0(x,p) = 0 \quad \text{for all} \quad x, p \in \mathbf{R}^3$$

Thus w_0 appears to be an eigenfunction of the operator A_0 with associated eigenvalue $|p|^2$. However, this cannot be the case as direct calculation shows that $w_0 \notin H(\mathbf{R}^3)$ and therefore w_0 must be a generalised eigenfunction of A_0. Consequently, the above results from the Fourier Plancherel theory are referred to collectively as a generalised eigenfunction expansion theorem.

We shall indicate, in the next few sections, how similar results can be obtained when dealing with scattering problems in stratified media.

8.3 Scattering in Plane Stratified Media

A medium is said to be **plane stratified** if the local sound speed, $c(x)$, and density, $\rho(x)$, are functions of a single Cartesian coordinate. This we will indicate by writing

$$c(x) = c(x_1, x_2, x_3) = c(x_3) =: c(z) \tag{8.44}$$

$$\rho(x) = \rho(x_1, x_2, x_3) = \rho(x_3) =: \rho(z) \tag{8.45}$$

where for ease of presentation we have written $x_3 = z$.

We shall investigate here acoustic wave phenomena in the particular case when

$$c(z) = \begin{cases} c_1, & 0 \leq z < h \\ c_2, & z \geqslant h \end{cases} \tag{8.46}$$

and

$$\rho(z) = \begin{cases} \rho_1, & 0 \leq z < h \\ \rho_2, & z \geqslant h \end{cases} \tag{8.47}$$

where c_1, c_2, ρ_1, ρ_2 and h are bounded constants.

If we assume that this particular stratified medium occupies the half space

$$\mathbf{R}^3_+ := \{x = (x_1, x_2, x_3) \in \mathbf{R}^3 : (x_1, x_2) \in \mathbf{R}^2 \text{ and } z > 0\} \tag{8.48}$$

then the IBVP governing acoustic wave phenomena in such a material, when $z = 0$ is a free surface, has the form (see (8.1) and (8.9))

$$\{\partial_t^2 + L(x)\}u(x, t) = 0, \quad x \in \mathbf{R}^3_+ \quad \text{and} \quad t > 0 \tag{8.49}$$

$$u(x, 0) = \varphi(x) \quad \text{and} \quad u_t(x, 0) = \psi(x), \quad x \in \mathbf{R}^3_+ \tag{8.50}$$

$$u(x, t) = 0, \quad z = 0, \quad t \geqslant 0 \tag{8.51}$$

where the differential expression $L(x)$ in (8.1) now assumes the modified form defined by

$$L(x)u(x) := -c^2(z)\rho(z)\nabla \cdot \left(\frac{1}{\rho(z)} \nabla u(x) \right) \tag{8.52}$$

The IVP (8.49)to (8.53) models acoustic wave phenomena in a medium comprising a layer of thickness h, sound speed c_1 and density ρ_1, lying underneath a layer having sound speed c_2 and density ρ_2. Acoustic wave processes in such media present a phenomena which we have so far not encountered. Specifically, waves can be trapped in the layer $0 \leq z \leq h$ as a consequence of total reflection at an interface between two media. We shall discuss these **trapped** or **guided waves** in more detail when we discuss generalised eigenfunction expansion theorems associated with (8.49) to (8.53).

As in Section 8.2, an IBVP such as (8.49) to (8.53) is conveniently analysed in a weighted L_2 setting. With this in mind we see that all the structures introduced in Section 8.2 are available here provided we take

$$\Omega = \mathbf{R}^3_+$$

and modify the weight function (8.19) to read

$$w(x)dx = c^{-2}(z)\rho^{-1}(z)dx \tag{8.53}$$

The solution of the IBVP (8.49) to (8.53) is based on being able to construct an operator

$$A : H(w) \supseteq D(A) \rightarrow H(w) := L_2(\mathbf{R}^3_+, w(x)dx) \tag{8.54}$$

which is a self-adjoint realisation in $H(w)$ of $L(x)$ that incorporates the boundary condition (8.51). A first step in this direction is given by (8.21) provided that the weight function and the boundary condition are modified as in (8.53) and (8.51) respectively. It then follows, as in Section 8.2, that the operator A introduced here is a positive, formally self-adjoint (symmetric) operator on its domain $D(A)$. It now remains to show that A, as defined here, is actually a self-adjoint operator on its domain. This will then ensure that we have the spectral theorem available for interpreting the solutions of abstract IVPs centred on the operator A. In establishing this result we shall use the following notations (see Chapter 3)

$$\mathcal{D}(\mathbf{R}^3_+) = \text{Schwartz space of infinitely differentiable functions on } \mathbf{R}^3_+ \text{ with compact support} \tag{8.55}$$

$$\mathcal{D}'(\mathbf{R}^3_+) = \text{dual space of all distributions on } \mathbf{R}^3_+ \tag{8.56}$$

We also introduce the Sobolev Hilbert spaces [1], [3]

$$H^m(\mathbf{R}^3_+) = L_2(\mathbf{R}^3_+) \bigcap \{u : D^\alpha u \in L_2(\mathbf{R}^3_+) \text{ for } |a| \leq m\}$$

where

$$\alpha = (\alpha_1, \alpha_2, \alpha_3, \ldots, \alpha_n) = \text{multi-index of non-negative integers}$$

$$|\alpha| = \sum_{k=1}^{n} \alpha_k = \text{order of } \alpha$$

$$f_\alpha(x) = f_{\alpha_1\alpha_2\alpha_3\ldots\alpha_n}(x)$$

$$D^\alpha f(x) = D_1^{\alpha_1} D_2^{\alpha_2} \ldots D_n^{\alpha_n}, \quad \text{where } D_k^m := \frac{\partial^m}{\partial x_k^m}$$

The space $H^m(\mathbf{R}^3_+)$ is a Hilbert space with inner product

$$(u, v)_m := \int_{\mathbf{R}^3_+} \sum_{|\alpha| \leq m} \overline{D^\alpha u(x)} D^\alpha v(x) dx \tag{8.57}$$

and $\mathcal{D}(\mathbf{R}^3_+)$ is a linear subset of $H^m(\mathbf{R}^3_+)$ for all m. Of particular use to us here is the fact that

$$H_0^1(\mathbf{R}^3_+) := \text{closure of } \mathcal{D}(\mathbf{R}^3_+) \text{ in } H^1(\mathbf{R}^3_+) \tag{8.58}$$

is a closed subspace of $H^1(\mathbf{R}^3_+)$. It is known [1] that all elements of $H_0^1(\mathbf{R}^3_+)$ satisfy the homogeneous Dirichlet condition required in (8.21).

In proving that the operator A is self-adjoint we shall make use of the following general result.

Theorem 8.1. *Let H be a Hilbert space. A symmetric operator $B : H \to H$ which is such that*

$$\text{Range of } B =: \text{Ran}B = H$$

is self-adjoint.

Proof. If $B : H \to H$ is a symmetric operator then evidently $B \subset B^*$. Consequently, it is sufficient to show that every element $g \in D(B^*)$ is also an element of $D(B)$. Therefore, let $g \in D(B^*)$ and $B^*g = g^*$. Then since $\text{Ran}B = H$ there exists an element $h \in D(B)$ such that $Bh = g^*$. Consequently, for each $f \in D(B)$ and because B has been assumed to be symmetric we obtain

$$(Bf, g) = (f, g^*) = (f, Bh) = (Bf, h)$$

Since $\text{Ran}B = H$ then we have $g = h$. Hence $g \in D(B)$ as required. ∎

With this preparation we now prove the following result.

Theorem 8.2. *Range of* $(I + A) = \text{Ran}\ (I + A) = H(w)$.

Proof. If $f \in \text{Ran}\ (I + A)$ then there exists an element $u \in D(A)$ such that $(I + A)u = f$.

We want to show that for *any* $f \in H$ there exists a $u \in D(A)$ such that $(I + A)u = f$.

Now, for all $u \in D(A)$ and $v \in H^1(\mathbf{R}^3_+)$ see that (8.16) and (8.20) imply

$$(Au, v) = -\int_{\mathbf{R}^3_+} \nabla \cdot \left(\frac{1}{\rho(z)} \nabla \bar{u}(x) \right) v(x) dx \tag{8.59}$$

$$= \int_{\mathbf{R}^3_+} (\nabla \bar{u} \cdot \nabla v)(x) \rho^{-1}(z) dx \tag{8.60}$$

$$= (f - u, v)_{H(w)} \quad \text{for all} \quad v \in H_0^1(\mathbf{R}^3_+) \tag{8.61}$$

Consequently, for all $v \in H_0^1(\mathbf{R}^3_+)$

$$(f, v)_{H(w)} = \int_{\mathbf{R}^3_+} \overline{f(x)} v(x) w(x) dx$$

$$= \int_{\mathbf{R}^3_+} \{(\nabla \bar{u} \cdot \nabla v)(x) \rho^{-1}(z) + \bar{u}(x) v(x)\} dx \tag{8.62}$$

If now we set

$$\{u, v\} = \int_{\mathbf{R}^3_+} \{(\nabla \bar{u} \cdot \nabla v)(x) \rho^{-1}(z) + \bar{u}(x) v(x)\} dx \tag{8.63}$$

then we see that (8.63) defines an inner product on $H_0^1(\mathbf{R}^3_+)$ which is equivalent to the inner product $(\cdot, \cdot)_1$, defined by (8.57). Furthermore, for all $f \in H(w)$ and $v \in H_0^1(\mathbf{R}^3_+)$

$$|(f, v)_{H(w)}| \leq \|f\|_{H(w)} \|v\|_{H(w)} \leq \|f\|_{H(w)} \{v, v\}^{1/2} \tag{8.64}$$

the last inequality following because

$$\|v\|_{H(w)} = \int_{\mathbf{R}^3_+} |v(x)|^2 w(x) dx \leq \{v, v\} \tag{8.65}$$

for all $v \in H_0^1(\mathbf{R}^3_+)$.

For the sake of illustration we now introduce, bearing (8.60) and (8.61) in mind, a linear functional Φ_f defined according to

$$\Phi_f(v) := (f, v)_{H(w)} = \{u, v\} \tag{8.66}$$

The expression (8.66) taken in conjunction with the results (8.62) to (8.65) provides an expression of the Riesz representation theorem. Therefore we can conclude that for each $f \in H(w)$ there is a $u \in H_0^1(\mathbf{R}^3_+)$ such that (8.62) holds for all $v \in H_0^1(\mathbf{R}^3_+)$.

Before completing the proof we first emphasise that the "divergence operator" appearing in the definitions of L and $D(A)$ (see (8.1) and (8.21)) must be understood in the sense of distributions. Consequently, if a vector $\mathbf{W} = (W_1, W_2, W_3)$ is such that $W_j \in L_2(\mathbf{R}^3_+)$, $j = 1, 2, 3$ then $(\nabla \cdot \mathbf{W}) \in L_2(\mathbf{R}^3_+)$ if and only if

$$\int_{\mathbf{R}^3_+} (\mathrm{W} \cdot \nabla \varphi)(x) dx = -\int_{\mathbf{R}^3_+} (\nabla \cdot \mathrm{W})(x) \varphi(x) dx \quad \text{for all} \quad \varphi \in \mathcal{D}(\mathbf{R}^3_+) \tag{8.67}$$

Now $u \in H_0^1(\mathbf{R}_+^3)$ by construction. Furthermore, (8.62) with $v \in \mathcal{D}(\mathbf{R}_+^3) \subset H_0^1(\mathbf{R}_+^3)$ taken together with (8.67) and (8.59) to (8.61) implies

$$-\nabla \cdot \left(\frac{1}{\rho(z)} \nabla u(x) \right)(x) = (f - u)(x) w(z) \tag{8.68}$$

each side of which defines an element in $L_2(\mathbf{R}_+^3)$. Hence we can conclude that $u \in D(A)$ and that $(I + A)u = f$, as required. ■

8.3.1 The Eigenfunctions of A

We have seen that A, defined in (8.21) is a positive, self-adjoint operator in $H(w)$. Hence, it has a spectral family $\{E_\lambda, \lambda \geqslant 0\}$ and corresponding spectral representation (see Section 5.2)

$$A = \int_0^\infty \lambda dE_\lambda$$

$$\Phi(A) = \int_0^\infty \Phi(\lambda) dE_\lambda$$

where Φ is a bounded, Lebesgue measurable function of λ. However, a practical difficulty centred on these spectral representations concerns the actual determination of the spectral family $\{E_\lambda\}$. The situation can be eased, as we have already mentioned in the previous chapter, by using generalised eigenfunction expansion theorems of a similar form to those indicated in Section 8.2 and Chapters 1, 6 and 7. Therefore one of our principal aims is to determine the generalised eigenfunctions of A that have a clear wave theoretic interpretation. Consequently, paralleling as far as possible the analysis of homogeneous media problems, we will require that in the present case the generalised eigenfunctions of A, denoted by $\psi(x)$, should be characterised by the following properties.

The generalised eigenfunctions in our present case are solutions of the differential equation

$$L(x)\psi(x) := -c^2(x)\rho(x)\nabla \cdot \left\{ \frac{1}{\rho(x)} \nabla \psi(x) \right\} = \lambda \psi(x), \quad \lambda \in \mathbf{R}_+ \tag{8.69}$$

Solutions of (8.69) will be expected to satisfy

$$\psi(x) = \text{locally in } D(A) \text{ (i.e. } \varphi\psi \in \mathcal{D}(A)) \text{ for each } \varphi \in \mathcal{D}(\mathbf{R}^3) \tag{8.70}$$

$$\psi(x) = \text{locally in } \mathbf{R}_+^3 \tag{8.71}$$

Furthermore, the $\psi(x)$ will be expected to satisfy certain boundary and interface conditions. Specifically, if we introduce the notation

$$x = (x_1, x_2, x_3) =: (y, z) \tag{8.72}$$

where $y = (x_1, x_2) \in \mathbf{R}^2$ and $x_3 = z$ then we will require that ψ satisfies the following.

Boundary Condition on $z = 0$

$$\psi(x)|_{z=0} = \psi(y,0) = 0, \quad y \in \mathbf{R}^2 \tag{8.73}$$

Interface Conditions $z = h$

$$\psi(y, h^+) = \psi(y, h^-) \tag{8.74}$$

$$\frac{1}{\rho_2}\frac{\partial \psi}{\partial z}(y, h^+) = \frac{1}{\rho_1}\frac{\partial \psi}{\partial z}(y, h^-) \tag{8.75}$$

where $h^+ = h + 0$ and $h^- = h - 0$

We shall look for solutions of the IBVP (8.69) to (8.75) in the form

$$\psi(x) = \psi(y, z) = \{\exp(ip{\cdot}y)\}\theta(z) \tag{8.76}$$

where $p := (p_1, p_2)$ and $p{\cdot}y = p_1x_1 + p_2x_2$.

The form (8.76) is suggested by taking the Fourier transform of the above IBVP with respect to the variables x_1 and x_2 with the aim of reducing the partial differential equation involved to an equivalent ordinary differential equation.

Substitution of (8.76) into (8.69), (8.73), (8.74) and (8.75) leads to the following ordinary differential equations and boundary conditions.

$$\left\{\frac{d^2}{dz^2} + \left(\frac{\lambda}{c_1^2} - |p|^2\right)\right\}\theta(z) = 0 \quad \text{for } 0 \le z < h \tag{8.77}$$

$$\left\{\frac{d^2}{dz^2} + \left(\frac{\lambda}{c_2^2} - |p|^2\right)\right\}\theta(z) = 0 \quad \text{for } z > h \tag{8.78}$$

$$\theta(0) = 0 \tag{8.79}$$

$$\theta(h^+) = \theta(h^-) \tag{8.80}$$

$$\frac{1}{\rho_2}\frac{\partial \theta}{\partial z}(h^+) = \frac{1}{\rho_1}\frac{\partial \psi}{\partial z}(h^-) \tag{8.81}$$

It is clear that $\psi(z)$ as defined in (8.76) is not an element of $H(w)$, for this reason it is referred to as a generalised (improper) eigenfunctionof $L(x)$ and hence A. The quantities $\psi(z)$ are referred to as **reduced** (generalised) eigenfunctions since they satisfy the reduced equations (8.77), (8.78).

In constructing solutions of (8.77) to (8.81) the following notations will be used

$$\xi := \left(\frac{\lambda}{c_2^2} - |p|^2\right)^{1/2} \quad \text{and} \quad \eta := \left(\frac{\lambda}{c_1^2} - |p|^2\right)^{1/2} \tag{8.82}$$

$$\xi \geqslant 0 \text{ for } \lambda \geqslant c_2^2|p|^2 \text{ and } \xi = i\xi' \text{ with } \xi' > 0 \text{ for } \lambda < c_2^2|p|^2 \tag{8.83}$$

$$\eta \geqslant 0 \text{ for } \lambda \geqslant c_1^2|p|^2 \text{ and } \eta = i\eta' \text{ with } \eta' > 0 \text{ for } \lambda < c_1^2|p|^2 \tag{8.84}$$

For any $p \in \mathbf{R}^2$ and $\lambda \in \mathbf{R}_+$ such that $\lambda \neq c_1^2|p|^2$ and $\lambda \neq c_2^2|p|^2$ the general solution of (8.77), (8.78) and (8.79) has the form

$$\begin{aligned} \psi(z) &= \alpha \sin \eta z, & 0 < z < \mathrm{h} \\ &= \beta \exp(i\xi z) + \gamma \exp(-i\xi z), & z > \mathrm{h} \end{aligned} \tag{8.85}$$

where α, β and γ are suitable constants.

This function will also satisfy (8.80), (8.81) if and only if

$$\begin{aligned} &\beta \exp(i\xi h) + \gamma \exp(-i\xi h) = \alpha \sin \eta h \\ &\beta \exp(i\xi h) + \gamma \exp(-i\xi h) = \alpha \frac{\rho_2}{\rho_1} \frac{\eta}{i\xi} \cos \eta h \end{aligned} \tag{8.86}$$

Solving (8.85) and (8.86) for β and γ we obtain

$$\beta = \frac{1}{2}\left(\alpha \sin \eta h - \alpha \frac{\rho_2}{\rho_1} \frac{i\eta}{\xi} \cos \eta h\right) \exp(-i\xi h) \tag{8.87}$$

$$\gamma = \frac{1}{2}\left(\alpha \sin \eta h + \alpha \frac{\rho_2}{\rho_1} \frac{\eta}{i\xi} \cos \eta h\right) \exp(i\xi h) \tag{8.88}$$

Hence the solutions of (8.77) to (8.79) are given by (8.85) to (8.88) with α an arbitrary constant.

It remains to determine the values of λ and p for which (8.77) to (8.79) give bounded functions of $z \in \mathbf{R}_+$. In this connection we first notice the following.

(i) If ξ is real (that is $\lambda > c_2^2|p|^2$) then $\theta(z)$ is bounded.

(ii) If $\xi = i\xi'$ is purely imaginary (that is $\lambda < c_2^2|p|^2$) then $\theta(z)$ is bounded if and only if $\gamma = 0$, that is if and only if

$$\xi' = -\frac{\rho_2}{\rho_1} \eta \cot \eta h \tag{8.89}$$

(iii) If $\xi = i\xi'$ is purely imaginary and η is real then (8.89) has solutions. If both $\xi = i\xi'$ and $\eta = i\eta'$ are purely imaginary then (8.89) has no solutions. To see this notice that (8.89) may be written

$$\xi' = -\frac{\rho_2}{\rho_1} \eta' \cot \eta' h \tag{8.90}$$

and this has no solutions with $\xi' > 0$ and $\eta' > 0$.

In the special case when $c_1 = c_2$ and $\rho_1 = \rho_2$ then L reduces to the negative Laplacian (with a multiply factor c) in $\mathbf{R}^3_+$ which has an associated Dirichlet condition. In this case we have seen in the earlier chapters that there is a complete set of improper eigenfunctions which can be defined in the form

$$\psi(x, p) = \exp\{i(x_1p_1 + x_2p_2 + x_3p_3)\} - \exp(i(x_1p_1 + x_2p_2 - x_3p_3)) \quad (8.91)$$

where $x = (x_1, x_2, x_3)$ and $p = (p_1, p_2, p_3)$.

In physical terms(8.91) describes a plane wave propagating towards the boundary $x_3 = 0$ of $\mathbf{R}^3_+$ together with a reflected wave. We discuss more general cases below. We shall see that if $c_1 > c_2$ in (8.46) then the improper eigenfunctions are much like those displayed in (8.91) and consist of an incident and a reflected wave. However if $c_1 < c_2$ another class of improper eigenfunctions arises. In addition to improper eigenfunctions like (8.91) we will now obtain functions localised near the region$(0 \leq x_3 \leq h)$. They correspond physically to waves that are **trapped** as a result of total reflection at an interface between the two media.

Bearing in mind the observations (i), (ii) and (iii) above, and the definitions (8.82), we can obtain the following results.

Case 1: $c_1 \geqslant c_2$

In this case it is obvious that $c_1^2|p|^2 \geqslant c_2^2|p|^2$. Hence there are improper eigenfunctions for every $\lambda > c_2^2|p|^2$ and none for $\lambda < c_2^2|p|^2$.

Case 2: $c_1 < c_2$

It is clear that here $c_1^2|p|^2 < c_2^2|p|^2$. Hence, there are improper eigenfunctions for every $\lambda > c_2^2|p|^2$ and none for $\lambda < c_1^2|p|^2$.

In addition there are improper eigenfunctions for those values of λ and p which satisfy

$$c_1^2|p|^2 < \lambda < c_2^2|p|^2 \quad (8.92)$$

Case 2 is the more interesting case. Consequently, we shall assume for the remainder of this chapter that

$$c_1 < c_2 \quad (8.93)$$

A detailed study of both cases can be found in [9]. However, as mentioned earlier, it is frequently the case that many of the proofs in this area are very technical and long. Since in this monograph our main interest is not so much in the detailed proof of a result as in its application a number of the main results will simply be stated and their applications reviewed.

8.3.2 The Wave Eigenfunctions of A

The required generalised eigenfunctions are characterised by (8.69), (8.72), (8.73) and (8.76). These eigenfunctions, as we have already mentioned, are of two main types will be called **free wave eigenfunctions** and **guided wave eigenfunc-**

tions. When discussing such eigenfunctions there are a number of cases to be considered.

Case 2a:

$$\lambda > c_2^2|p|^2 > c_1^2|p|^2 \quad \text{where} \quad |p|^2 = p_1^2 + p_2^2 \tag{8.94}$$

In this case functions satisfying (8.69) to (8.73) and (8.76) exist. They are denoted by ψ_0, referred to as free wave eigenfunctions and, with a slight abuse of notation, can be expressed in the form

$$\psi_0(x, p, \lambda) \equiv \psi_0(y, z, p, \lambda) = (2\pi)^{-1}\{\exp i(p\cdot y)\}\psi_0(z, p, \lambda) \tag{8.95}$$

where

$$\psi_0(z, p, \lambda) = a(p, \lambda)\begin{cases} \sin \eta z \\ 0 < z < h \\ \beta(\xi, \eta)\exp(i\xi(z-h)) + \gamma(\xi, \eta)\exp(-i\xi(z-h)) \\ z > h \end{cases} \tag{8.96}$$

It can be shown that the positive normalising constant $a(p, \lambda)$ can be conveniently taken to be

$$a(p, \lambda) = \left(\frac{\rho_2}{4(\pi\xi)|\beta(\xi, \eta)|}\right)^{1/2}$$

In physical terms the eigenfunction $\psi_0(y, z, p, \lambda)$ characterises an acoustic field with time dependence $\exp(-it\lambda^{1/2})$ (see (8.49) and (8.69)). With (8.85) in mind this field can be interpreted as a plane wave which propagates in the region $z > h$, is refracted at this interface $z = h$, (totally) reflected at the boundary $z = 0$ and refracted again at the interface $z = h$.

Case 2b:

$$c_1^2|p|^2 < \lambda < c_2^2|p|^2 > \quad \text{where} \quad |p|^2 = p_1^2 + p_2^2 \tag{8.97}$$

For those values of λ which satisfy (8.97) the functions $\psi_0(y, z, p, \lambda)$ defined by (8.95) and (8.96) together with (8.82), (8.77) and (8.78) still satisfy (8.69) to (8.73), and (8.76). However, in this case (8.97) implies that ξ in (8.82) is pure imaginary thus, recalling (8.83), we write

$$\xi = i\xi' \quad \text{where} \quad \xi' = \left(|p|^2 - \frac{\lambda}{c_2^2}\right)^{1/2} > 0 \tag{8.98}$$

In this case (8.82) indicates that η is real and positive. It follows that the boundedness condition is satisfied by $\psi_0(y, z, p, \lambda)$ in (8.95) and satisfies the boundedness condition (8.89) if and only if

$$\gamma(i\xi', \eta) = 0 \tag{8.99}$$

or, recalling (8.88), if and only if

$$\xi' = -\frac{\rho_2}{\rho_1}\eta \cot \eta h \tag{8.100}$$

When λ and $|p|$ are such that (8.97) holds then (8.100) is equivalent to the following sequence of relations

$$h\eta = \left(k - \frac{1}{2}\right)\pi + \tan^{-1}\frac{\rho_1 \xi'}{\rho_2 \eta} \tag{8.101}$$

For values of $k = 1, 2, \ldots$, the relation (8.101) defines a functional relation between $|p|$ and

$$\omega = \lambda^{1/2} \tag{8.102}$$

Solutions of this relationship will be denoted

$$\lambda = \lambda_k(|p|), \quad \omega = \omega(|p|) = \lambda_k(|p|)^{1/2} \tag{8.103}$$

each one of which represents a relation between the wave number $|p|$ of the plane wave in $\psi_0(y, z, p, \lambda)$ and the corresponding frequencies ω. In the general theory of wave motions relations such as those just described are known as **dispersion relations**.

The functions ψ_k defined by

$$\psi k(y,z,p) = \psi_0(y,z,p,\lambda_k(|p|)), \quad k = 1, 2, \ldots \tag{8.104}$$

satisfy, by construction, (8.69) to (8.71) and (8.76) for all eigenvalues λ which satisfy (8.97). These functions have the specific form, bearing in mind (8.76) and (8.82),

$$\psi_k(y, z, p) = (2\pi)^{-1}\{\exp i(p\cdot y)\}\psi_k(z, p) \tag{8.105}$$

where

$$\psi_k(z, p) = a_k(p)\begin{cases}\sin \eta_k(|p|)z, & 0 < z < h. \\ \sin \eta_k(|p|)h \exp(-\xi_k'(|p|)(z-h)), & z > h\end{cases} \tag{8.106}$$

$$\eta_k(|p|) = \left(\frac{\lambda_k(|p|)}{c_1^2} - |p|^2\right)^{1/2} \tag{8.107}$$

$$\xi_k'(|p|) = \left(|p|^2 - \frac{\lambda_k(|p|)}{c_2^2}\right)^{1/2} \tag{8.108}$$

In (8.106) the quantity $a_k(p)$ is a positive constant which can be conveniently determined by the normalising condition

$$\int_0 |\psi_k(z,p)|^2 w(z)dz := \int_0^\infty |\psi_k(z,p)|^2 c^{-2}(z)\rho^{-1}(z)dz = 1 \tag{8.109}$$

The eigenfunctions $\psi k(z, p)$ represent an acoustic field having a time dependence $\exp(-i\omega_k(|p|))$which characterises a plane wave which is trapped in the layer $0 \leq z \leq h$ by reflection at $z = 0$ and total internal reflection at the interface $z = h$. In the region $z > h$ the acoustic field is exponentially damped in the z direction and propagates strictly in the horizontal direction defined by the vector $p = (p_1, p_2)$.

8.3.3 Generalised Eigenfunction Expansions

The generalised eigenfunctions introduced in the previous subsections have different wave theoretic interpretations depending on the relative magnitudes of λ, $|p|$, c_1 and c_2. We saw that there were two classes of wave eigenfunctions namely the free and the guided. These two classes can be conveniently characterised by defining

$$p_1 = \frac{\pi c_1}{2h(c_2^2 - c_1^2)^{1/2}}, \quad p_k = (2k-1)p_1 \tag{8.110}$$

$$\Omega_0 := \{(p, \lambda) : p \in \mathbf{R}^2 \text{ and } c_2^2|p|^2 < \lambda\} \subset \mathbf{R}^3 \tag{8.111}$$

$$\Omega_k = \{p : |p| > p_k, k = 1, 2, \ldots\} \tag{8.112}$$

The required expansion theorems can be obtained by following the very precise but lengthy analytical techniques developed in Wilcox [7]. Before stating the results we recall and just gather together, purely for convenience, the modified notations we use when dealing with this particular stratified medium.

Notation Summary

$$x = (x_1, x_2, x_3) =: (y, z) \in \mathbf{R}^3, \; y := (x_1, x_2) \in \mathbf{R}^2, \quad z = x_3$$

$$\mathbf{R}_+^3 := \{x \in \mathbf{R}^3 : y \in \mathbf{R}^2, z > 0\}, \quad \text{see (8.48)}$$

$$w(x)dx = c^{-2}(z)\rho^{-1}(z)dx = c^{-2}(z)\rho^{-1}(z)dydz = w(z)dx, \quad \text{see (8.53)}$$

$$H(w) = L_2(\mathbf{R}_+^3, w(x)dx), \quad \text{see (8.54)}$$

With these particular notations the operator A in (8.21) will be modified accordingly.

Theorem 8.3. *For every $f \in H(w)$ the limits*

$$(F_0 f)(p,\lambda) = \hat{f}_0(p,\lambda) = L_2(\Omega_0) - \lim_{M\to\infty} \int_0^M \int_{|y|\le M} \overline{\psi_0(y,z,p,\lambda)} f(y,z)\omega(z)dx \tag{8.113}$$

$$(F_k f)(p,\lambda) = \hat{f}_k(p,\lambda) = L_2(\Omega_k) - \lim_{M\to\infty} \int_0^M \int_{|y|\le M} \overline{\psi_k(y,z,p)} f(y,z)w(z)dx \tag{8.114}$$

where $k = 1, 2, \ldots$ exist and satisfy the Parseval relation

$$\|f\|^2_{H(w)} = \sum_{k=0}^{\infty} \|f_k\|^2_{L_2(\Omega_k)} \tag{8.115}$$

Furthermore if we define

$$\Omega_0^M := \{(p,\lambda) : p \in \mathbf{R}^2 \quad \text{and} \quad c_2^2|p|^2 < \lambda < M\} \tag{8.116}$$

$$\Omega_k^M := \{p : p < |p| < M \quad \text{for} \quad k = 1, 2, \ldots\} \tag{8.117}$$

then it can be shown that the limits

$$f_0(y,z) = H(w) - \lim_{M\to\infty} \int_{\Omega_0^M} \psi_0(y,z,p,\lambda)\hat{f}_0(p,\lambda)dpd\lambda \tag{8.118}$$

$$f_k(y,z) = H(w) - \lim_{M\to\infty} \int_{\Omega_k^M} \psi_k(y,z,p)\hat{f}_k(p)dp \ldots, \quad k = 1, 2, \ldots \tag{8.119}$$

exist and moreover

$$f(y,z) = H(w) - \lim_{M\to\infty} \sum_{k=0}^{\infty} f_k(y,z) \tag{8.120}$$

With the understanding we adopted in Chapters 1, 6 and 7 the above results can be written in the following more concise symbolic form

$$\hat{f}_0(p,\lambda) = \int_{\mathbf{R}^3_+} \overline{\psi_0(y,z,p,\lambda)} f(y,z)w(z)dx \tag{8.121}$$

$$f_0(y,z) = \int_{\Omega_0} \psi_0(y,z,p,\lambda)\hat{f}_0(p,\lambda)dpd\lambda \tag{8.122}$$

$$\hat{f}_k(p,\lambda) = \int_{\mathbf{R}^3_+} \overline{\psi_k(y,z,p)} f(y,z)w(z)dx \tag{8.123}$$

$$f_k(y,z) = \int_{\psi_k} (y,z,p)\hat{f}_k(p)dp, \quad k = 1, 2, \ldots \tag{8.124}$$

$$f(y,z) = \sum_{k=0}^{\infty} f_k(y,z) \tag{8.125}$$

The relations (8.121) to (8.125) are the required generalised eigenfunction expansions. They provide a spectral representation associated with the operator A in the sense that for all $f \in D(A)$

$$(F_0(Af))(p,\lambda) = \widehat{(Af)}_0(p,\lambda) = \lambda(F_0 f)(p,\lambda) = \lambda \hat{f}_0(p,\lambda) \tag{8.126}$$

$$(F_k(Af))(p) = \widehat{(Af)}_k(p) = \lambda_k(|p|)(F_k, f)(p) = \lambda_k(|p|)\hat{f}_k(p) \tag{8.127}$$

where $k = 1, 2, \ldots$. The relations (8.121) to (8.125) define a **modal decomposition** appropriate for the model we are investigating in this section.

Once these modal decompositions are established then we can obtain the required physical solutions from abstract solution forms obtained when dealing with the operator A introduced in (8.54). Specifically, if the IBVP (8.49) to (8.52) is realised in the weighted L_2-space $H(w)$ then paralleling the analysis for homogeneous media but now using A rather than A_0 we can represent the solutions of (8.49) to (8.52) in the form

$$u(x, t) = u(y, z, t) = \mathrm{Re}\{v(x, t)\} = \mathrm{Re}\{v(y, z, t)\} \tag{8.128}$$

where $v(x, t)$ is the complex form of the required solution and is defined as an abstract quantity in the form

$$v(\cdot\,, t) \equiv v(\cdot\,, \cdot\,, t) = (\exp\{-itA^{1/2}\})f \tag{8.129}$$

where

$$f := \varphi + iA^{-1/2}\psi \in H(w) \tag{8.130}$$

it being assumed that $\varphi \in H(w)$ and $\psi \in D(A^{-1/2})$.

By applying the generalised eigenfunction expansion theorem (8.121) to (8.124) a physical form for the required solution can be obtained from (8.129). This yields the modal decomposition

$$v(x,t) = v(y,z,t) = \sum_{k=0}^{\infty} v_k(y,z,t) \tag{8.131}$$

where

$$v_0(y,z,t) = \int_{\Omega_0} \psi_0(y,z,p,\lambda)\{\exp(-it\lambda^{1/2})\}\hat{f}_0(p,\lambda)dpd\lambda \tag{8.132}$$

$$v_k(y,z,t) = \int_{\Omega_k} \psi_k(y,z,p)\{\exp(-itw_k(|p|))\}\hat{f}_k(p)dp \tag{8.133}$$

where $k = 1, 2, \ldots$.

8.3.4 Some Remarks about Asymptotic Wave Functions

If the representation (8.76) is substituted into the modal decomposition (8.133) then the spectral integrals for the associated guided modes, $k \geqslant 1$, take the form

$$v_k(y,z,t)=\frac{1}{2\pi}\int_{\Omega_k}\{\exp(-i(x\cdot p-t\omega_k(|p|)))\}\psi_k(z,p)\hat{f}_k(p)dp, \quad k=1,2,\dots \tag{8.134}$$

where $\psi_k(z,p)$ is defined in (8.105).

The behaviour for large t of these integrals can be calculated by the method of stationary phase. As might be suspected this is a lengthy procedure. However, the manner in which this is done is worked through in considerable detail in [7]. The following convergence theorem can be obtained using this technique.

Theorem 8.4. *For each $k \geqslant 1$*

(i) *there exists an element $v_k^{\infty}(\cdot,\cdot,t) \in H(w)$ for all $t > 0$ and the mapping $t \to v_k^{\infty}$ is continuous*

(ii) *the element $v_k^{\infty}(\cdot,\cdot,t) \in H(w)$ is an asymptotic wave function for the modal wave vk$(\cdot,\cdot,t)$ in the sense that*

$$\lim_{t\to\infty}\left\|v_k(\cdot,\cdot,t)-v_k^{\infty}(\cdot,\cdot,t)\right\|=0 \tag{8.135}$$

To investigate the asymptotic behaviour of the free mode we begin by substituting (8.95) into (8.132) to obtain

$$v_0(y,z,t)=\frac{1}{2\pi}\int_{\Omega_0}\{\exp(i(y\cdot p-t\lambda^{1/2}))\}\psi_0(z,p,\lambda)\hat{f}_0(p,\lambda)dpd\lambda$$

The representation (8.96) for ψ_0 implies that

$$v_0(y,z,t) = v_0^{+}(y,z-h,t)+ v_0^{-}(y,z-h,t) \tag{8.136}$$

where (see (8.82), (8.87), (8.88))

$$v_0^{+}(y,z,t)=\frac{1}{2\pi}\int_{\Omega_0}\left\{\exp\left(i\left(y\cdot p+z\xi-t\lambda^{1/2}\right)\right)\right\}a(p,\lambda)\beta(\xi,\eta)\hat{f}_0(p,\lambda)dpd\lambda \tag{8.137}$$

$$v_0^{-}(y,z,t)=\frac{1}{2\pi}\int_{\Omega_0}\left\{\exp\left(i\left(y\cdot p-z\xi-t\lambda^{1/2}\right)\right)\right\}a(p,\lambda)\gamma(\xi,\eta)\hat{f}_0(p,\lambda)dpd\lambda \tag{8.138}$$

If we make the change of variable

$$(p, \lambda) \rightarrow (p, g) \quad \text{where} \quad q = \xi = \left(\frac{\lambda}{c_2^2} - |p|^2 \right)^{1/2}$$

then (8.137) can be written (see (8.96) and (8.130) to (8.133))

$$v_0^+(y, z, t) = \frac{1}{(2\pi)^{3/2}} \int_{q \geqslant 0} \{\exp(i(y \cdot p + zq - t\omega(p, q)))\} \hat{f}(p, \lambda) dp dq \tag{8.139}$$

where

$$\hat{f}(p, \lambda) - c_2^2 \rho_2^{1/2} (2|q|)^{1/2} \left\{ \frac{\beta(\xi, \eta)}{|\beta(\xi, \eta)|} \right\} \hat{f}_0(p, \lambda) \tag{8.140}$$

and

$$\lambda = \lambda(p, q) = \omega(p, q)^2 = c_2^2 (|p|^2 + q^2) \tag{8.141}$$

Similarly by making the change of variable in (8.138)

$$(p, \lambda) \rightarrow (p, q) \quad \text{where} \quad q = -\xi = \left(-\frac{\lambda}{c_2^2} + |p|^2 \right)^{1/2}$$

we obtain

$$v_0^-(y, z, t) = \frac{1}{(2\pi)^{3/2}} \int_{q \leq 0} \{\exp(i(y \cdot p + zq - t\omega(p, q)))\} \hat{f}(p, \lambda) dp dq \tag{8.142}$$

where $\hat{f}$ is defined in (8.140).

If we now add (8.139) and (8.142), bearing in mind (8.136) then we obtain

$$v_0(y, z + h, t) = \frac{1}{(2\pi)^{3/2}} \int_{\mathbf{R}^3} \{\exp(i(y \cdot p + zq - t\omega(p, q)))\} \hat{f}(p, \lambda) dp dq \tag{8.143}$$

for all $z \geqslant 0$. Hence (8.143) and (8.141) imply that in the half space $z \geqslant h$ the quantity $v_0(y, z, t)$ coincides with a solution, in $L_2(\mathbf{R}^3)$, of the d'Alembert equation with propagation constant c_2.

The discussions in Chapter 1 and in [3, Chapter 8], indicate that the right-hand side of (8.143) defines an asymptotic wave function v^∞ in $L_2(\mathbf{R}^3)$ of the typical form

$$v^\infty(y, z, t) = \frac{G(r - c_2 t, \theta)}{r} \tag{8.144}$$

where

$$r^2 = |y|^2 + z^2 \quad \text{and} \quad \theta = \frac{(y,z)}{r}$$

It then follows that if

$$v_0^\infty(y,z,t) = \begin{cases} v^\infty(y, z-h, t), & z \geqslant h \\ 0 & 0 \leq z \leq h \end{cases} \tag{8.145}$$

then

$$\lim_{t\to\infty} \left\| v_0(\cdot,\cdot,t) - v_0^\infty(\cdot,\cdot,t) \right\|_{H(w)} = 0 \tag{8.146}$$

References and Further Reading

[1] R.A. Adams: *Sobolev Spaces*, Academic Press, New York, 1975.

[2] G.F. Roach: *Greens Functions* (2nd Edn), Cambridge Univ. Press, London, 1970/1982.

[3] G.F. Roach: *An Introduction to Linear and Nonlinear Scattering Theory*, Pitman Monographs and Surveys in Pure and Applied Mathematics, Vol. 78, Longman, Essex, 1995.

[4] E. Skudrzyk: *The Foundations of Acoustics*, Springer, New York, 1971.

[5] E.C. Titchmarsh: *Introduction to the Theory of Fourier Integrals*, Oxford University Press, 1937.

[6] H.F. Weinberger, *A First Course in Partial Differential Equations*, Blaisdell, Waltham, MA, 1965.

[7] C.H. Wilcox: *Scattering Theory for the d'Alembert Equation in Exterior Domains*, Lecture Notes in Mathematics, No. 442, Springer, Berlin, 1975.

[8] C.H. Wilcox: Transient Electromagnetic Wave Propagation in a Dielectric Waveguide, Proc. Conf. on the Mathematical Theory of Electromagnetism, Instituto Nazionale di Alto Mathematica, Rome, 1974.

[9] C.H. Wilcox: Spectral Analysis of the Perkeris Operator in the Theory of Acoustic Wave Propagation in Shallow Water, *Arch. Rat. Mech. Anal.* **60**, 1976, 259–300.

9

Scattering in Spatially Periodic Media

9.1 Introduction

In the previous chapter we discussed acoustic wave scattering processes in a nonhomogeneous medium, specifically in a stratified medium. There we saw that the stratification could give rise to trapped wave phenomena.

In this chapter we outline some additional effects that can occur when the stratification of the medium is periodic. Such media and associated scattering effects arise in many areas of practical interest, for instance, in non-destructive testing, ultrasonic medical diagnosis and radar and sonar problems to name but a few.

For ease of presentation we shall in this chapter only be concerned with problems in one space dimension, that is, with string problems. Additional studies in more than one space dimension are referred to in the Commentary. This will allow us to avoid having to investigate trapped wave phenomena and so be able to concentrate on periodic media effects on acoustic wave propagation. We will also take the opportunity here of writing some of the results in terms of an associated spectral family rather than immediately in terms of generalised eigenfunctions (see Chapter 5).

9.2 The Mathematical Model

In this chapter we will investigate the initial value problem

$$\{\partial_t^2 - L(x)\}u(x, t) = f(x)\exp(-i\omega t), \quad (x, t) \in \mathbf{R} \times \mathbf{R}_+ \tag{9.1}$$

$$u(x, 0) = \varphi(x), \quad u_t(x, 0) = \psi(x) \tag{9.2}$$

where

$$L(x) = -c^2(x)\left\{\rho(x)\frac{\partial}{\partial x}\left(\frac{1}{\rho(x)}\frac{\partial}{\partial x}\right) - q(x)\right\} \tag{9.3}$$

$$c, \rho, q \text{ are given functions of } x \text{ with period } p \tag{9.4}$$

$$\omega \text{ is a frequency parameter satisfying } \omega > 0 \tag{9.5}$$

The source term f and the initial values $\varphi(x)$ and $\psi(x)$ are assumed to vanish outside a bounded interval of **R** (see also Chapter 8).

The initial value problem (9.1), (9.2) describes the vibrations of a periodically nonhomogeneous infinite string generated by a time harmonic source term. We would emphasise that the periodicity terms ω and p are not necessarily the same.

Again for the sake of simplicity we shall assume that all data functions are sufficiently "nice" in the sense that we will assume

$$c, \rho, q \in C(\mathbf{R}) \quad \text{and} \quad \varphi, \psi, f \in C_0^\infty(\mathbf{R}) \tag{9.6}$$

As mentioned in Chapter 8, regularity conditions such as (9.6) can always be relaxed provided that the equation (9.1) can be interpreted meaningfully.

The analysis of this problem follows along the lines used in Chapter 8. Consequently, we introduce the weighted Hilbert space

$$H(w) = L_2(\mathbf{R}, c^{-2}(x)\rho^{-1}(x)dx) \tag{9.7}$$

which is endowed with the inner product $(\cdot\,,\cdot)_{H(w)}$ defined by

$$(u, v)_{H(w)} := \int_{\mathbf{R}} \overline{u(x)}v(x)w(x)dx \tag{9.8}$$

where

$$w(x) = c^{-2}(x)\rho^{-1}(x)$$

The initial value problem (9.1), (9.2) can be represented in $H(w)$ by introducing an operator A as follows

$$\mathrm{A} : H(w) \to H(w) \tag{9.9}$$

$$Au = Lu \quad \text{for all} \quad u \in D(A)$$

$$D(A) = \{u \in H(w) \colon Lu \in H(w)\}$$

Then, as in Chapter 8, we can show that A is a positive, formally self-adjoint operator on $D(A) \subseteq H(w)$.

With this preparation we see that the given IVP (9.1), (9.2) involving a partial differential equation can be realised, in $H(w)$, as an abstract IVP for an ordinary differential equation in the form

$$\{d_t^2 + A\}u(t) = f\exp(-i\omega t) \tag{9.10}$$

$$u(0) = \varphi, \quad u_t(0) = \psi \tag{9.11}$$

where, as in the previous chapter, we understand

$$u = u(\cdot\,,\cdot): t \to u(,\cdot\, t) = u(t) \in H(w) \tag{9.12}$$

The solution of (9.10), (9.11) can be written in the form

$$u(t) = \left(\cos\left(tA^{1/2}\right)\right)\varphi + A^{-1/2}\left(\sin t\left(A^{1/2}\right)\right)\psi + \int_0^t G(t,s)g(s)ds \tag{9.13}$$

where $g(t) = f\exp(-i\omega t)$ and where $G(t, s)$ is the Green's function for the IVP (9.10), (9.11) [8] which satisfies

$$LG(t, s) = \delta(\mathrm{t} - s) \tag{9.14}$$

$$G(0, s) = 0, \quad Gt(0, s) = 0 \tag{9.15}$$

In this case $G(t, s)$ has the form [8]

$$G(t,s) = \frac{\theta_1(t)\theta_2(s) - \theta_1(s)\theta_2(t)}{W(\theta_1,\theta_2)} \tag{9.16}$$

where θ_1, θ_2 are any two linearly independent solutions of $L\theta = 0$ and

$$W(\theta_1, \theta_2) = \theta_1(s)\theta_2'(t) - \theta_1'(s)\theta_2(t) \tag{9.17}$$

If we choose

$$\theta_1(t) = \sin tA^{1/2}, \quad \theta_2(t) = \cos tA^{1/2} \tag{9.18}$$

then

$$W(\theta_1, \theta_2) = -A^{1/2} \tag{9.19}$$

Furthermore, after a completely straight forward but rather lengthy calculation, we obtain

$$\int_0^t G(t,s)g(s)ds = \frac{f}{(A-\omega^2)}\left\{\exp(-i\omega t) + \frac{i\omega}{A^{1/2}}\sin\left(tA^{1/2}\right) - \cos\left(tA^{1/2}\right)\right\} \tag{9.20}$$

Applying the functional calculus for self-adjoint operators indicated in (5.41) we can now write (9.13) in the following form

$$u(\cdot,t)=\int_0^\infty \cos(\lambda t)d(E_\lambda\varphi)+\int_0^\infty \frac{\sin\sqrt{\lambda}t}{\sqrt{\lambda}}d(E_\lambda\psi)+\int_0^\infty \xi(\lambda,t)d(E_\lambda f) \tag{9.21}$$

where

$$\xi(\lambda,t)=\frac{1}{(\lambda-\omega^2)}\left\{\exp(-i\omega t)+\frac{i\omega}{\sqrt{\lambda}}\sin\left(t\sqrt{\lambda}\right)-\cos\left(t\sqrt{\lambda}\right)\right\} \tag{9.22}$$

and $\{E_\lambda\}$ denotes the spectral family of A.

The result (9.21) indicates that the asymptotic behaviour of $u(x, t)$ as $t \to \infty$ is, as might be expected, closely related to the properties of $\sigma(A)$, the spectrum of A, and of $\{E_\lambda\}$ the spectral family of A.

The properties of $\sigma(A)$ are, as we have seen in Chapter5, governed by

$$R(\lambda) := (\mathrm{A} - \lambda)^{-1} \tag{9.23}$$

the **resolvent** of A.

The relationship between the spectral family, $\{E_\lambda\}$, and the resolvent $R(\lambda)$ is given by Stone's formula [9],[7] which in our present case assumes the form

$$\begin{aligned}&((E_{\beta+0}+E_{\beta-0})f,g)_{H(w)}-((E_{\alpha+0}+E_{\alpha-0})f,g)_{H(w)}\\&=\frac{1}{\pi i}\lim_{\tau\downarrow 0}\int_\alpha^\beta((R(\varepsilon+i\tau)-R(\varepsilon-i\tau))f,g)_{H(w)}d\varepsilon\end{aligned} \tag{9.24}$$

where $f, g \in H(w)$.

Since A is positive and has no eigenvalues (see Theorem 9.8 for another proof of this result) then $E_0 = 0$ and $E_{\beta+0} = E_{\beta-0}$ so that (9.24) reduces to

$$(E_\beta f,g)=\frac{1}{2\pi i}\lim_{\tau\downarrow 0}\int_0^\beta((R(\varepsilon+i\tau)-R(\varepsilon-i\tau))f,g)_{H(w)}d\varepsilon \tag{9.25}$$

Thus we see that the computation of $\{E_\lambda\}$ can be reduced to examining the limiting behaviour of $R(z)f$ as $\operatorname{Im} z \downarrow 0$ and $\operatorname{Im} z \uparrow 0$.

If we write

$$U(x, z) = R(z)f(x) \tag{9.26}$$

then we see that $U(x, z)$ satisfies

$$c^2(x)\left\{\rho(x)\frac{\partial}{\partial x}\left(\frac{1}{\rho(x)}\frac{\partial}{\partial x}\right)-q(x)\right\}U(x,z)+zU(x,z)=f \tag{9.27}$$

where the coefficients are periodic. The equation (9.27) is an example of Hill's equation, the classical theory of which can be applied to an investigation of the resolvent.

An analysis of Hill's equation can be made using results of Floquet's theory of periodic differential equations. For convenience we give in the next section a short account of Floquet theory and display a number of results that are particularly relevant for our present study.

9.3 Elements of Floquet Theory

In this section we shall be concerned with the general second order equation

$$a_0(x)y''(x) + a_1(x)y'(x) + a_2(x)y(x) = 0 \tag{9.28}$$

in which the coefficients $a_r(x)$ are complex valued, piecewise continuous and periodic, all with the same period p. Consequently,

$$a_r(\mathrm{x} + p) = a_r(x), \quad 0 \leq r \leq 2 \tag{9.29}$$

where p is a non-zero constant. In investigating (9.28) we shall assume that the usual theory of linear differential equations without singular points applies [3], [4], [2].

We notice that if $\zeta(x)$ is a solution of (9.28), since the equation is unaltered if x is replaced by $(x + p)$, then $\zeta(x + p)$ also is a solution. However, in general (9.28) need not have any non-trivial solutions with period p. Nevertheless the following result holds.

Theorem 9.1. *There is a non-zero constant β and a nontrivial solution $\zeta(x)$ of* (9.28) *such that*

$$\zeta(x + p) = \beta\zeta(x) \tag{9.30}$$

Proof. Let $\theta_k(x)$, $k = 1, 2$ be linearly independent solutions of (9.28) which satisfy

$$\theta_1(0) = 1, \quad \theta_1'(0) = 0, \quad \theta_2(0) = 0, \quad \theta_2'(0) = 1 \tag{9.31}$$

Since $\theta_k(x + p)$, $k = 1, 2$ also linearly independent solutions of (9.28) there are constants C_{ij}, $i, j \leq$ such that

$$\theta_1(x + p) = C_{11}\theta_1(x) + C_{12}\theta_2(x) \tag{9.32}$$

$$\theta_2(x + p) = C_{21}\theta_1(x) + C_{22}\theta_2(x) \tag{9.33}$$

where the matrix $\mathrm{C} := [C_{ij}]$ is non-singular.

Every solution $\zeta(x)$ of (9.28) has the form

$$\zeta(x) = \alpha_1\theta_1(x) + \alpha_2\theta_2(x) \tag{9.34}$$

where α_1, α_2 are constants. Now, recognising (9.32) and (9.33) we see that (9.30) holds provided

$$\begin{vmatrix} C_{11}-\beta & C_{12} \\ C_{21} & C_{22}-\beta \end{vmatrix} = 0$$

that is, provided

$$\beta^2 - (C_{11} + C_{22})\beta + \det C = 0 \tag{9.35}$$

This quadratic equation for β has at least one non-trivial solution since $\det C \neq 0$.

An alternative form of (9.35) can be obtained as follows. Using (9.32) to (9.34) it follows that

$$C_{11} = \theta_1(p), \quad C_{12} = \theta_1'(p), \quad C_{21} = \theta_2(p), \quad C_{22} = \theta_2'(p) \tag{9.36}$$

Hence, using Liouvilles' formula for the Wronskian [2], [4]

$$\det C = W(\theta_1, \theta_2)(p) = \exp\left\{-\int_0^p \left\{\frac{a_1(\eta)}{a_0(\eta)}\right\} d\eta\right\} \tag{9.37}$$

We remark that we have also used the fact that $W(\theta_1, \theta_2)(0) = 1$. Thus (9.35) can be written

$$\beta^2 - (C_{11} + C_{22})\beta + \exp\left\{-\int_0^p \left\{\frac{a_1(\eta)}{a_0(\eta)}\right\} d\eta\right\} = 0 \tag{9.38}$$

One of the main results of the Floquet theory is the following.

Theorem 9.2. *There are linearly independent solutions $\zeta_1(x)$ and $\zeta_2(x)$ of* (9.28) *such that either*

(i)

$$\zeta_1(x) = \{\exp(m_1 x)\} v_1(x)$$
$$\zeta_2(x) = \{\exp(m_2 x)\} v_2(x)$$

where m_1, m_2 are constants, not necessarily distinct, and v_1, v_2 are periodic functions of x of period p, or

(ii)

$$\zeta_1(x) = \{\exp(mx)\} v_1(x)$$
$$\zeta_2(x) = \{\exp(mx)\}\{x v_1(x) + v_2(x)\}$$

where m is a constant and v_1, v_2 are periodic functions of x of period p.

Proof. Assume that (9.35) has distinct solutions β_1 and β_2 then, by Theorem 9.1 there are non-trivial solutions $\zeta_1(x)$, $\zeta_2(x)$ of (9.28) such that

$$\zeta_k(x+p) = \beta_k \zeta_k(x), \quad k = 1, 2 \tag{9.39}$$

and these solutions are linearly independent.

Since $\beta_k, k = 1, 2$ are non-zero we can define m_k, k $= 1, 2$ so that

$$\exp(pm_k) = \beta_k \tag{9.40}$$

We then define

$$v_k(x) = \{\exp(-m_k x)\}\zeta_k(x) \tag{9.41}$$

and it follows from (9.39) and (9.40) that

$$v_k(x+p) = \{\exp(-m_k(x+p))\}\beta_k\zeta_k(x) = v_k(x)$$

Hence by (9.41)

$$\zeta_k(x) = \{\exp(m_k x)\}v_k(x)$$

where the v_k have period p.

When (9.35) has a repeated root β then we define m so that

$$\exp(pm) = \beta$$

From Theorem 9.1 it follows that there is a non-trivial solution $\Psi_1(x)$ of (9.28) such that

$$\Psi_1(x+p) = \beta\Psi_1(x) \tag{9.42}$$

Let $\Psi_2(x)$ be a solution of (9.28) so that $\Psi_1(x)$, $\Psi_2(x)$ are not linearly dependent.

Since $\Psi_2(x+p)$ also satisfies (9.28) there are constants d_1, d_2 such that

$$\Psi_2(x+p) = d_1\Psi_1(x) + d_2\Psi_2(x) \tag{9.43}$$

From (9.42) and (9.43) we obtain

$$W(\Psi_1, \Psi_2)(x+p) = \beta d_2 W(\Psi_1, \Psi_2)(x)$$

Therefore, using Liouville's formula for the Wronsksian

$$\beta d_2 = \exp\left\{-\int_x^{x+p}\left\{\frac{a_1(\eta)}{a_0(\eta)}\right\}d\eta\right\} = \exp\left\{-\int_0^p\left\{\frac{a_1(\eta)}{a_0(\eta)}\right\}d\eta\right\}$$

which follows since the integrand has period p. Furthermore the right-hand form in this expression is equal to β^2 since β is also a solution of (9.38). Hence $d_2 = \beta$. From (9.43) we now obtain

$$\Psi_2(x+p) = d_1\Psi_1(x) + \beta\Psi_2(x) \tag{9.44}$$

There are now two possibilities to consider.
Possibility 1: $d_1 = 0$
In this case (9.44) implies

$$\Psi_2(x+p) = \beta\Psi_2(x)$$

This, together with (9.42), indicates that we have the same situation as in (9.39) but with $\beta_1 = \beta_2 = \beta$. Consequently, this case is now covered by part (i) of the theorem.

Possibility 2: $d_1 \neq 0$
In this case we define

$$V_1(x) = \{\exp(-mx)\}\Psi_1(x)$$

and

$$V_2(x) = \{\exp(-mx)\}\Psi_2(x) - \left(\frac{d_1}{p\beta}\right)xV_1(x)$$

Then by (9.42) and (9.44) $V_1(x)$ and $V_2(x)$ have period p. Therefore since

$$\Psi_1(x) = \{\exp(mx)\}V_1(x)$$
$$\Psi_2(x) = \{\exp(mx)\}\left\{\left(\frac{d_1}{p\beta}\right)xV_1(x) + V_2(x)\right\}$$

then part (ii) is covered with $\zeta_1(x) = \Psi_1(x)$ and $\xi_2(x) = (p\beta/d_1)\Psi_2(x)$.

Definition 9.3. The solutions $\beta_k, k = 1, 2$ of (9.35) or (9.38) are called the characteristic multipliers of (9.28) and the $m_k, k = 1, 2$ are called the characteristic exponents of (9.28). ■

9.3.1 Hill's Equation

A particular form of (9.28) which frequently occurs in practice is the equation

$$\{P(x)y'(x)\}' + Q(x)y(x) = 0 \tag{9.45}$$

where P and Q are real-valued, periodic functions with the same period p. Furthermore, we shall assume that P is continuous and non-vanishing and that P' and Q are piecewise continuous. Thus, (9.45) is a particular form of (9.28) and is referred to as **Hill's equation**.

If we assume that

$$\int_0^p \left\{\frac{a_1(\eta)}{a_0(\eta)}\right\} d\eta = 0 \tag{9.46}$$

and set

$$B(x) = \int_0^x \left\{\frac{a_1(\eta)}{a_0(\eta)}\right\} d\eta$$

then multiplying (9.28) by $\{a_0(x)\}^{-1}\exp(B(x))$ we obtain an equation of the form (9.45) with

$$P(x) = \exp(B(x)), \quad Q(x) = \left\{\frac{a_2(x)}{a_0(x)}\right\}\exp(B(x))$$

Using (9.46) we see that B, P and Q all have period p.

For the equation (9.45) the quadratic equation (9.38) becomes, bearing in mind (9.36),

$$\beta^2 - (\theta_1(p) + \theta_2'(p))\beta + 1 = 0 \tag{9.47}$$

From the familiar properties of quadratic equations we see that the characteristic multipliers β_1, β_2 satisfy

$$\beta_1\beta_2 = 1 \tag{9.48}$$

The solutions $\theta_k(x)$ of (9.45) are real valued because $P(x)$ and $Q(x)$ are real valued.

The properties of the solution $\theta_k(x), k = 1, 2$ of (9.45) depend on the properties of the solutions $\beta_k, k = 1, 2$ of (9.47). In this connection we introduce the real number D, called the **discriminant** of (9.45), defined by

$$D := \theta_1(p) + \theta_2'(p) \tag{9.49}$$

Using again familiar properties of quadratic equations we see that when discussing the roots of (9.47) there are five cases to be considered.

Case 1: $D > 2$

It follows from (9.47) that the $\beta_k, k = 1, 2$ are real and distinct. Furthermore, they are positive but not equal to unity. Consequently, using (9.48) it follows that there is a real number m such that

$$\exp(mp) = \beta_1 \quad \text{and} \quad \exp(-mp) = \beta_2$$

This follows from (9.40). Therefore, by part (i) of Theorem 9.2 we obtain

$$\zeta_1(x) = \{\exp(mx)\}v_1(x) \quad \text{and} \quad \zeta_2(x) = \{\exp(-mx)\}v_2(x) \tag{9.50}$$

where v_1 and v_2 each have period p.

Case 2: $D < -2$

This case is similar to Case 1 except that here β_1 and β_2 are now negative but they are not equal to (-1). Hence the m in (9.50) must be replaced by $(m + i\pi/p)$.

Case 3: $-2 < D < 2$

It follows from (9.47) that the $\beta_k, k = 1, 2$ are not real but distinct. However, they are complex conjugates and (9.48) indicates that they have unit modulus. Hence there exists a real number α such that either $0 < \alpha p < \pi$ or $-\pi < \alpha p < 0$ and

$$\{\exp(i\alpha p)\} = \beta_1 \quad \text{and} \quad \{\exp(-i\alpha p)\} = \beta_2$$

Therefore, using part (i) of Theorem 9.2 we can write

$$\zeta_1(x) = \{\exp(i\alpha x)\}v_1(x) \quad \text{and} \quad \zeta_2(x) = \{\exp(-i\alpha x)\}v_2(x) \tag{9.51}$$

where the $v_k, k = 1, 2$ each have period p.

Case 4: $D = 2$

In this case (9.47) has equal roots $\beta_k = 1, \mathrm{k} = 1, 2$. Consequently, we have to decide which part of Theorem 9.2 applies. The standard theory of algebraic equations indicates, bearing in mind that $\beta_k, k = 1, 2$, that part (i) of Theorem 9.2 applies if rank $(C - I) = 0$ whilst part (ii) applies if rank $(C - I) = 1$. There are two possibilities to consider.

Possibility 1: $\theta_2(p) = \theta_1'(p) = 0$

Since

$$W(\theta_1, \theta_2)(p) = W(\theta_1, \theta_2)(0) = 1$$

then we have

$$\theta_1(p)\theta_2'(p) = 1$$

$$D = \theta_1(p) + \theta_2'(p) = 2$$

Consequently

$$\theta_1(p) = \theta_2'(p) = 1$$

Hence rank $(C - I) = 0$ and part (ii) of Theorem 9.2 applies.

Since $\beta_k = 1$, $k = 1, 2$ then the characteristic exponents $m_k, k = 1, 2$ must be zero. Consequently, Theorem 9.2 implies that

$$\zeta_k(x) = v_k(x), \quad k = 1, 2$$

where each of the v_k have period p.

Possibility 2: $\theta_2(p)$ and $\theta_1'(p)$ not both zero

In this case rank $(C - I) \neq 0$ and part (ii) of Theorem 9.2 applies with $m = 0$. Hence

$$\zeta_1(x) = v_1(x) \quad \text{and} \quad \zeta_2(x) = xv_1(x) + v_2(x)$$

where the $v_k, k = 1, 2$ each have period p.

Case 5: $D = -2$

In this case (9.47) has equal roots $\beta_k = -1$, $k = 1, 2$. We proceed as in Case 4 and consider two possibilities.

Possibility 1: $\theta_2(p) = \theta_1'(p) = 0$

Arguing as in Case 4 we have that rank $(C + 1) = 0$ and part (i) of Theorem 9.2 applies with $m_1 = m_2 = \dfrac{i\pi}{p}$.

Hence

$$\xi_1(x) = \left\{\exp\left(\frac{i\pi x}{p}\right)\right\} v_1(x) \quad \text{and} \quad \xi_2(x) = \left\{\exp\left(\frac{i\pi x}{p}\right)\right\} v_2(x) \tag{9.52}$$

where v_1 and v_2 each have period p. It now follows from (9.52) that

$$\zeta_k(x + p) = -\zeta_k(x), \quad k = 1, 2$$

and hence that *all* solutions of (9.45) satisfy

$$\zeta(x + p) = -\zeta(x)$$

Possibility 2: $\theta_2(p)$ and $\theta_1'(p)$ are not both zero.

Here rank $(C + 1) \neq 0$ and part (ii) of Theorem 9.2 holds with $m = \dfrac{i\pi}{p}$. Hence

$$\zeta_1(x) = u_1(x) \text{ and } \zeta_2(x) = xu_1(x) + u_2(x)$$

where $u_k(x) = \left\{\exp\left(\dfrac{i\pi x}{p}\right)\right\} v_k(x), \quad k = 1, 2$

Thus the $u_k, k = 1, 2$ satisfy

$$u_k(x + p) = -u_k(x), \quad k = 1, 2$$

Definition 9.4. A function f with the property $f(x+p) = -f(x)$ for all x is said to be **semi-periodic** with **semi-period** p.

It follows from Definition 9.4 that a function which is semi-periodic with semi-period p is periodic with period $2p$.

We shall need the following results in later sections.

Theorem 9.5. (i) *If* $|D| > 2$ *then all non-trivial solutions of* (9.45) *are unbounded on* $(-\infty, \infty)$.
(ii) *If* $|D| < 2$ *then all non-trivial solutions of* (9.45) *are bounded on* $(-\infty, \infty)$.

Proof. If $D > 2$ then we have Case 1 and (9.50) holds. It is clear that any linear combination of $\zeta_1(x)$ and $\zeta_2(x)$ is unbounded either as $x \to \infty$ or as $x \to -\infty$. The argument for $D < -2$ (which is Case 2 in the above) is similar. Part (i) is proved.

If $|D| < 2$ then Case 3 applies and (9.51) holds. Hence

$$|\zeta_k(x)| = |v_k(x)|, \quad k = 1, 2$$

Since v_k,k $= 1, 2$ are periodic in $(-\infty, \infty)$ they are also bounded there. Hence the $\zeta_k, k = 1, 2$ are bounded in $(-\infty, \infty)$. This establishes part (ii). ■

Definition 9.6. The equation (9.45) is said to be

(a) **unstable** if all non-trivial solutions are unbounded on $(-\infty, \infty)$.

(b) **conditionally stable** if there exists a non-trivial solution which is bounded in $(-\infty, \infty)$.

(c) **stable** if all solutions are bounded in $(-\infty, \infty)$.

Using Cases 4 and 5 the following theorem can be established.

Theorem 9.7 *The equation* (9.45) *has non-trivial solutions with period p if and only if* $D = 2$ *and with semi-period p if and only if* $D = -2$. *All solutions of* (9.45) *have either period p or semi-period p if, in addition,* $\theta_2(p) = \theta_1'(p) = 0$.

Proof. This is left as an exercise.

9.4 Solutions of the Mathematical Model

If we set

$$V(x, z) = \frac{U(x, z)}{\sqrt{\rho(x)}} \tag{9.53}$$

$$q_0(x) = q(x) - \sqrt{\rho(x)}\left\{\frac{1}{\rho(x)}\left(\sqrt{\rho(x)}\right)'\right\} \tag{9.54}$$

$$f_1 = \frac{f(x)}{\sqrt{\rho(x)}} \tag{9.55}$$

then (9.27) can be written in the form

$$c^2(x)\{d_x^2 - q_0(x)\}V(x, z) = -f_1(x) \tag{9.56}$$

The homogeneous form of (9.56) is a form of (9.28) in the particular case when

$$a_0 = 1, \quad a_1 = 0, \quad a_2 = q_0 \tag{9.57}$$

Following the analysis of equations of the form (9.28) given in the last section, let $\theta_1(x, z)$ and $\theta_2(x, z)$ be solutions of the homogeneous equations

$$c^2(x)\{d_x^2 - q_0(x)\}\theta_k(x, z) = 0, \quad k = 1, 2 \tag{9.58}$$

which satisfy

$$\theta_1(0, z) = 1, \quad \theta_1'(0, z) = 0 \tag{9.59}$$

$$\theta_2(0, z) = 0, \quad \theta_2'(0, z) = 1 \tag{9.60}$$

where the primes denote differentiation with respect to x.

Using (9.58) to (9.60) we find that the Wronskian of θ_1 and θ_2 has the constant value one. Furthermore, since c and q_0 are p-periodic then

$$\theta_k(x + p, z) = \theta_k(p, z)\theta_1(x, z) + \theta_k'(p, z)\theta_2(x, z), \quad k = 1, 2 \tag{9.61}$$

This follows from the fact that both sides of (9.61) are solutions of (9.58) with the same initial conditions.

We can now use results from Floquet theory introduced in the previous section to investigate solutions of (9.58) that have the form

$$\xi(p, z) = \alpha_1\theta_1(p, z) + \alpha_2\theta_2(p, z) \tag{9.62}$$

and which have the property

$$\xi(x + p, z) = \beta\xi(x, z) \tag{9.63}$$

Bearing in mind (9.61) we find that we must have

$$(\theta_1(p, z) - \beta)\alpha_1 + \theta_2(p, z)\alpha_2 = 0 \tag{9.64}$$

$$\theta_1'(p, z)\alpha_1 + (\theta_2'(p, z) - \beta)\alpha_2 = 0 \tag{9.65}$$

This system has non-trivial solutions α_1 and α_2 if and only if

$$\beta^2 - D(p, z)\beta + 1 = 0 \tag{9.66}$$

where

$$D(p, z) = \theta_1(p, z) + \theta_2'(p, z) \tag{9.67}$$

We remark that in obtaining (9.66) we have used the fact that the Wronskian $W(\theta_1, \theta_2) = 1$.

We notice that every solution of (9.58) that satisfies (9.63) with $\beta := \exp(I\alpha p)$ has the form

$$\zeta(x, z) = \{\exp(i\alpha x)\}V(x, z)$$

with $V(x, z) = \{\exp(-i\alpha x)\}\zeta(x, z)$ being p-periodic.

It now follows that the properties, in particular the growth properties, of the solutions of (9.58) are determined by the location of the roots of (9.66). Just as in the previous section there are now a number of possible cases to consider. For convenience we give them for the particular case of interest in this section.

Case 1: $-2 < D(p, z) < 2$

In this case (9.66) has no real roots β_1 and β_2 satisfying $\beta_2 = \overline{\beta}_1$ and $|\beta_1| = 1$. It follows that (9.58) has a fundamental system of solutions of the form

$$\xi_1(x, z) = \{\exp(i\alpha x)\}V_1(x, z), \quad \xi_2(x, z) = \{\exp(i\alpha x)\}V_2(x, z) \tag{9.68}$$

with $0 < \alpha < \dfrac{\pi}{p}$ and p-periodic functions V_1 and V_2.

Case 2: $D(p, z) = 2,\ \theta_2(p, z) = \theta_1'(p, z) = 0$

In this case every solution of (9.58) is p-periodic. This follows since the Wronskian $W(\theta_1, \theta_2) = 1$ will now imply that $\theta_2'(p, z)\theta_1(p, z) = 1$ and hence $\theta_2'(p, z) = \theta_1(p, z) = 1$ since

$$D(p, z) = \theta_1(p, z) + \theta_2(p, z) = 2$$

It follows that $\theta_1, \theta_1', \theta_2, \theta_2',$ have the same values at $x = 0$ and $x = p$ and so θ_1 and θ_2 are p-periodic.

Case 3: $D(p, z) = 2,\ \theta_2(p, z) \neq 0,\ \theta_1'(p, z) \neq 0$

The equation (9.58) has a non-trivial p-periodic solution since $\beta = 1$ is a root of (9.66). Furthermore, since $\theta_2(p, z)$ and $\theta_1'(p, z)$ are not both zero then the conditions (9.59), (9.60) indicate that at least one of the functions θ_1 and θ_2 is not periodic. In this case (9.58) has a fundamental system of the form

$$\zeta_1(x, z) = V_1(x, z), \quad \zeta_2(x, z) = xV_1(x) + V_2(x, z) \tag{9.69}$$

where V_1 and V_2 are p-periodic.

Case 4: $D(p, z) = -2,\ \theta_2(p, z) = \theta_1'(p, z) = 0$

In this case every solution of (9.58) satisfies

$$\zeta(x + p, z) = -\zeta(x, z) \tag{9.70}$$

This follows since, arguing as in Case 2, we find that θ_1, θ_1', θ_2, θ_2' have opposite signs at $x = 0$ and $x = p$.

Case 5: $D(p, z) = -2$, $\theta_2(p, z) \neq 0$, $\theta_1'(p, z) = 0$

Arguing as in Case 2, we find that in this case (9.58) has a fundamental system of the form (9.43) where V_1 and V_2 satisfy (9.70).

Case 6: $D(p, z) \in C - [-2, 2]$

In this case the roots β_1 and β_2 of (9.66) do not lie on the unit circle. If they did lie on the unit circle then they would be of the form $\beta = \exp(i\alpha p)$ with a real. This being so then on solving the quadratic equation (9.66) we would obtain

$$D(p,z) \pm \sqrt{D(p,z)^2 - 4} = 2\exp(i\alpha)$$

which implies

$$D(p, z)^2 - 4 = (2\exp(i\alpha) - D(p, z))^2$$

and hence

$$D(p, z) = \{\exp(-i\alpha)\}(\{\exp(2i\alpha)\} + 1) = 2\cos\alpha \in [-2, 2]$$

Since $\beta_1\beta_2 = 1$ then the system (9.64), (9.65) has exactly one root β with $|\beta| > 1$. Consequently, on setting $\beta = \exp(\alpha p)$ we obtain the fundamental system

$$\xi_1(x, z) = \{\exp(\alpha x)\}V_1(x, z), \quad \xi_2(x, z) = \{\exp(-\alpha x)\}V_2(x, z) \qquad (9.71)$$

with Re $\alpha > 0$ and $-\pi/p < \operatorname{Im}\alpha < \pi/p$ together with p-periodic functions V_k, $k = 1, 2$.

We can now obtain the following result.

Theorem 9.8. *Let A denote the $H(w)$ realisation of the differential expression defined in* (9.3).

The spectrum of A, denoted $\sigma(A)$, is purely continuous.

Proof. If A has an eigenvalue z and an associated eigenvector ξ then we have

$$A\xi = z\xi \qquad (9.72)$$

(see also (9.27)).

We have seen that A is a positive, self-adjoint operator on its domain $D(A) \subseteq H(w)$. Consequently, we require non-trivial solutions of (9.72) such that

$$\int_{-\infty}^{\infty} |\xi(x)|^2 w(x)dx < \infty \qquad (9.73)$$

The equation (9.72) is a particular form of (9.45) for which $P(\mathrm{x}) = 1$ and $Q(x) = (\mathrm{z} - q(x))$. Consequently the analysis of the previous sections is available;

in particular the Cases 1 to 5 and Theorems 9.5 and 9.7. Therefore, we can conclude

(i) for $|D| > 2$ non-trivial solutions, $\zeta(x)$, of (9.72) are unbounded on $(-\infty, \infty)$

(ii) for $|D| < 2$ non-trivial solutions, $\zeta(x)$, of (9.72) are bounded on $(-\infty, \infty)$.

In the case of conclusion (ii) it follows from Case 3 that we must have non-trivial solutions such that

$$\zeta(x+p) = p\zeta(x)$$

$$|\zeta(x+p)| = |\zeta(x)| = 1$$

However, this being the case (9.73) would not be convergent. Hence, there is no non-trivial ζ satisfying (9.72). Thus A has no isolated eigenvalues and we conclude that $\sigma(A)$ can only be continuous.

It follows directly from the above that

$$z \in \sigma(A) \text{ if and only if } D(p, z) \in [-2, 2] \tag{9.74}$$

The above discussion indicates that scalars μ and v which ensure that $D(p, \mu) = 2$ and $D(p, v) = -2$, respectively, are of special interest. Recalling (9.27) and the boundary value problems (9.58) to (9.60) the previous discussions indicate that $D(p, \mu) = 2$ if and only if μ is an eigenvalue of the problem

$$c^2(x)\{d_x^2 - q_0(x)\}\varphi(x, \mu) + \mu\varphi(x, \mu) = 0 \tag{9.75}$$

$$\varphi(0, \mu) = \varphi(p, \mu), \ \varphi'(0, \mu) = \varphi'(p, \mu) \tag{9.76}$$

Similarly $D(p, \mu) = -2$ if and only if v is an eigenvalue of the problem

$$c^2(x)\{d_x^2 - q_0(x)\}\varphi(x, v) + v\varphi(x, v) = 0 \tag{9.77}$$

$$\varphi(0, v) = -\varphi(p, v), \ \varphi'(0, v) = \varphi'(p, v) \tag{9.78}$$

By the standard Sturm–Liouville theory [3] the eigenvalues of (9.75), (9.76) and of (9.77), (9.78) can be arranged as monotonic sequences $\{\mu_k\}_{k=1}^{\infty}$ and $\{v_k\}_{k=1}^{\infty}$ with μ_k and v_k tending to infinity, double eigenvalues being counted twice.

The location of the eigenvalues μ_k, v_k, $k = 1, 2$ can be shown to be of the form [3], [4]

$$\mu_1 < v_1 \leq v_2 < \mu_2 \leq \mu_3 < v_3 \leq v_4 \leq \ldots \tag{9.79}$$

In particular the boundary of $\sigma(A)$, denoted $\partial\sigma(A)$, consists of simple eigenvalues $\lambda_1, \lambda_2, \ldots$ of both problems (9.75), (9.76) and (9.77), (9.78). Further details on the properties of these eigenvalues can be found in [4] and [8].

After this preparation we can indicate the behaviour of $R(\lambda + i\tau)$ as $\tau \downarrow 0$. There are two principal results that can be obtained in this connection. However, as on a number of previous occasions, the proofs of these results are highly technical and long. Consequently, we will simply state the results here. Full details of the associated proofs can be found in the references cited in the Commentary. In this connection we would particularly mention [4] and [8, Section XIII].

The first of the two important results is the following.

Theorem 9.9. *If A is the operator defined in (9.9) and if $f \in C_0^\infty(\mathbf{R})$ then $R(z)f = (A - zI)^{-1}f$ depends analytically on $z \in \rho(A)$, the resolvent set of A. Moreover, $R(z)f$ can be extended analytically from both above and below $\sigma(A)^0$, the interior of $\sigma(A)$, The quantity $R(z)f$, together with its analytic continuations, depend continuously on (x, z) for $x \in \mathbf{R}$.*

If $\lambda \in \partial\sigma(A)$, the boundary of $\sigma(A)$, then there exists a non-trivial, p-periodic solution, $y(x, \lambda)$, of the homogeneous form of equation (9.27) with the properties

$$y(x + p, \lambda) = y(x, \lambda), \qquad \lambda = \mu_k \tag{9.80}$$

$$y(x + p, \lambda) = -y(x, \lambda), \quad \lambda = \nu_k \tag{9.81}$$

and a solution, $U(x, \lambda)$ of (9.27) *such that*

$$(R(z)f)(x) = y(x,\lambda)\left\{\int_{-\infty}^{\infty} y(s,\lambda)w(s)f(s)ds\right\}\{\varepsilon(z-\lambda)\}^{1/2} + U(x,\lambda) + O\left(|z-\lambda|^{1/2}\right) \quad \text{as} \quad z \to \lambda \tag{9.82}$$

with either $z \in \rho(A)$ or $z = \eta + i0$ and $\eta \in \sigma(A)^0$. Furthermore, $w(x) = c^{-2}(x)\rho^{-1}(x)$ is the weight function introduced earlier and ε is defined by

$$\varepsilon = \begin{cases} 1 & \text{if} \quad \lambda = \mu_{2k} \quad \text{or} \quad \lambda = \nu_{2k-1} \\ -1 & \text{if} \quad \lambda = \mu_{2k+1} \quad \text{or} \quad \lambda = \nu_{2k} \end{cases} \tag{9.83}$$

The estimate (9.82) holds uniformly with respect to $x \in \mathbf{R}$.

The quantity $U(x, \lambda)$ is bounded as $|x| \to \infty$ if and only if

$$\int_{-\infty}^{\infty} y(s,\lambda)w(s)f(s)ds = 0 \tag{9.84}$$

or equivalently if and only if

$$U(x, z) = R(z)f(x) \to U(x, \lambda) \quad \text{as} \quad z \to \lambda$$

Finally, we state the following result.

Theorem 9.10. *Let $y(x, \lambda)$ and $U(x, \lambda)$ be the solutions of the homogeneous forms of the equation* (9.27) *with $z = \lambda$ which appear in the estimate* (9.82) *for the resolvent $R(z)$. Then the solution $u(x, t)$ of the initial value problem* (9.1), (9.2) *satisfies the following estimates uniformly on bounded subsets of* $\mathbf{R}$.

(i) *If $q \neq 0$ and $\omega^2 \in \partial\sigma(A)$ then*

$$u(x,t) = \frac{1 - i\varepsilon}{(\pi\omega)^{1/2}} \sqrt{t} \left\{ \exp(-i\omega t) y(x, \omega^2) \int_{-\infty}^{x} w(s) y(s, \omega^2) f(s) ds \right\} + U(x, \omega^2) \exp(-i\omega t) + o(1) \quad \text{as} \quad t \to \infty \tag{9.85}$$

where ε is defined in (9.83) *and $\lambda = \omega^2$.*

(ii) *If $q \neq 0$ and $\omega^2 \notin \partial\sigma(A)$ then the principle of limiting amplitude holds in the sense that*

$$u(x, t) = (R(\omega^2 + i0)f)(x) \exp(-i\omega t) + o(1) \text{ as t} \to \infty \tag{9.86}$$

(see [5], [6], [9] and the Commentary).

(iii) *If $q = 0$ and $\omega = 0$ then zero belongs to $\sigma(A)$ and $y(x, 0)$ is constant and*

$$u(x,t) = y^2(x,0) t \int_{-\infty}^{\infty} f(s) w(s) ds + U(x,0) + o(1) \quad \text{as} \quad t \to \infty \tag{9.87}$$

(iv) *If $q = 0$ and $\omega^2 > 0$ then the (static) term*

$$y^2(x,0) \int_{-\infty}^{\infty} w(s) \left\{ \psi(s) + \frac{1}{i\omega} f(s) \right\} ds \tag{9.88}$$

must be added to the right-hand side of (9.85) and (9.86).

References and Further Reading

[1] R.A. Adams: *Sobolev Spaces*, Academic Press, New York, 1975.

[2] G. Birkhoff and G.-C. Rota: Ordinary Differential Equations (3rd Edn), John Wiley, New York, 1978.

[3] E. Coddington and Levinson: *Theory of Ordinary Differential Equations*. McGraw-Hill, New York, 1955.

[4] M.S.P. Eastham: *Theory of Ordinary Differential Equations*, Van Nostrand, London, 1970.

[5] D.M. Eidus: The principle of limiting absorption, *Math. Sb.*, **57**(99), 1962 and *AMS Transl.*, **47**(2), 1965, 157–191.

[6] D.M. Eidus: The principle of limiting amplitude, *Uspekhi Mat. Nauk.* **24**(3), 1969, 91–156 and *Russ. Math. Surv.* **24**(3), 1969, 97–167.

[7] R. Leis: *Initial Boundary Value Problems in Mathematical Physics*, John Wiley, Chichester, 1986.

[8] M. Reed and B. Simon: *Methods of Mathematical Physics, Vols 1–4*, Academic Press, New York, 1972–1979.

[9] G.F. Roach: *Greens Functions* (2nd Edn), Cambridge Univ. Press, London, 1970/1982.

[10] G.F. Roach: *An Introduction to Linear and Nonlinear Scattering Theory*, Pitman Monographs and Surveys in Pure and Applied Mathematics, Vol. 78, Longman, Essex, 1995.

10

Inverse Scattering Problems

10.1 Introduction

The determination of properties of a medium from a knowledge of wave scattering processes in that medium is an inverse scattering problem. Perhaps the simplest illustration of such a problem, which at the same time distinguishes it from the majority of scattering problems we have discussed so far, is provided by the following "tennis ball" problem. Assume that the system of interest consists of a light bulb, a tennis ball and a screen. When the light is switched on it shines on the tennis ball which in turn casts a shadow on the screen. If we know the details of the light source and the tennis ball then the problem of determining the details of the shadow is a **direct scattering problem**. If, however, we only know the details of the light source and the shadow and we want to determine the details of the tennis ball then this is an **inverse scattering problem**.

In the earlier chapters we have concentrated on direct scattering problems and have seen that a knowledge of scattered waves (the shadow)in the far field (of the bulb and ball) can lead to reliable constructive solution methods. For the inverse problems a knowledge of the scattered waves in the far field is an essential ingredient from the outset. This is due almost entirely to the fact that in practice the majority of measurements are made in the far field.

In more general terms the inverse scattering problem can be thought of as the determination of an impurity in an otherwise homogeneous region from the measurements available of a field scattered by the inhomogeneity.

Such problems frequently arise when analysing, for example, various ultrasonic diagnostic techniques and other non-destructive testing processes. Typical areas include remote sensing problems associated with radar, sonar, geophysics and medical diagnosis.

Since in the majority of practical problems measurements of the scattered field can only be made in the far field of the transmitter and scatterer we can expect that scattering theory can be used to provide a satisfactory means of investigating the sensing problem mentioned above. To see this recall that when developing a scattering theory we always try to work with self-adjoint operators A_0, A_1 on a Hilbert space H. Associated with these operators are Evolution operators:

$$U_0(t) := \exp(-itA_0), \quad U_1(t) := \exp(-itA_1)$$

Wave operators:

$$W_{\pm} := \lim_{t\to\pm\infty} U_0^*(t)U_1(t)$$

Scattering operator:

$$S := W_+^* W_-$$

If the operators A_0, A_1 characterise initial value problems with solutions u_0, u_1 respectively which are required to satisfy given initial data f_0, f_1 respectively then we have seen that the solutions can be written in the form

$$u_0(t) = U_0(t)f_0, \quad u_1(t) = U_1(t)f_1$$

In practice f_1 is given from the outset and u_1 is regarded as the solution of a Perturbed Problem (PP). When the intention is to use scattering theory techniques to approach solutions to this problem then the aim is to determine initial data for an associated Free Problem (FP), which has a reasonably easily obtainable solution u_0, which will ensure that u_0 and u_1 are asymptotically equal (AE) as $t \to \pm\infty$. To this end we introduce initial data $f_0^{\pm}$ for the FP which generates FP solutions $u_{\pm}$ such that $(u_1 - u_{\pm}) \to 0$ in a suitable energy norm as $t \to \pm\infty$ respectively. The initial data $f_0^{\pm}$ are related by means of the scattering operator S in the form

$$Sf_0^- = f_0^+$$

With the above understanding we can describe the two types of scattering problems mentioned above as follows:

Direct Scattering Problems: Knowing U_0, U_1 and f_1 determine S and $f_0^{\pm}$.
Inverse Scattering Problems: Knowing S and $U_0(t)$ determine $U_1(t)$.

In this chapter, with practical problems such as those mentioned above in mind, we shall reduce considerably the generality of these abstract problems by restricting attention to a study of a perturbed wave equation of the form

$$\{\partial_t^2 - \Delta + q(x)\}u(x, t) = 0, \quad (x, t) \in \mathbf{R}^n \times \mathbf{R} \tag{10.1}$$

where $1 \leq n \leq \infty$. This equation is also referred to as the **plasma wave equation** and as such is intimately connected with potential scattering problems. It arises in the modelling of many physical systems ranging, for example, from the study of the electron density in the atmosphere of the earth to the vibrations of an elastically braced string. We concentrate here on indicating methods which are used to determine the potential term, q, when u, the state of the system, is known (in the far field).

When the potential term is time independent a number of methods are available for tackling inverse scattering problems. In particular, taking the Fourier transform with respect to time of the plasma wave equation yields a stationary Schrödinger equation. The potential is then recovered by means of the celebrated Marchenko equations in one or other of its many forms in both the frequency and time domains (see Commentary). A comprehensive and self-contained account of the derivation of the two basic types of Marchenko equations is given in [8]. Essentially, the required potential term is obtained in terms of a certain functional of a reflection coefficient. There are other methods of recovering the potential and these may be conveniently called high energy limit methods. Before introducing this method some preparation is required. This is given in the next section where we concentrate on the plasma wave equation rather than on the classical wave equation as we have been doing in previous chapters.

10.2 Some Asymptotic Formulae for the Plasma Wave Equation

We consider the plasma wave equation (10.1) in which the potential term $q(x)$ satisfies the following.

Assumption 10.1. *The potential function q is defined in $\boldsymbol{R}^3$ and is a non-negative, bounded function with compact support, that is, q is a measurable function on $\boldsymbol{R}^3$ satisfying*

$$0 < q(x) \leq c_0, \quad x \in \mathbf{R}^3$$

$$q(x) = 0, \quad |x| \geqslant R_0$$

where c_0 and R_0 are positive constants.

We introduce the operator

$$\mathrm{A} : L_2(\mathbf{R}^3) \to L_2(\mathbf{R}^3) \equiv H \tag{10.2}$$

$$A_u = -\Delta u + q(x), \quad \mathrm{u} \in D(A)$$

$$D(A) = \{\mathrm{u} \in H : (-\Delta u + qu) \in H\}$$

which is a self-adjoint realisation of $(-\Delta + q(x))$ in H.

We also introduce a real-valued functions $s \in C^2$ defined on $\mathbf{R}$ such that

$$\text{supp}\,(s) \subset [a, b]$$

with $-\infty < a < b < \infty$. We denote by θ a unit vector in $\mathbf{R}^3$ and set

$$u_0(x, t) \equiv u_0(x, t, \theta, s) = s(x \cdot \theta - t) \tag{10.3}$$

where $x{\cdot}\theta$ denotes the scalar product in $\mathbf{R}^3$. The form of the argument of s indicates that the right-hand side of (10.3) represents a plane wave and that consequently u_0 satisfies the free equation (see [12, Chapter 7])

$$(\partial_t^2 - \Delta)u_0(x, t) = 0 \tag{10.4}$$

As in earlier chapters s characterises the **signal profile**. Furthermore, since

$$q(x)u_0(x, t, \theta) = 0$$

with $(x, t) \in \mathbf{R}^3 \times (-\infty, t_0)$ and $(x, t) \in \mathbf{R}^3 \times (t_1, \infty)$ where

$$t_0 := -\mathrm{b} - R_0 \quad \text{and} \quad t_1 := -a + R_0$$

then $u_0(x, t, \theta)$ satisfies the (free) plasma wave equation for $t \leq t_0$ and $t \geqslant t_1$.

Let $u(x, t) \equiv u(x, t, \theta)$ denote the solution of (10.1), the perturbed wave equation, that is the plasma wave equation, which satisfies the initial conditions

$$u(x, t_0, \theta) = u_0(x, t, \theta) \quad \text{and} \quad \partial_t u(x, t_0, \theta) = \partial_t u_0(x, t, \theta) \tag{10.5}$$

We shall refer to $u(x, t)$ as the **total field**. We denote the scattered wave by $u_{sc}(x, t) = u_{sc}(x, t, \theta)$ where

$$u_{sc}(x, t, \theta) = u(x, t, \theta) - u_0(x, t, \theta)$$

and the scattered field is assumed to satisfy the IVP

$$\{\partial_t^2 - \Delta + q(x)\}u_{sc}(x, t) = -q(x)u_0(x, t, \theta) \tag{10.6}$$

$$u_{sc}(x, t, \theta) = \partial_t u_{sc}(x, t, \theta) = 0, \quad t \leq t_0, \quad x \in \mathbf{R}^3 \tag{10.7}$$

where we have recognised (10.5).

In practical cases it is usually u_{sc} that is measured and, moreover, measured in the far field of the transmitter and receiver. Consequently we need to study the asymptotic behaviour of $u_{sc}(x, t, \theta)$ as $t \to \infty$. In order to do this we need some preparation. Consequently, we shall introduce the notion of a **scattering amplitude** denoted by $F\,(k, \omega, \omega')$ where

$$F(k,\omega,\omega')=\frac{1}{4\pi}\int_{\mathbf{R}^3}\varphi(y,-k\omega)q(y)\{\exp(iky\omega')\}dy \tag{10.8}$$

with $k>0$ and ω, $\omega' \in S^2$ the unit sphere in three dimensions. Here $\varphi(x,\xi)$, with $x, \chi \in \mathbf{R}^3$ is the unique solution of the celebrated Lippmann–Schwinger equation [1], [3], [6], [11]

$$\varphi(x,\xi)=\exp(ix\cdot\xi)-\frac{1}{4\pi}\int_{\mathbf{R}^3}\frac{\exp(i|\xi||x-y|)}{|x-y|}q(y)\varphi(y,\xi)dy \tag{10.9}$$

The solutions $\varphi(x,\varepsilon)$ are the **distorted plane waves** associated with the plasma wave equation (10.1).

We now introduce the **far field solution** u_{sc}^{∞} defined by

$$u_{sc}^{\infty}(x,t)\equiv u_{sc}^{\infty}(x,t,\theta,s)=|x|^{-1}K(|x|-t,\omega_x,\theta,s) \tag{10.10}$$

where $\omega_x = x/|x|$,

$$K(v,\omega,\theta,s)=\frac{1}{(2\pi)^{1/2}}\int_{-\infty}^{\infty}\{\exp(iv\rho)\}\hat{s}(\rho)F(\rho,\omega,\theta)d\rho \tag{10.11}$$

is the **asymptotic wave function (profile)** (see Chapter 6) and $\hat{s}(\rho)$ is the usual one-dimensional Fourier transform of $s(\tau)$ given by

$$\hat{s}(\rho)=\frac{1}{(2\pi)^{1/2}}\int_{-\infty}^{\infty}\{\exp(-i\rho\tau)\}s(\tau)d\tau$$

The following result can be obtained [17].

Theorem 10.2. *If the requirements of Assumption 10.1 hold then*

$$u_{sc}(x,t,\theta,s) = u_{sc}^{\infty}(x,t,\theta,s) \quad \text{as} \quad t\to\infty$$

in the sense that

$$\lim_{t\to\infty}\left\|u_{sc}(\cdot,t,\theta,s)-u_{sc}^{\infty}(\cdot,t,\theta,s)\right\|_H=0$$

A proof of this result for the plasma wave equation rather than the classical wave equation is given in Appendix A10.1.

10.3 The Scattering Matrix

In this section we introduce a representation of a Hilbert space H which is in terms of a given self-adjoint operator $A : H \to H$. Such a representation will enable us to define a so-called **scattering matrix** associated with A. As we shall see, the scattering matrix plays a central role in the development of what have become known as high energy limit methods for solving inverse scattering problems.

10.3.1 Decomposable Operators

We start with the Hilbert space $H := L_2(\mathbf{R}^n)$ which as we have seen consists of (equivalence classes) of square-integrable functions defined on $\mathbf{R}^n$ and taking values in $\mathbf{C}$. A generalisation of this structure would be to consider functions on $\mathbf{R}^n$ taking values in a more general vector space than $\mathbf{C}$; in a fixed Hilbert space H_0 (say). With this in mind let $\Lambda \subseteq \mathbf{R}$ and suppose that f is a vector-valued function on Λ which takes values in a fixed Hilbert space H_0. We shall write f_λ to signify the value, in H_0, of the function f at the point $\lambda \in \Lambda$. We will then denote the function f by the collection $\{f_\lambda\}$ and often write

$$f = \{f_\lambda\}_{\lambda \in \Lambda} \equiv \{f_\lambda\} \tag{10.12}$$

The function f in (10.12) is measurable if for every element $g \in H_0$ the complex-valued function $(g, f_0)_0$ is measurable where $(\cdot, \cdot)_0$ denotes the inner product in H_0.

Measure is a generalisation of the familiar concept of length defined on intervals of $\mathbf{R}$.

In an abstract setting we introduce

Definition 10.3. Let X denote a set and Θ a class of subsets of X. A **positive measure** on the pair (X, Θ) is a mapping m with the properties

(i) $m(\phi) = 0$, $\phi =$ the empty set.

(ii) For $\{\theta_k\}$ a countable collection of disjoint elements of Θ (that is $\theta_k \cap \theta_n = \phi$ for $k \neq n$)

$$m\left[\bigcup_{k=1}^{\infty} \theta_k\right] = \sum_{k=1}^{\infty} m(\theta_k)$$

When such a measure exists then the pair (X, Θ) is called a **measurable space** and the elements of Θ are called **measurable sets**.

A more mathematically rigorous account of measure and measurability can be found in such texts as [9], [10], [12], [11].

Definition 10.4. Let X be a space and $f : X \to \mathbf{R}$. The function f is said to be **measurable** if for all $\alpha \in \mathbf{R}$ the set

$$\{x \in X : f(x) > \alpha\}$$

is measurable.

In a similar manner to that adopted for $L_2(\mathbf{R}^n)$ we shall denote by $L_2(\Lambda, \mathbf{R}^n)$ the set of all (equivalence classes) of (measurable) functions f which satisfy

$$\|f\|^2 = \int_\Lambda \|f_\lambda\|_0^2 \, d\lambda < \infty \tag{10.13}$$

The operations on $L_2(\Lambda, \mathbf{R}^n)$ of vector addition, multiplication by a scalar and definition of an inner product are

$$f + g = \{f_\lambda + g_\lambda\}, \quad \alpha f = \{\alpha f_\lambda\} \tag{10.14}$$

$$(f, g) = \int_\Lambda (f_\lambda, g_\lambda)_0 \, d\lambda \tag{10.15}$$

where $f = \{f_\lambda\}$ and $g = \{g_\lambda\}$.

Furthermore all the Hilbert space axioms including that of completeness can be shown to hold for $L_2(\Lambda, \mathbf{R}^n)$ in just the same manner as for $L_2(\mathbf{R}^n)$ [10], [11].

A further generalisation is possible which is centred on an operator of interest. To this end let $\lambda \in \Lambda \subseteq \mathbf{R}$ and let $A(\lambda) \in B(H_0)$.

An operator-valued function A is said to be measurable if for every $f \in L_2(\Lambda, H_0)$ the vector-valued function defined as $\{A(\lambda)f_\lambda\}$ is measurable. Then for this $B(H_0)$-valued function we define an operator in $L_2(\Lambda, H_0)$ by

$$D(A) - \left\{f \in L_2(\Lambda, H_0) : \int_\Lambda \|A(\lambda)f_\lambda\|_0^2 \, d\lambda < \infty\right\} \tag{10.16}$$

$$Af = \{A(\lambda)f_\lambda\} \tag{10.17}$$

An operator A defined as in (10.16), (10.17) is said to be **decomposable**. Furthermore, if $A(\lambda) = \varphi(\lambda)I_0$ where φ is a measurable function and I_0 is the identity element in H_0 then A is said to be **diagonalisable**.

Two important results in this area are given in the following theorems.

Theorem 10.5. *Let F be a bounded operator in $L_2(\Lambda, H_0)$ which commutes with the self-adjoint operator*

$$A = \{\lambda I_0\}$$

Then F is a decomposable operator given by $F = \{F(\lambda)\}$ and

$$\|F\| = \operatorname*{ess\,sup}_{\lambda \in \Lambda} \|F(\lambda)\| < \infty$$

Theorem 10.6. *A bounded decomposable operator $A = \{A(\lambda)\}$ is bounded invertible (respectively, self-adjoint, unitary) if and only if $A(\lambda)$ is bounded invertible (respectively, self-adjoint, unitary) almost everywhere.*

A proof of these theorems can be found in such standard texts as, for example, [9], [10], [11], [13].

10.3.2 Some Algebraic Properties of $W_\pm$ and S

We begin by recalling, for convenience, a number of definitions and notations introduced in Chapter 6.

Let A_0 and A_1 denote the free and perturbed operators respectively, each being defined on a Hilbert space H. We have introduced Evolution operators, $U_0(t)$, $U_1(t)$

$$U_0(t)=\exp\{-itA_0^{1/2}\},\quad U_1(t)=\exp\{-itA_1^{1/2}\} \tag{1.23}$$

Wave operators, $W_\pm$

$$W_\pm := \lim_{t\to\pm\infty} U_0^*(t)U_1(t) \tag{6.67}$$

Scattering operator S

$$S := W_+W_-^* \tag{6.105}$$

A number of useful properties of these quantities can be established; for example

$$\begin{aligned}U_1(T)W_\pm^* &= U_1(T)\lim_{t\to\pm\infty} U_1^*(t)U_0(t)\\ &= \lim_{t\to\pm\infty} U_1(T)U_1^*(t)U_0(t)\\ &= \lim_{t'\to\infty} U_1(T)U_1^*(t'+T)U_0(t'+T)\\ &= \lim_{t'\to\pm\infty} U_1^*(t')U_0(t')U_0(T)\\ &= W_\pm^*U_0(T)\end{aligned}$$

Similarly

$$U_0(T)W_\pm = W_\pm U_1(T)$$

The relations

$$U_0(T)W_\pm = W_\pm U_1(T) \tag{10.18}$$

$$U_1(T)W_\pm^* = W_\pm^*U_0(T) \tag{10.19}$$

are known as the **intertwining relations**.

The intertwining relations (10.18), (10.19) lead to the following result

$$SU_0(T) = W_+W_-^*U_0(T) = W_+U_1(T)W_-^* = U_0(T)W_+W_-^* = U_0(T)S \tag{10.20}$$

Finally, in this subsection, we set $W = W_{\pm}$ to denote one or other of the two wave operators and notice

$$\|iT^{-1}W^*(U_0(T)-I)f - W^*A_0^{1/2}f\| \leq \|W^*\|\|iT^{-1}(U_0(T)-I)f - A_0^{1/2}f\|$$

As $T \to 0$ the right-hand side of this inequality tends to zero by virtue of the properties of the infinitesimal generator of $U_0(T)$ (see Chapter 5). Furthermore, in the limit as $T \to 0$ the left-hand side indicates that

$$i\frac{d}{dT}(W^*U_0(T)f) = W^*A_0^{1/2}f \tag{10.21}$$

However, using (10.19) we also have

$$\begin{aligned} i\frac{d}{dT}(U_1(T)W^*f) &= i\frac{d}{dT}(W^*U_0(T)f) \\ &= iW^*\frac{d}{dT}(U_0(T)f) \\ &= iW^*(-iA_0^{1/2}U_0(T)f) \qquad (10.22) \\ &\to WA_0^{1/2} \text{ as } T \to 0 \qquad (10.23) \end{aligned}$$

But

$$\begin{aligned} i\frac{d}{dT}(U_1(T)W^*f) &= i(-A_1^{1/2}U_1(T)W^*f) \\ &\to A_1^{1/2}W^*f \text{ as } T \to 0 \qquad (10.24) \end{aligned}$$

Thus, combining (10.22) to (10.24) we conclude that

$$A_1^{1/2}W^* = W^*A_0^{1/2} \tag{10.25}$$

We have seen in (10.20) that S and $U_0(T)$ commute. Therefore Theorems 10.5 and 10.6 indicate that S is a bounded, decomposable operator and, hence, we can write S in the form

$$S = \{S(\lambda)\} \tag{10.26}$$

The quantity $S(\lambda)$ is referred to as the S-**matrix of energy** λ associated with the scattering operator S.

We also have that

$$U_0SU_0^{-1} = SU_0U_0^{-1} = S = \{S(\lambda)\} \tag{10.27}$$

If we now use the spectral theorem together with the Fourier transform, $\mathcal{F}$, defined in $\mathbf{R}^3$ in the form

$$(\mathcal{F}u)(\xi) = (2\pi)^{-3/2} \int_{\mathbf{R}^3} \{\exp(-ix\xi)\} u(x) dx \tag{10.28}$$

$$(\mathcal{F}^* v)(x) = (2\pi)^{-3/2} \int_{\mathbf{R}^3} \{\exp(ix\xi)\} v(\xi) d\xi \tag{10.29}$$

then (10.27) can be written in the equivalent form

$$(\mathcal{F} S \mathcal{F}^* f) = \{S(|\xi|) f(|\xi| \cdot)\} \tilde{\xi} \tag{10.30}$$

where

$$f \in C_0^\infty(\mathbf{R}^3), \quad \xi \in \mathbf{R}^3, \quad \tilde{\xi} = \xi / |\xi|$$

The details of obtaining this alternative representation are left as an exercise. We remark that (10.30) is sometimes taken as a relation which defines the scattering matrix.

10.4 The Inverse Scattering Problem

The inverse scattering problem is a non-linear, ill-posed problem. When dealing with classical wave scattering phenomena there are essentially two types of such problems. One arises as a consequence of target scattering phenomena whilst the other centres on potential scattering effects. In this chapter we are dealing with the plasma wave equation which is typical of the type of equation which characterises potential scattering. The associated inverse problem is the determination, or reconstruction as it is sometimes called, of the potential, q, from the far field behaviour of the scattered field. Details of inverse scattering problems associated with target scattering are given in the Commentary.

In the particular case of an inverse scattering problem associated with the plasma wave equation there are three basic questions that have to be addressed.

1. Does a potential term q exist which is compatible with measured data?
2. Is the potential term unique?
3. How can the potential term be constructed from measured data?

In Section 10.2 we can see a possible strategy for attacking the inverse problem associated with the plasma wave equation. Specifically, we could take the following steps.

- Launch the incident pulses $u_0(x, t, \theta, s)$ characterised by (10.3).
- Evaluate $u_{sc}(x, t, \theta, s)$ from experimental data in practice, and recover the asymptotic wave function $K(\upsilon, \omega, \theta, s)$ via (10.8). Here we acknowledge the results of Theorem 10.2.

- Obtain the scattering amplitude using (10.11).
- Construct the potential function q by solving the first kind Volterra integral equation (10.8).

The above set of steps can involve a great deal of hard, technical work [2], [8].

Newton [8] addressed these questions by investigating an extension of the Marchenko equations in $\mathbf{R}^1$. However, the potential term was required to satisfy quite strict conditions. Furthermore, either the scattering data or the associated spectral measure was required on the whole real line and for all energy numbers $k \geqslant 0$.

An alternative approach which offers good prospects for reconstructing the potential, q, is the so-called high energy limit method.

10.5 A High Energy Limit Method

In order to explain this method we first recall that the scattering operator S can be represented in the form $S = \{S(\lambda)\}$, where $S(\lambda)$ is the S-matrix associated with S (see Subsection 10.3.1). We then define

$$F(k) = -2\pi i k^{-1}(S(k) - I) \tag{10.31}$$

where I is the identity operator on $L_2(S^2)$. The scattering operator is a decomposable operator and as a consequence $F(k)$ is a Hilbert Schmidt operator on $L_2(S^2)$ with Hilbert Schmidt kernel which we denote by $F(k, \omega, \omega')$. That is

$$(\mathcal{F}(k)f)(\omega) = \int_{S^2} \mathcal{F}(k, \omega, \omega')f(\psi')d\omega' \tag{10.32}$$

$$\int_{S^2}\int_{S^2} |\mathcal{F}(k, \omega, \omega'|^2 d\omega d\omega') < \infty \tag{10.33}$$

for $f \in L_2(S^2)$. The quantity $F(k, \omega, w')$ is referred to as the **scattering amplitude**. If the potential term, $q(x)$, in the plasma wave equation satisfies

$$|q(x)| \leq C_0(1+|x|)^{-\beta}, \quad x \in \mathbf{R}^3, \quad \beta \in \mathbf{R} \tag{10.34}$$

then Faddeev [3] showed that provided $\beta > 3$ in (10.34) then

$$\lim_{k\to\infty} \mathcal{F}(k, \omega, \omega') = \frac{1}{4\pi}\int_{\mathbf{R}^3} \{\exp(-i\xi x)\}q(x)dx \tag{10.35}$$

Here the limit is taken so that $k \to \infty$ maintaining the relation $\xi = k(\omega - \omega')$ for a given $\xi \in \mathbf{R}^3$. When (10.35) holds then q can be recovered by using the usual Fourier inversion formula. Faddeev's work [3] would appear to be the first which gives a rigorous proof for the fact that a high energy limit of the scattering amplitude is the Fourier transform of the potential.

Instead of approaching the inverse scattering problem by considering, as did Faddeev, the high energy limit of the scattering amplitude, another profitable approach is to examine a high energy limit of the scattering matrix. One such approach was introduced by Saito [15]. This begins by setting

$$\psi_{x,k}(\omega) = \exp\{-ikx\omega\}, \quad \omega \in S^2 \tag{10.36}$$

where $x \in \mathbf{R}^3$ and $k > 0$. We then define

$$f(x,k) = k(F(k)\psi_{x,k}, \psi_{x,k})_{S^2} \tag{10.37}$$

here $(\cdot\,,\cdot)_{S^2}$ is the inner product in $L_2(S^2)$. We regard $\psi_{x,k}$ as a function defined on S^2 with parameters $k \geqslant 0$ and $x \in \mathbf{R}^3$. Here $x\omega$ is the inner product in $\mathbf{R}^3$.

It can be shown, [15], [16], that provided $\beta > 1$ in (10.34) then the following limit exists

$$f(x,\infty) = \lim_{k\to\infty} f(x,k) = -2\pi \int_{\mathbf{R}^3} \frac{q(y)}{|y-x|^2}\,dy \tag{10.38}$$

The equation (10.38) is an integral equation of the first kind for $q(x)$ and the following result can be obtained

$$q(x) = -(4\pi^3)^{-1}(\mathcal{F}^*(|\xi|\mathcal{F}f(.,\infty)))(x) \tag{10.39}$$

where $\mathcal{F}$ denotes the usual Fourier transform (see below).

A proof of (10.39) is sketched in the next section.

In the methods of Faddeev and of Saito we do not need any low energy scattering data to recover q but we do need scattering data in the neighbourhood of $k = \infty$; hence the name of the method.

10.5.1 The Solution of an Integral Equation

Bearing in mind (10.38) we see that the high energy limit method for solving inverse scattering problems for the plasma wave equation centres on the ability to solve an integral equation of the form

$$-2\pi \int_{\mathbf{R}^3} \frac{q(y)}{|y-x|^{n-1}}\,dy = g(x) \tag{10.40}$$

where g is a known function and it is required to determine the function q.

Since the equation (10.40) has been defined in $\mathbf{R}^n$ we will find it necessary to use Fourier transforms in $\mathbf{R}^n$. These are defined by

$$(\mathcal{F}f)(\xi) = (2\pi)^{-n/2} \int_{\mathbf{R}^3} \{\exp(-i\xi y)\} f(y) dy \tag{10.41}$$

$$(\bar{\mathcal{F}}f)(\xi) = (\mathcal{F}f)(-\xi) \tag{10.42}$$

$$(\mathcal{F}^* f)(y) = (2\pi)^{-n/2} \int_{\mathbf{R}^3} \{\exp(i\xi y)\} f(\xi) d\xi \tag{10.43}$$

$$(\bar{\mathcal{F}}^* f)(y) = (\mathcal{F}^* f)(-\xi) \tag{10.44}$$

where ξy, denotes the inner product, in $\mathbf{R}^3$, of ξ and y.

We introduce the notations

$$M_\varepsilon := \left\{ f \in C(\mathbf{R}^n) : f(y) = O(|y|^{-\varepsilon}) \text{ as } y \to \infty \right\} \tag{10.45}$$

that is $f \in M_\varepsilon$ implies that f satisfies the estimate

$$|f(y)| \le c(1 + |y|)^{-\varepsilon}, \quad y \in \mathbf{R}^n \tag{10.46}$$

$S \equiv S(\mathbf{R}^n) =$ all rapidly decreasing functions on $\mathbf{R}^n$

$S' \equiv S'(\mathbf{R}^n) =$ all linear, continuous functionals on S.

We shall denote the (duality) pairing between S and S' by $\langle \cdot\,, \cdot \rangle$.

Definition 10.7. For $g \in M_\varepsilon$, $\varepsilon > 0$ and $s > 0$ define a linear functional $\Lambda^s g$ on $S_\xi \equiv S(\mathbf{R}^n_\xi)$ by

$$\langle \Lambda^s g, G \rangle := \int_{\mathbf{R}^n} g(y) \{ \bar{\mathcal{F}}^* (|\xi|^s G) \}(y) dy \tag{10.47}$$

where $G \in S_\xi$ and $\bar{\mathcal{F}}^*$ is defined as in (10.43).

It can be shown [15] that $\Lambda^s g \in S'_\xi$.

Let $q \in M_\mu$ with $1 < \mu < n$ and let g be defined by (10.40). Then it can be seen that $g \in M_{\mu-1}$ and consequently $\Lambda^s g$ is well defined for any $s > 0$.

If we take the Fourier transform throughout (10.40) and make use of the formula for the convolution of two functions

$$\mathcal{F}(f * g) = (2\pi)^{n/2} (\mathcal{F}f) \times (\mathcal{F}g) \tag{10.48}$$

then we obtain

$$\mathcal{F}g = -(2\pi)^{n/2+1} \mathcal{F}(|y|^{-(n-1)}) \times \mathcal{F}q \tag{10.49}$$

We also notice that we have

$$\mathcal{F}(|y|^{-t})(\xi) = B_t |\xi|^{-(n-t)}, \quad 0 < t < n \tag{10.50}$$

in S' where

$$B_t = 2^{n/2+1} \frac{\Gamma\left(\frac{n-t}{2}\right)}{\Gamma\left(\frac{t}{2}\right)} \tag{10.51}$$

when Γ denotes the gamma function [4].

It follows from (10.49) and (10.50) that with $t = n - 1$ we obtain

$$\mathcal{F}g = \frac{-4\pi^{(n+3)/2}}{\Gamma\left(\frac{n-1}{2}\right)} |\xi|^{-1} \mathcal{F}q \tag{10.52}$$

Hence

$$\mathcal{F}q = -B_n |\xi| \mathcal{F}g = -B_n \Lambda g, \quad \text{in } S \tag{10.53}$$

where $\Lambda g = \Lambda^1 g$ and

$$B_n = \frac{\Gamma\left(\frac{(n-1)}{2}\right)}{4\pi \frac{n+3}{2}} \tag{10.54}$$

Therefore, using Fourier inversion we obtain from (10.53)

$$q(y) = -B_n \mathcal{F}^*(|\xi| \mathcal{F}g)(y) = -B_n(\mathcal{F}^* \Lambda g)(y) \tag{10.55}$$

This whole process can be reversed. Consequently, we see that (10.55) gives the unique solution of (10.40) provided $\mathcal{F}^* \Lambda g \in M_\mu$. Thus we have the following result.

Theorem 10.8. *The integral equation (10.40) has a unique solution $q \in M_\mu$ with $1 < \mu < n$ if and only if*

$$g \in M_{\mu-1} \quad \text{and} \quad \mathcal{F}^* \Lambda g \in M_\mu \tag{10.56}$$

The solution $q(y)$ is given by (10.55).

In this section we are particularly interested in problems in $\mathbf{R}^3$. In this case we recognise the properties of the gamma function and obtain

$$B_3 = \frac{1}{4\pi^3}$$

$$q(y) = -\frac{1}{4\pi^3}\mathcal{F}^*(|\xi|\mathcal{F}g)(y) = -\frac{1}{4\pi^3}(\mathcal{F}^*\Lambda g)(y)$$

We are interested in the case when $g(x)$ is replaced by the $f(x, \infty)$ which is defined in (10.38). Consequently, if we now define

$$q(x,k) := -\frac{1}{4\pi^3}\mathcal{F}^*(|\xi|\mathcal{F}f(\cdot,k))(x) \tag{10.57}$$

then the following approximation result for q can be obtained [15], [16].

Theorem 10.9. *Let*
(i) $q(x) \leq c(1 + |x|)^{-2}$, $c > 0$
(ii) $q \in C^2(\mathbf{R}^3)$ *satisfying*

$$|D^\alpha q(x)| \leq c_1(1+|x|)^{-\beta}, \quad x \in \mathbf{R}^3, \quad |\alpha| = 1,2$$

with constants $c_1 > 0$ and $\beta > 5/2$. Here $\alpha = (\alpha_1, \alpha_2, \alpha_3)$ is a multi-index of non-negative integers such that

$$|\alpha| = \alpha_1 + \alpha_2 + \alpha_3.$$

Furthermore, we have written

$$D^\alpha = D_1^{\alpha_1} D_2^{\alpha_2} D_3^{\alpha_3}, \quad D_j = \frac{\partial}{\partial x_j}, \quad j = 1, 2, 3$$

Then

$$\|q(\cdot) - q(\cdot, k)\|_{L_2(\mathbf{R}^3)} \leq c_2(k)^{-1} \quad \text{as} \quad k \to \infty$$

where c_2 depends on c_1, β and $\max_{x \in \mathbf{R}^3} |q(x)|$ *but it does not depend on k.*

These last two results indicate that we need only work with measurements taken at sufficiently high values of k.

Appendix A10.1 Proof of Theorem 10.2

The scattered wave, $u_{sc}(x, t)$ satisfies the IVP (10.6), (10.7).

We recall the following notation

$$u_0(x, t) \equiv u_0(x, t, \theta, s) = s(x\theta - t) \tag{10.3}$$

where s characterises the incident waveform.

Using the Duhamel integral we see that the scattered field can be expressed in the form

$$\begin{aligned} u_{sc}(x,t,\theta) &= \operatorname{Re}\left\{iA_1^{1/2}\int \left(\exp\left[-i(t-\tau)A_1^{1/2}\right]\right)Q(x,\tau,\theta,s)\right\}d\tau \\ &= \operatorname{Re}\left\{\left[\exp\left(-itA_1^{1/2}\right)\right]h\right\} \end{aligned} \tag{A1.1}$$

where

$$A_1 = \text{realisation of } (\Delta + q) \text{ in } L_2(\mathbf{R}^3)$$

$$Q(x, t, \theta) = -q(x)u_0(x, t, \theta) \equiv Q(x, t, \theta, s) = -q(x)u_0(x, t, \theta, s)$$

$$h = h(x,\theta,s) = iA_1^{1/2}\int_{-\infty}^{\infty}\left(\exp\left[i\tau A_1^{1/2}\right]\right)Q(x,\tau,\theta,s)$$

(See also [17].)

Let A_0 denote the self-adjoint realisation of $(-\Delta)$ in $L_2(\mathbf{R}^3)$, that is

$$A_0 u = -\Delta u, \quad u \in H^2(\mathbf{R}^3)$$

where $H^2(\mathbf{R}^3)$ denotes the Sobolev Hilbert of second order "derivatives" (see [1], [13]).

Introduce, as in [17] and Chapter 6, the complex valued, scattered wave v_{sc} defined by

$$v_{sc}(x, t, \theta, s) \equiv v_{sc} = \{\exp(-itA^{1/2})\}h = U_1(t)h \tag{A1.2}$$

where h is defined as in (A1.1).

We notice that (6.67) implies

$$v_{sc} = U_0(t)U_0^*(t)U_1(t)h = U_0(t)W_{\pm}h \tag{A1.3}$$

and, in particular, as $t \to \infty$

$$v_{sc}(\cdot, t, \theta, s) = U_0(t)W_{+}h(\cdot\,, \theta, s) \tag{A1.4}$$

The relation (A1.4) is to be understood in the sense

$$\lim_{t\to\infty} \|v_{sc}(\cdot, t, \theta, s) - U_0(t)W_{+}H(\cdot, \theta, s)\| = 0 \tag{A1.5}$$

Let $g \in L_2(\mathbf{R}^3)$ and set

$$w = U_0(t)g \tag{A1.6}$$

It is shown in [17] that ω satisfies the asymptotic relation

$$\omega(x,t) = |x|^{-1} G_0(|x| - t, \omega_x, g), \quad t \to \infty \tag{A1.7}$$

where $\omega_x = x/|x|$ and

$$G_0(v, \omega, g) \frac{1}{2\pi^{3/2}} \int_0^{\infty} \{\exp(iv\rho)\}\{\mathcal{F}g\}(\xi)|_{\xi=p\omega}(-i\rho)d\rho \tag{A1.8}$$

where $\mathcal{F}$ denotes the usual Fourier transform taken in $L_2(\mathbf{R}^3)$ with $\rho > 0$ and $\omega \in S^2$.

If we now set $g = W_+ h$ in (A1.7) then we get

$$U_0 W_+ h = |x|^{-1} G_0(|x| - t, \omega_x, W_+ h) \tag{A1.9}$$

as $t \to \infty$.

Define

$$G(v, \omega, \theta, s) := G_0(v, \omega, W_+ h(\cdot, \theta, s)) \tag{A1.10}$$

and set

$$v_{sc}^{\infty}(x, t, \theta, s) = |x|^{-1} G(|x| - t, \omega, \theta, s) \tag{A1.11}$$

It now follows from (A1.2) and (A1.8) to (A1.11) that

$$v_{sc}(x, t, \theta, s) = v_{sc}^{\infty}(x, t, \theta, s) \quad \text{as} \quad t \to \infty$$

that is

$$\lim_{t \to \infty} \left\| v_{sc}(\cdot, t, \theta, s) - v_{sc}^{\infty}(x, t, \theta, s) \right\|_{L_2(\mathbf{R}^3)} = 0$$

References and Further Reading

[1] R.A. Adams: *Sobolev Spaces*, Academic Press, New York, 1975.

[2] W.O. Amrein, J.M. Jauch and K.B. Sinha: *Scattering Theory in Quantum Mechanics*. Lecture Notes and Supplements in Physics, Benjamin, Reading, 1977.

[3] L.D. Faddeev: The Uniqueness of Solutions for the Scattering Inverse Problem, *Vestnik Leningrad Univ.* 7, 1956, 126–130.

[4] I.M. Gel'fand and G. Shilov: *Generalised Functions Vol. I*, Academic Press, New York, 1964.

[5] T. Ikebe: Eigenfunction expansions associated with the Schrodinger Operators and their application to scattering theory, *Arch. Rat. Mech. Anal.* 5, 1960, 2–33.

[6] S.T. Kuroda: *An Introduction to Scattering Theory*, Lecture Notes No 51, Aarhus Univ. 1978.

[7] N.F. Mott and M.S.W. Massey: *The Theory of Atomic Collisions*, Oxford, 1949.

[8] R.G. Newton: *Scattering Theory of Waves and Particles*, McGraw-Hill, New York, 1966.

[9] A. Olek: *Inverse Scattering Problems for Moving, Penetrable Bodies*. PhD Thesis, University of Strathclyde, Glasgow, 1997.

[10] D.B. Pearson: *Quantum Scattering and Spectral Theory*, Academic Press, London, 1988.

[11] F. Riesz and B. Sz-Nagy: *Functional Analysis*, Ungar, New York, 1955.

[12] M. Reed and B. Simon: *Methods of Mathematical Physics, Vols 1–4*, Academic Press, New York, 1972–1979.

[13] G.F. Roach: *An Introduction to Linear and Nonlinear Scattering Theory*, Pitman Monographs and Surveys in Pure and Applied Mathematics, Vol. 78, Longman, Essex, 1995.

[14] W. Rudin: *Principles of Mathematical Analysis* (3rd Edn), McGraw-Hill, New York, 1976.

[15] Y. Saito: An inverse problem in potential theory and the inverse scattering problem. *J. Math. Kyoto Univ.* 22, 1982, 307–321.

[16] Y. Saito: Some properties of the scattering amplitude and the inverse scattering problem, *Osaka J. of Math.* 19(8), 1982, 527–547.

[17] C.H. Wilcox: *Scattering Theory for the d'Alembert Equation in Exterior Domains*, Lecture Notes in Mathematics, No.442, Springer, Berlin, 1975.

11

Scattering in Other Wave Systems

11.1 Introduction

So far in this monograph we have been concerned with acoustic wave scattering problems. In dealing with such problems we have adopted the Wilcox theory of acoustic scattering introduced in [8]. The main reasons for doing this were, on the one hand, that the Wilcox theory uses quite elementary results from functional analysis, the spectral theory of self-adjoint operators on Hilbert spaces and semigroup theory and, on the other it leads quite readily to the development of constructive methods based on generalised eigenfunction expansion theorems. Furthermore, unlike the Lax–Phillips theory [2] the Wilcox theory applies to scattering problems in both even and odd space dimensions.

In the following sections we indicate how scattering problems associated with electromagnetic waves and with elastic waves can be placed in a similar framework to that used when studying acoustic wave scattering problems. Indeed we will see that the electromagnetic and the elastic wave problems can be given the same symbolic form as acoustic wave problems. Consequently, the constructive methods developed for acoustic wave problems can, in principle, become available for electromagnetic and for elastic wave problems.

11.2 Scattering of Electromagnetic Waves

Each step in the analysis of acoustic wave scattering problems has its analogue in the analysis of electromagnetic wave scattering problems. In particular the scalar d'Alembert equation which arises when studying the acoustic field is replaced by a vector wave equation when investigating the electromagnetic wave field. However, the vector nature of the electromagnetic problems does lead to more demanding algebraic manipulations and calculations.

We consider here electromagnetic problems in $\mathbf{R}^3$ and understand that an element $x \in \mathbf{R}^3$ can be written in component form as $x = (x_1, x_2, x_3)$. Furthermore, we write $(x, t)=(x_1, x_2, x_3, t)$ to denote the space-time coordinates. When studying target scattering we let $B \subset \mathbf{R}^3$ denote a closed bounded set such that $\Omega = \mathbf{R}^3 - B$ is connected. The set B represents the scattering body. For potential type scattering problems there is no need to introduce the set B. In this case the scattering arises as a result of perturbations of coefficients in the governing field equations.

We assume that the medium filling Ω is characterised by a dielectric constant ε and a magnetic permeability μ.

Electromagnetic phenomena are governed by the celebrated **Maxwell's equations** which we write in the form

$$\{\varepsilon \mathbf{E}_t - \nabla \times \mathbf{H}\}(x, t) = \mathbf{J}(x, t) \tag{11.1}$$

$$\{\mu \mathbf{H}_t + \nabla \times \mathbf{E}\}(x, t) = \mathbf{K}(x, t) \tag{11.2}$$

where $\mathbf{E}$ and $\mathbf{H}$ represent the electrical and magnetic field vectors respectively, whilst $\mathbf{J}$ and $\mathbf{K}$ denote the electric and magnetic currents respectively. Furthermore, we have introduced the symbol ∇, referred to as nabla, to denote the vector differential expression

$$\nabla := \mathbf{i}\frac{\partial}{\partial x_1} + \mathbf{j}\frac{\partial}{\partial x_2} + \mathbf{k}\frac{\partial}{\partial x_3}$$

where $\mathbf{i}$, $\mathbf{j}$, $\mathbf{k}$ are a triad of unit vectors used to characterise $\mathbf{R}^3$. In terms of this symbol we define

$$\text{grad } \varphi = \nabla \varphi$$

$$\text{div } \mathbf{v} = \nabla \cdot \mathbf{v}$$

$$\text{curl } \mathbf{v} = \nabla \times \mathbf{v}$$

where the dot denotes the usual scalar product and $\times$ the usual vector product [4].

Maxwell's equations (11.1), (11.2) can be written more conveniently in the form

$$\begin{bmatrix} \mathbf{E} \\ \mathbf{H} \end{bmatrix}_t (x,t) + \begin{bmatrix} 0 & -\frac{1}{\varepsilon}\nabla \times \\ \frac{1}{\mu}\nabla \times & 0 \end{bmatrix} \begin{bmatrix} \mathbf{E} \\ \mathbf{H} \end{bmatrix} (x,t) = \begin{bmatrix} \frac{1}{\varepsilon}\mathbf{J} \\ \frac{1}{\mu}\mathbf{K} \end{bmatrix} (x,t) \tag{11.3}$$

This in turn can be written

$$(\mathbf{U}_t - i\mathbf{G}\mathbf{U})(x, t) = \mathbf{f}(x, t) \tag{11.4}$$

where

$$\mathbf{U}(x,t) = \langle \mathbf{E}, \mathbf{H} \rangle (x,t), \quad \mathbf{f}(x,t) = \left\langle \frac{1}{\varepsilon}\mathbf{J}, \frac{1}{\mu}\mathbf{K} \right\rangle (x,t)$$

$$-i\mathbf{G} = \mathbf{M}^{-1}\begin{bmatrix} 0 & -\nabla\times \\ \nabla\times & 0 \end{bmatrix}, \quad \mathbf{M} = \begin{bmatrix} \varepsilon & 0 \\ 0 & \mu \end{bmatrix}$$

and, as usual $\langle \cdot, \cdot \rangle$ denotes the transpose of the vector $\begin{bmatrix} \cdot \\ \cdot \end{bmatrix}$.

Equation (11.4) has the same symbolic form as that discussed in Chapter 7 (see (7.17)). However, here the unknown $\mathbf{U}$ has a more complicated structure since it has the general form $\mathbf{U} = \langle \mathbf{u}_1, \mathbf{u}_2 \rangle$ where in our particular case $\mathbf{u}_1 = \mathbf{E}$ and $\mathbf{u}_2 = \mathbf{H}$.

We shall use the notation

$$\mathbf{v}(x, t) = \langle v_1, v_2, v_3 \rangle \, (x, t), \quad v_j \in \mathbf{C}, \quad j = 1, 2, 3$$

to denote a vector in $\mathbf{R}^3$ with complex-valued components. Throughout we shall be interested in vectors $\mathbf{v}$ having the properties

$$\mathbf{v} \in (L_2(\Omega))^3 := \{\mathbf{v} = \langle v_1, v_2, v_3 \rangle : v_j \in L_2(\Omega), j = 1, 2, 3\} \tag{11.5}$$

Occasionally we will emphasise matters by using the notation

$$L_2(\Omega, \mathbf{C}^3) = (L_2(\Omega))^3 \cap \left\{ v_j : \|v_j\|^2 < \infty, j = 1, 2, 3 \right\}$$

where $\|\cdot\|$ denotes the usual $L_2(\Omega)$ norm.

The collection $L_2(\Omega, \mathbf{C}^3)$ is a Hilbert space, which we shall denote by H, with respect to the inner product and norm structure

$$(\mathbf{u}, \mathbf{v})_H = (u_1, v_1) + (u_2, v_2) + (u_3, v_3) \tag{11.6}$$

$$\|\mathbf{u}\|_H^2 = \|u_1\|^2 + \|u_2\|^2 + \|u_3\|^2 \tag{11.7}$$

where $(\cdot, \cdot)$ and $\|\cdot\|$ denote the usual $L_2(\Omega)$ inner product and norm.

We also introduce

$$\mathbb{D}(\Omega) := \{\mathbf{u} \in (L_2(\Omega))^3 : \operatorname{div} \mathbf{u} \in L_2(\Omega)\} \tag{11.8}$$

$$\mathbb{R}(\Omega) := \{\mathbf{u} \in (L_2(\Omega))^3 : \operatorname{curl} \mathbf{u} \in (L_2(\Omega))^3\} \tag{11.9}$$

with structure

$$\|\mathbf{u}\|_{\mathbb{D}}^2 := \|\mathbf{u}\|_{\mathrm{H}}^2 + \|\operatorname{div} \mathbf{u}\|^2 \tag{11.10}$$

$$\|\mathbf{u}\|_{\mathbb{R}}^2 := \|\mathbf{u}\|_{\mathrm{H}}^2 + \|\operatorname{curl} \mathbf{u}\|_{\mathrm{H}}^2$$

Furthermore, we define

$$\mathbb{R}_0(\Omega) := \{\mathbf{u} \in \mathbb{R}(\Omega) : (\nabla \times \mathbf{u}, \mathbf{v}) = (\mathbf{u}, \nabla \times \mathbf{v})\} \tag{11.11}$$

With this preparation it would therefore seem natural to look for solutions **U** of (11.4) which belong to the class

$$\mathrm{L} := (L_2(\Omega))^3 \times (L_2(\Omega))^3 \tag{11.12}$$

On this class we will find it convenient to define a product of the two elements $\mathbf{W} = \langle \mathbf{w}_1, \mathbf{w}_2 \rangle$ and $\mathbf{V} = \langle \mathbf{v}_1, \mathbf{v}_2 \rangle$ to be

$$\begin{aligned}(\mathbf{W}, \mathbf{V})_{\mathbf{L}} &:= \left(\begin{bmatrix} \mathbf{w}_1 \\ \mathbf{w}_2 \end{bmatrix}, \begin{bmatrix} \mathbf{v}_1 \\ \mathbf{v}_2 \end{bmatrix} \right) \\ &= (\mathbf{w}_1, \mathbf{v}_1)_H + (\mathbf{w}_2, \mathbf{v}_2)_H \end{aligned} \tag{11.13}$$

When dealing with Maxwell's equations (11.1), (11.2) we find it convenient to use, instead of **L** as defined above, the class

$$\mathcal{H} = (L_2(\Omega))^3 \times (L_2(\Omega))^3$$

which is a Hilbert space with respect to a weighted inner product defined by

$$(\mathbf{W}, \mathbf{V})_{\mathcal{H}} = (\mathbf{W}, \mathbf{MV})_{\mathbf{L}} \tag{11.14}$$

where the weight matrix **M** is defined in (11.4).

If we now introduce the operator

$$\begin{aligned} &\mathbf{G} : \mathcal{H} \to \mathcal{H} \\ &-i\mathbf{GU} = \mathbf{M}^{-1} \begin{bmatrix} 0 & -\nabla\times \\ \nabla\times & 0 \end{bmatrix} \begin{bmatrix} \mathbf{u}_1 \\ \mathbf{u}_2 \end{bmatrix} \end{aligned} \tag{11.15}$$

where

$$\mathbf{U} = \langle \mathbf{u}_1, \mathbf{u}_2 \rangle \in D(\mathbf{G})$$

$$D(\mathbf{G}) = \mathbb{R}_0(\Omega) \times \mathbb{R}(\Omega) \subset \mathcal{H}$$

then we see that (11.1), (11.2) have the following representation in $\mathcal{H}$

$$\mathbf{U}_t(t) - i\mathbf{GU}(t) = \mathbf{f}(t), \quad \mathbf{U}(0) = \mathbf{U}_0 \in \mathcal{H} \tag{11.16}$$

As in previous chapters we understand

$$\mathbf{U}(\cdot, \cdot) : t \to \mathbf{U}(\cdot, t) =: \mathrm{U}(t)$$

Implicit in this development are the assumptions of

Initial conditions:

$$\mathbf{E}(0) = \mathbf{E}_0, \quad \mathbf{H}(0) = \mathbf{H}_0$$

Boundary conditions:
Total reflection at the boundary $\partial\Omega$ of Ω that is

$$(\mathbf{n} \times \mathbf{E}) = 0 \quad \text{on } \partial\Omega$$

where $\mathbf{n}$ is the outward drawn unit vector normal to $\partial\Omega$. Other boundary conditions can be imposed as required [3].

We require solutions of (11.5) that have finite energy. With the above structure we have that the energy, ε, of the electromagnetic field can be expressed in the form

$$\varepsilon := \|\mathbf{U}\|_{\mathcal{H}}^2 = (\mathbf{u}_1, \varepsilon\,\mathbf{u}_1)_H + (\mathbf{u}_2, \mu\mathbf{u}_2)_H \tag{11.17}$$

Once we have a Hilbert space representation of (11.1), (11.2) the matter of existence and uniqueness of solutions to (11.16) arises. These results can be obtained using the methods outlined in Chapter 5. To this end the following results can be obtained in much the same manner as for acoustic equations but now they have to be specialised bearing in mind the structure of $\mathcal{H}$. Specifically, the following theorem can be established [3].

Theorem 11.1. *If* $\mathbf{G}$ *is defined as in* (11.15) *then*
(i) $\mathbf{G}$ *is a self-adjoint operator*
(ii) $\mathbf{C}\backslash\mathbf{R} \subset \rho(\mathbf{G})$
(iii) $\|(\mathbf{G} + i\lambda)^{-1}\| \leq 1/|\lambda|$.

The results in Theorem 8.1 when combined with Stone's theorem yield, as outlined in Chapter 5, the following important result.

Theorem 11.2. *The initial boundary value problem*

$$\mathbf{U}_t(t) - i\mathbf{G}\mathbf{U}(t) = 0, \quad \mathbf{U}(0) = \mathbf{U}_0 \tag{11.18}$$

is uniquely solvable in $\mathcal{H}$. *Moreover, the solution* $\boldsymbol{U}$ *has finite energy.*

Once questions of existence and uniqueness have been settled then, as in previous chapters, we can take the first component (say) of the solution U to provide details solely of the electric field $\mathbf{E}$. However, once it has been established that Theorem 11.2 holds, that is, that (11.1), (11.2) is a well-posed system, then an attractive alternative approach is to deal with the equations (11.1), (11.2) directly and this we shall do.

If we recall the vector identity

$$\text{curl curl } \mathbf{v} = \text{grad div } \mathbf{v} - \Delta\mathbf{v} \tag{11.19}$$

then we can eliminate $\mathbf{H}$ in (11.1), (11.2) and obtain the second order vector equation

$$\{\partial_t^2 + \text{curl curl}\}\mathbf{E}(x, t) = -\partial_t\mathbf{J}(x, t) \tag{11.20}$$

Whilst we shall use (11.20) in this section we would point out that a similar vector wave equation to (11.20) can be obtained for $\mathbf{H}$ by eliminating $\mathbf{E}$ in (11.1), (11.2).

The equation (11.20) is equivalent to a system of three second order partial differential equations for the components of the electric field. Once these are solved then the magnetic field can be found from (11.20).

In the following it will be convenient to recall some of the more detailed notations and conventions of matrix algebra.

If the electric field vector $\mathbf{E}$ is written in the form

$$\mathbf{E} = E_1\mathbf{i}_1 + E_2\mathbf{i}_2 + E_3\mathbf{i}_3 \tag{11.21}$$

where $(\mathbf{i}_1, \mathbf{i}_2, \mathbf{i}_3)$ denotes the orthonormal basis associated with the coordinate system (x_1, x_2, x_3) then (11.21) defines a one-to-one correspondence

$$\mathbf{E} \leftrightarrow u = (E_1, E_2, E_3)^T \tag{11.22}$$

where T denotes the transpose of a matrix. Consequently for two vectors $\mathbf{a}$ and $\mathbf{b}$ with associated correspondences

$$\mathbf{a} = a_1\mathbf{i}_1 + a_2\mathbf{i}_2 + a_3\mathbf{i}_3 \leftrightarrow a = (a_1, a_2, a_3)^T \tag{11.23}$$

$$\mathbf{b} = b_1\mathbf{i}_1 + b_2\mathbf{i}_2 + b_3\mathbf{i}_3 \leftrightarrow b = (b_1, b_2, b_3)^T \tag{11.24}$$

the associated scalar and vector products of vector algebra correspond to the following matrix operations

$$\mathbf{b}\cdot\mathbf{a} \leftrightarrow b^T a = b\cdot a = b_1 a_1 + b_2 a_2 + b_3 a_3 \tag{11.25}$$

$$\mathbf{b} \times \mathbf{a} \leftrightarrow b \times a = \mathcal{M}(b)a \tag{11.26}$$

where the matrix $\mathcal{M}(b)$ is defined by

$$\mathcal{M}(b) = \begin{bmatrix} 0 & -b_3 & b_2 \\ b_3 & 0 & -b_1 \\ -b_2 & b_1 & 0 \end{bmatrix} \tag{11.27}$$

Similarly, the vector differential operator

$$\text{curl} = \nabla \times = \left(\mathbf{i}_1 \frac{\partial}{\partial x_1} + \mathbf{i}_2 \frac{\partial}{\partial x_2} + \mathbf{i}_3 \frac{\partial}{\partial x_3} \right) \tag{11.28}$$

has a matrix representation which has the typical form

$$\text{curl } \mathbf{E} = \nabla \times \mathbf{E} \leftrightarrow \nabla \times u = \mathcal{M}(\partial) u \tag{11.29}$$

where

$$\partial = (\partial_1, \partial_2, \partial_3) = \left(\frac{\partial}{\partial x_1}, \frac{\partial}{\partial x_2}, \frac{\partial}{\partial x_3} \right)$$

Using (11.29) repeatedly we can obtain

$$\text{curl curl } \mathbf{E} = \nabla \times \nabla \times \mathbf{E} \leftrightarrow \nabla \times \nabla \times u = \mathcal{A}(\partial) u \tag{11.30}$$

where

$$\mathcal{A}(b) = \mathcal{M}(b)^2 = bb - |b|^2 I \tag{11.31}$$

The term bb denotes a (3×3) matrix whose jkth element is $b_j b_k$ and I denotes the (3×3) identity matrix. Hence the identity (11.30) can be written

$$\nabla \times \nabla \times u = (\nabla \nabla \cdot - \Delta I) u \tag{11.32}$$

where Δ denotes the three-dimensional Laplacian

$$\Delta = \partial_1^2 + \partial_2^2 + \partial_3^2 \tag{11.33}$$

We remark that (11.19) and (11.32) are corresponding expressions. However, care must be exercised when dealing with the vector identity (11.19) in coordinate systems other than Cartesian. This is particularly the case when interpreting the vector Laplacian Δ (see [4]).

With these various notations in mind we see that the electromagnetic field generated in Ω is characterised by a function u of the form

$$u(x, t) = (u_1(x, t), u_2(x, t), u_3(x, t))^T, \quad x \in \Omega, \quad t \in \mathbf{R} \tag{11.34}$$

where $u_j = E_j, j = 1, 2, 3$ are the components of the electric field. The quantity $u(x, t)$ is a solution of the inhomogeneous vector equation

$$\{\partial_t^2 + \nabla \times \nabla \times\} u(x, t) = -\partial_t J(x, t) =: f(x, t), \quad (x, t) \in \Omega \times \mathbf{R} \tag{11.35}$$

and $J(x, t)$ is the electric current density which generates the field. We notice that if we are dealing with divergence-free fields then the first term on the right-hand side of (11.32) vanishes and (11.35) reduces to a vector form of the familiar scalar wave equation.

The FP associated with (11.35) has a field which we shall denote by $u_0(x, t)$. This is the field generated by the sources (transmitters) in the medium when no scatterers are present. It can be represented, as in the scalar case, in terms of a retarded potential (see Appendix A12.5). To see this we apply the divergence operator to both sides of (11.35) and use the well-known vector identity $\nabla \cdot \nabla \times u = 0$ to obtain

$$\partial_t^2 \nabla \cdot u(x, t) = -\partial_t J(x, t) \tag{11.36}$$

We now integrate (11.36) twice over the interval $t_0 \leq \tau \leq t$ and use the initial condition

$$u(x, t) = 0 \quad \text{for} \quad t < t_0, \quad x \in \Omega$$

to obtain

$$\nabla \cdot u(x, t) = -\int_{t_0}^{t} \nabla . J(x, \tau) d\tau, \quad x \in \Omega, \quad t \in \mathbf{R}. \tag{11.37}$$

Using (11.35), (11.37) and the identity (11.32) we see that u satisfies

$$\{\partial_t^2 - \Delta\} u(x, t) = \nabla \nabla \cdot \int_{t_0}^{t} J(x, \tau) d\tau - \partial_t J(x, t) \tag{11.38}$$

Since

$$\partial_t^2 \int_{t_0}^{t} J(x, \tau) d\tau = \partial_t J(x, t)$$

we see, on replacing u by u_0 in (11.38) that the free field is determined by the inhomogeneous wave equation

$$\{\partial_t^2 - \Delta\} u_0(x, t) = \{\nabla \nabla - \partial_t^2 \mathbf{I}\} \cdot \int_{t_0}^{t} J(x, \tau) d\tau \tag{11.39}$$

for $(x, t) \in \Omega \times \mathbf{R}_+$ and the initial condition $u_0(x, t) = 0$ for all $t \leq t_0$.

Equation (11.39) is equivalent to three scalar wave equations and as in the acoustic case they can be integrated by the retarded potential formula. For ease of presentation we define

$$IJ(x, t) = \int_{t_0}^{t} J(x, \tau) d\tau$$

Consequently, the function V defined by

$$V(x,t) = \frac{1}{4\pi}\int_{|x'-x_0|\le\delta_0} \frac{IJ(x',t-|x-x'|)}{|x-x'|}dx' \tag{11.40}$$

satisfies

$$\{\partial_t^2 - \nabla\}V(x,t) = IJ(x,t), \quad (x,t) \in \mathbf{R}^3 \times \mathbf{R} \tag{11.41}$$

and the initial condition $V(x, t) = 0$ for $t \le t_0$. It then follows, by the linearity of the wave equation, that the free field is given by

$$u_0(x,t) = \{\nabla\nabla - \partial_t^2\mathbf{I}\}\cdot V(x,t) \tag{11.42}$$

We have now reached the stage where we are in a position to take the same steps for analysing electromagnetic scattering problems as we did when dealing with acoustic scattering problems in Chapters 6 and 7.

For convenience we gather together in the next section the salient features of the analysis of the scattering of acoustic waves.

11.3 Overview of Acoustic Wave Scattering Analysis

The governing equation in this case is

$$\{\partial_t^2 - \Delta\}u(x,t) = f(x,t), \quad (x,t) \in \mathbf{R}^3 \times \mathbf{R} \tag{11.43}$$

where u characterises the acoustic field.

It is assumed that the transmitter is localised near the point x_0. It will also be assumed that the transmitter emits a signal at time t_0 which, in the first instance, is in the form of a pulse of time duration T. Consequently, we will have

$$\text{supp}\, f \subset \{(x,t): t_0 \le t \le t_0 + T\}, \quad |x - x_0| \le \delta_0$$

where δ_0 is a given constant.

If the pulse is scattered by a body B then we will assume that

$$\mathrm{B} \subset \{x : |x| \le \delta, \delta = \text{constant}\}$$

It will be assumed that the transmitter and the scattering body are very far apart and disjoint; this introduces the so-called far field approximation

$$|x_0| >> \delta_0 + \delta$$

The acoustic field for the primary, or Free Problem, that is, the acoustic field which obtains when there are no scatterers present in the medium, is given in terms of a retarded potential in the form (see Appendix A12.5)

$$u_0(x,t) = \frac{1}{4\pi}\int_{|x'-x_0|\leq\delta_0} \frac{f(x', t-|x-x'|)}{|x-x'|}dx' \tag{11.44}$$

for $(x, t) \in \mathbf{R}^3 \times \mathbf{R}$.

If θ_0 denotes a unit vector defined by

$$x_0 = -|x_0|\,\theta_0 \tag{11.45}$$

then expanding $|x - x'|$ in powers of $|x_0|$ we find

$$u_0(x,t) = \frac{s(\theta_0, x\cdot\theta_0 - t + |x_0|)}{|x_0|} + O\left(\frac{1}{|x_0|^2}\right)$$

uniformly for $t \in \mathbf{R}$, $|x| \leq \delta$ where

$$s(\theta_0, \tau) = \frac{1}{4\pi}\int_{|x'-x_0|\leq\delta_0} f(x', \theta_0\cdot(x'-x_0)-\tau)dx', \quad \tau \in \mathbf{R}$$

is the signal wave form.

When the error term is dropped in the above then the primary field is a plane wave propagating in the direction of unit vector θ_0.

When the primary acoustic field is a plane wave

$$u_0(x, t) = s(\theta_0, x\cdot\theta_0 - t), \quad \operatorname{supp} s(\theta_0,\cdot) \subset [a, b] \tag{11.46}$$

which is scattered by B then the resulting acoustic field $u(x, t)$ is the solution of the IBVP

$$\{\partial_t^2 - \Delta\}u(x, t) = 0, \qquad (x, t) \in \Omega \times \mathbf{R} \tag{11.47}$$

$$u(x, t) \in (bc) \tag{11.48}$$

$$u(x, t) \equiv u_0(x, t), \quad x \in \Omega,\ t + b + \delta < 0 \tag{11.49}$$

The echo or scattered field is then defined to be

$$u_{sc}(x, t) = u(x, t) - u_0(x, t), \quad (x, t \in \Omega \times \mathbf{R}) \tag{11.50}$$

It can then be shown [8] that in the far field

$$u_{sc}(x,t) \approx \frac{e(\theta_0, \theta, |\mathrm{x}| - t)}{|x|}, \quad x = |x|\theta \tag{11.51}$$

The quantity $e(\theta_0, \theta, t)$ is the echo wave form.

A main aim in practical scattering theory is to calculate the relationship between the signal and echo wave forms. A way of doing this is as follows.

Assume the source function f has the form

$$f(x,t) = g_1(x)\cos\omega t + g_2(x)\sin\omega t = \mathrm{Re}\{g(x)\exp(-i\omega t)\} \tag{11.52}$$

where $g(x) = g_1(x) + ig_2(x)$.

The associated wave field, $u(x, t)$, will have the same time dependence and have the typical form

$$u(x,t) = w_1(x)\cos\omega t + w_2(x)\sin\omega t = \mathrm{Re}\{w(x)\exp(-i\omega t)\} \tag{11.53}$$

where $w(x) = w_1(x) + iw_2(x)$.

The wave field $u(x, t)$ must satisfy the d'Alembert equation (11.43) and the imposed boundary conditions when the source field has the form (11.46).

The boundary value problem for the complex wave function w is

$$\{\partial_t^2 - \Delta\}w(x) = -g(x), \quad x \in \Omega \tag{11.54}$$

$$w(x) \in (bc) \tag{11.55}$$

$$\left(\frac{\partial w}{\partial |x|} - i\omega w\right)(x) = O\left(\frac{1}{|x|^2}\right) \quad \text{as } |x| \to \infty \tag{11.56}$$

The radiation condition (11.56) ensures that the wave field characterised by u will be outgoing at infinity. Furthermore it guarantees the uniqueness of the wave field u [8].

The field $u_0(x, t)$ generated by $f(x, t)$ in the absence of scatterers is characterised by the complex wave function w_0 which is given by

$$w_0(x) = \frac{1}{4\pi}\int_{|x'-x_0|\le\delta_0} \frac{\exp(i\omega|x-x'|)}{|x-x'|} g(x')dx', \quad x \in \mathbf{R}^3 \tag{11.57}$$

Expanding $|x - x'|$ in powers $|x_0|$, as before, we obtain

$$w_0(x) = \frac{T(\omega\theta_0)}{|x_0|}\exp(i\omega\theta_0 \cdot x) + O\left(\frac{1}{|x_0|^2}\right) \quad \text{as } |x_0| \to \infty \tag{11.58}$$

uniformly for $|x| \le \delta$ where

$$T(\omega\theta_0) = \frac{1}{4\pi}\int_{|x'-cx_0|\le\delta_0} \{\exp-(i\omega\theta_0 \cdot x')g(x')dx'\} \tag{11.59}$$

With (11.53) in mind, together with the familiar forms of solutions of the d'Alembert equation, it will be convenient to express the primary wave function w_0 in the form

$$w_0(x, \omega\theta_0) = (2\pi)^{-3/2}\exp(i\omega\theta_0 \cdot x)$$

The wave field which is produced when $w_0(x, \omega\theta_0)$ is scattered by B is denoted $w^+(x, \omega\theta_0)$. It is taken to have the form

$$w^+(x, \omega\theta_0) = w_0(x, \omega\theta_0) + w_{sc}^+(x, \omega\theta_0)$$

and expected to satisfy the boundary value problem

$$(\Delta + \omega^2)w^+(x, \omega\theta_0) = 0$$

$$w^+(x, \omega\theta_0) \in (bc)$$

$$\left(\frac{\partial}{\partial|x|} - i\omega\right) w_{sc}^+(x, \omega\theta_0) = O\left(\frac{1}{|x|^2}\right) \quad \text{as} \quad |x| \to \infty$$

It can then be shown (see Chapter 6, [8] and [5]) that in the far field of B the scattered, or echo, field is a diverging spherical wave with asymptotic form

$$w_{sc}^+(x, \omega\theta_0) \approx \frac{\exp(i\omega|x|)}{4\pi|x|} T_+(\omega\theta, \omega\theta_0), \quad x = |x|\theta \tag{11.60}$$

The coefficient $T_+(\omega\theta, \omega\theta_0)$ is the scattering amplitude of B. It determines the amplitude and phase of the scattered field in the direction θ due to a primary field in the direction θ_0.

It can be shown that the echo wave profile e can be expressed in terms of known quantities in the form (see Chapter 6, [8] and [5])

$$e(\theta, \theta_0, \tau) = \mathrm{Re}\left\{\int_0^\infty \{\exp(i\omega\tau)\} T_+(\omega\theta, \omega\theta_0)\hat{s}(\omega, \theta_0)d\omega\right\} \tag{11.61}$$

where $\hat{s}$ denotes the Fourier transform of the signal wave form, that is

$$\hat{s}(\omega, \theta_0) := \frac{1}{(2\pi)^{1/2}} \int_{-\infty}^{\infty} \{\exp(-i\omega\tau)\} s(\tau, \theta_0)d\tau \tag{11.62}$$

11.4 More about Electromagnetic Wave Scattering

Quite simply we follow the various steps outlined in the previous section but now we must be careful to recognise that we will be dealing with matrix-valued coefficients, vectors and tensors [1], [6], [7].

The defining equation we must now consider is the inhomogeneous vector wave equation

$$\{\partial_t^2 + \nabla \times \nabla\times\}u(x, t) = f(x, t), \quad (x, t) \in \Omega \times \mathbf{R} \tag{11.63}$$

The far field approximations can be obtained in a similar manner as in the acoustic case. The first result we can obtain is a representation of $s(\tau, \theta_0)$ the signal wave profile in the form

$$s(\tau,\theta_0)=Q(\theta_0)\frac{1}{4\pi}\int_{|x'-x_0|\leq\delta_0} f(x',\theta_0\cdot(x'-x_0)-\tau)dx' \tag{11.64}$$

where Q is a tensor defined by

$$Q(\theta) = I - \theta\theta \tag{11.65}$$

and is the projection onto the plane through the origin with normal in the direction θ.

We assume that an electric current density of the form

$$J(x, t) = \mathrm{Re}\{J(x)\exp(-i\omega t)\} \tag{11.66}$$

generates wave fields

$$u(x, t) = \mathrm{Re}\{w(x)\exp(-i\omega t)\} \tag{11.67}$$

where $w(x) = \langle w_1, w_2, w_3\rangle (x)$ has complex-valued components. The wave $u(x, t)$ must satisfy (11.63) with f defined by (11.66) and (10.35) and the imposed boundary and initial conditions.

In the electromagnetic case the Sommerfeld radiation conditions used in connection with acoustic problems are not adequate; they have to be replaced by the Silver–Muller radiation conditions [3]. Consequently, the boundary value problem for the vector quantity in (11.67) is

$$\begin{aligned} \{\nabla\times\nabla\times-\omega^2\}w(x)&=f(x), && x\in\Omega \\ w(x)&\in(bc), && x\in\partial\Omega \\ \{\theta\times\nabla\times+i\omega\}w(x)&=O\left(\frac{1}{|x|^2}\right), && |x|\to\infty \end{aligned} \tag{11.68}$$

where $x = |x|\ \theta$ and

$$f(x) = i\omega J(x)$$

An application of the far field assumptions yields the following estimate for $w_0(x)$, the incident or primary field,

$$w_0(x)=Q(\theta_0)\left\{\frac{T(\omega\theta_0)}{|x_0|}\exp(-i\omega\theta_0\cdot x)\right\}+O\left(\frac{1}{|x_0|^2}\right) \tag{11.69}$$

where

$$T(\omega\theta_0)=\frac{1}{4\pi}\int_{|x'-x_0|\le\delta_0}\{\exp(-i\omega\theta_0\cdot x')\}f(x')dx' \tag{11.70}$$

When the error term in (11.69) is dropped then w_0 represents a plane wave which is represented in the form

$$w_0(x,\,\omega\theta_0) = \{\exp(-i\omega\theta_0{\cdot}x)\}Q(\theta_0){\cdot}a \tag{11.71}$$

where a is a constant vector.

It will be convenient to introduce the matrix plane wave

$$\Psi_0(x,\,\omega\theta_0) = (2\pi)^{-3/2}\{\exp(-i\omega\theta_0{\cdot}x)\}Q(\theta_0) \tag{11.72}$$

where the columns of Ψ_0 are plane waves of the form given in (11.71).

The electric field produced when the primary wave $\Psi_0(x,\,\omega\theta_0)$ is scattered by B will be denoted by $\Psi^+(x,\,\omega\theta_0)$. It will be assumed to have the form

$$\Psi^+(x,\,\omega\theta_0) = \Psi_0(x,\,\omega\theta_0) + \Psi^+_{sc}(x,\,\omega\theta),\quad x\in\Omega$$

and to be a solution of the vector boundary value problem

$$\begin{aligned}\{\nabla\times\nabla\times-\omega^2\}\Psi^+(x,\omega\theta_0)&=0, && x\in\Omega\\ \Psi^+(x,\omega\theta_0)&\in(bc), && x\in\partial\Omega\\ \{\theta\times\nabla\times+i\omega\}\Psi^+_{sc}(x,\omega\theta_0)&=O\left(\frac{1}{|x|^2}\right), && |x|\to\infty\end{aligned}$$

where as in the acoustic case the plus sign indicates "outgoing wave".

It can be shown [8] that $\Psi^+_{sc}(x,\,\omega\theta_0)$ characterises a diverging spherical wave with asymptotic form

$$\Psi^+_{sc}(x,\omega\theta_0)\approx\frac{\exp(i\omega|x|)}{4\pi|x|}T_+(\omega\theta,\omega\theta_0),\quad x=|x|\theta$$

We would emphasise that here $T_+(\omega\theta,\,\omega\theta_0)$ is a matrix-valued coefficient.

With these various modifications of the acoustic case and using the same notations it can be shown that

$$e(\theta,\theta_0,\tau)=\mathrm{Re}\left\{\int_0^\infty\{\exp(i\omega\tau)\}T_+(\omega\theta,\omega\theta_0)\hat{s}(\omega,\theta_0)d\omega\right\} \tag{11.73}$$

where, as in the acoustic case, $\hat{s}$ denotes the Fourier transform of the signal profile, namely

$$\hat{s}(\omega,\theta_0):=\frac{1}{(2\pi)^{1/2}}\int_{-\infty}^{\infty}\{\exp(-i\omega t)\}s(\tau,\theta_0)d\tau \tag{11.74}$$

Although (11.73) and (11.74) have much the same symbolic form as for the acoustic case it must be emphasised that $T_{+}(\omega\theta, \omega\theta_0)$ is matrix valued, that $\hat{s}\ (\omega, \theta_0)$ is vector valued and that the order of the factors in (11.73) must be maintained.

11.5 Potential Scattering in Chiral Media

Chiral materials are those which exhibit optical activity in the sense that the plane of vibration of linearly polarized light is rotated on passage through the material. Consequently, chiral phenomena in a medium can be investigated analytically by introducing into the classical Maxwell's equations those constitutive relations indicating the coupling of the electric and magnetic fields which involve a so-called **chirality measure**. There are a number of such relations. Here, for the purposes of illustration we shall use the Drude–Born–Fedorov (DBF) relations since they are symmetric under time reversality and duality transforms.

We remark that references to work supporting the material outlined in this section are given and discussed in the Commentary.

11.5.1 Formulation of the Problem

We consider electromagnetic waves propagating in a homogeneous, three-dimensional, unbounded chiral material. The electric field $\mathbf{E}(x, t)$ and magnetic field $\mathbf{H}(x, t)$, where $x = (x_1, x_2, x_3)$, are taken to satisfy the Maxwell's equations in the form

$$\text{curl } \mathbf{E}(x, t) = -\mathbf{B}_t(x, t) \tag{11.75}$$

$$\text{curl } \mathbf{H}(x, t) = \mathbf{D}_t(x, t) \tag{11.76}$$

$$\text{div } \mathbf{B}(x, t) = 0 \text{ and div } \mathbf{D}(x, t) = 0 \tag{11.77}$$

where $\mathbf{D}(x, t)$ and $\mathbf{B}(x, t)$ are electric and magnetic flux densities.

We introduce constitutive relations which relate the material fields $\mathbf{H}(x, t)$ and $\mathbf{D}(x, t)$ to the primitive fields $\mathbf{E}(x, t)$ and $\mathbf{B}(x, t)$ by using the DBF relations, namely

$$\mathbf{D}(x, t) = \varepsilon(\mathbf{I} + \beta\text{curl})\mathbf{E}(x, t) \tag{11.78}$$

$$\mathbf{B}(x, t) = \mu(\mathbf{I} + \beta\text{curl})\mathbf{H}(x, t) \tag{11.79}$$

where $\mathbf{I}$ is the identity matrix operator, ε denotes the electric permittivity, μ the magnetic permeability and β is the chirality measure. It is clear that if $\beta = 0$ then (11.78), (11.79) reduce to the classical relations for the achiral case.

We consider a bounded obstacle, B, immersed in a chiral material through which an incident electromagnetic wave propagates. We are interested in solutions of (11.75), (11.76) which satisfy initial conditions of the form

$$\mathbf{E}(x, 0) = \mathbf{E}_0(x) \quad \text{and} \quad \mathbf{H}(x, 0) = \mathbf{H}_0(x) \tag{11.80}$$

and boundary conditions which are dependent on the physical properties of the obstacle B.

From (11.75), (11.76) we get

$$\text{curl } \mathbf{E}(x,t) = -\mu \frac{\partial}{\partial t}(\mathbf{I} + \beta \text{curl})\mathbf{H}(x,t) \tag{11.81}$$

Since $\mathbf{E}$ is solenoidal then there will exist a vector function $\mathcal{M}$ such that

$$\mathbf{E}(x, t) = \text{curl } \mathcal{M}(x, t)$$

If in addition we assume

$$\text{div } \mathcal{M}(x, t) = 0 \tag{11.82}$$

then using the vector identity

$$\text{curl curl } \mathbf{v} = \{\text{grad div } - \Delta\}\mathbf{v} \tag{11.83}$$

we obtain

$$\Delta \mathcal{M}(x,t) = \mu \frac{\partial}{\partial t}(\mathbf{I} + \beta \text{curl})\mathbf{H}(x,t) \tag{11.84}$$

Furthermore if we assume (the **magnetic field assumption**)

$$\frac{\partial}{\partial t}(\mathbf{I} + \beta \text{curl})\mathbf{H}(x,t) = \varepsilon \left\{ \frac{\partial^2}{\partial t^2} + \beta \right\} \mathcal{M}(x,t) \tag{11.85}$$

then (11.84) yields

$$\left\{ \frac{\partial^2}{\partial t^2} - \frac{1}{\varepsilon\mu}\Delta + \beta \right\} \mathcal{M}(x,t) = 0 \tag{11.86}$$

Consequently, knowing any solution $\mathcal{M}(x, t)$ of (11.86) then $\mathbf{H}(x, t)$ and $\mathbf{E}(x, t)$ can be determined via the relations (11.84) and (11.81).

We remark that (11.78) and (11.79) indicate

$$\text{curl } \mathbf{H}(x,t) = \varepsilon \frac{\partial}{\partial t}(\mathbf{I} + \beta \text{curl})\mathbf{E}(x,t) \tag{11.87}$$

Since $\mathbf{H}$ is solenoidal then there will exist a vector function $\mathcal{N}$ such that

$$\mathbf{H}(x, t) = \text{curl } \mathcal{N}(x, t) \tag{11.88}$$

and if in addition we assume

$$\text{div}\mathcal{N}(x, t) = 0 \tag{11.89}$$

then we obtain

$$-\Delta N(x,t) = \varepsilon \frac{\partial}{\partial t}(\mathbf{I} + \beta \text{curl})\mathbf{E}(x,t) \tag{11.90}$$

Furthermore, if we assume (the **electric field assumption**)

$$\frac{\partial}{\partial t}(\mathbf{I} + \beta \text{curl})\mathbf{E}(x,t) = \varepsilon \left\{ \frac{\partial^2}{\partial t^2} + \beta \right\} \mathcal{N}(x, \text{t}) \tag{11.91}$$

then (11.90) yields

$$\left\{ \frac{\partial^2}{\partial t^2} - \frac{1}{\varepsilon\mu}\Delta + \beta \right\} \mathcal{N}(x,t) = 0 \tag{11.92}$$

Hence we have reduced the scattering problem for a class of electromagnetic waves in chiral materials to an initial boundary value problem of the form

$$\left\{ \frac{\partial^2}{\partial t^2} - \frac{1}{\varepsilon\mu}\Delta + \beta \right\} \mathbf{u}(x,t) = 0 \tag{11.93}$$

$$\mathbf{u}(x, 0) = \mathbf{u}_0(x), \quad \mathbf{u}_t(x, 0) = \mathbf{u}_1(x) \tag{11.94}$$

$$\mathbf{u} \text{ satisfies either (11.85) or (11.91)} \tag{11.95}$$

$$\mathbf{u} \in \text{(bc)}, \quad \mathbf{u} \in \text{(rc)} \tag{11.96}$$

where the notations (bc) and (rc) denote that the solution $\mathbf{u}(x, t)$ is required to satisfy boundary conditions on ∂B, the boundary of B, and a radiation condition as $|x| \to \infty$ respectively.

11.6 Scattering of Elastic Waves

The strategy outlined in this monograph for investigating wave scattering problems indicates that our first task is to represent the given physical problem as an operator equation problem in a suitable Hilbert space where ideally the operator involved is self-adjoint. Following the approach in previous chapters we will require that the Hilbert space representation of the given physical problem should be in the form of an IVP for a first order differential equation.

11.6.1 An Approach to Elastic Wave Scattering

In $\mathbf{R}^3$ elastic wave phenomena are governed by IBVPs of the following typical form

$$\{\partial_t^2 - \Delta^*\}\mathbf{u}(x, t) = \mathbf{f}(x, t), \quad x \in \Omega \subseteq \mathbf{R}^3, \qquad t \in \mathbf{R}^+ \tag{11.97}$$

$$\mathbf{u}(x, 0) = \varphi(x), \quad \mathbf{u}_t(x, 0) = \psi(x), \quad x \in \mathbf{R}^3 \tag{11.98}$$

$$\mathbf{u}(x, t) \in (bc), \quad (x, t) \in \partial\Omega \times \mathbf{R}^+ \tag{11.99}$$

where

x $= (x_1, x_2, x_3)$
φ, ψ = given vector functions characterising initial conditions
Δ^* $= -\mu\text{curl curl} + \lambda\text{grad div}$ is the Lamé operator
λ, μ = Lamé constants
$\Omega \subset \mathbf{R}^3$ = connected, open region exterior to the scattering target B
$\partial\Omega$ = closed, smooth boundary of Ω
$\mathbf{u}(x, t)$ = vector quantity characterising the elastic wave field
$= \langle u_1, u_2, u_3\rangle\ (x, t)$
$\mathbf{f}(x, t)$ = vector quantity characterising the signal emitted by the transmitter.

Other quantities of interest are

$$c_1 = \sqrt{\lambda + 2\mu} = \text{longitudinal wave speed}$$

$$c_2 = \sqrt{\mu} = \text{shear wave speed}$$

$$\begin{aligned}\mathbf{T}(\mathbf{u}, \mathbf{n}) &= 2\mu\left(\frac{\partial \mathbf{u}}{\partial \mathbf{n}}\right) + \lambda\mathbf{n}\,\text{div}\,\mathbf{u} + \mathbf{n}\times\text{curl}\,\mathbf{u} \\ &= \text{vector traction at a boundary point where the normal is } \mathbf{n}.\end{aligned} \tag{11.100}$$

For $\mathbf{u}(x, t) = \langle u_1, u_2, u_3\rangle\ (x, t)$ we shall write

$$\mathbf{u} \in (L_2(\mathbf{R}^3))^3 \quad \text{whenever} \quad u_j \in L_2(\mathbf{R}^3), j = 1, 2, 3$$

As in the acoustic and electromagnetic cases we begin an investigation of elastic waves by examining the FP. To this end we introduce the following notations and function spaces.

$$X(\Omega) = \text{space of scalar functions defined on a region } \Omega$$

$$\mathbf{X}(\Omega) = \text{space of vector functions defined on a region } \Omega.$$

For example if $\Omega \subseteq \mathbf{R}^3$ and

$$X(\Omega) = L_2(\Omega) := \left\{ u : \int_\Omega |u(x)|^2 \, dx < \infty \right\}$$

$$\begin{aligned}\mathbf{X}(\Omega) = \mathbf{L}_2(\Omega) &:= \{\mathbf{u} = \langle u_1, u_2, u_3 \rangle : u_j \in L_2(\Omega), j = 1, 2, 3\} \\ &= \left\{ \mathbf{u} : \mathbf{u} \in (L_2(\Omega))^3 \right\} \\ &= (X(\Omega))^3\end{aligned}$$

then

$L_2(\Omega)$ is a Hilbert space with inner product

$$(u, v) = \int_\Omega u(x)\overline{v(x)} dx \tag{11.101}$$

$\mathbf{L}_2(\Omega)$ is a Hilbert space with inner product

$$(\mathbf{u}, \mathbf{v}) = (u_1, v_1) + (u_2, v_2) + (u_3, v_3) \tag{11.102}$$

We also introduce

$$\mathbf{L}_2(\Delta^*, \Omega) := \mathbf{L}_2(\Omega) \cap \{\mathbf{u} : \Delta^* \mathbf{u} \in \mathbf{L}_2(\Omega)\} \tag{11.103}$$

$\mathbf{L}_2(\Delta^*, \Omega)$ is a Hilbert space with inner product

$$(\mathbf{u}, \mathbf{v})_{\Delta^*} := (\mathbf{u}, \mathbf{v}) + (\Delta^* \mathbf{u}, \Delta^* \mathbf{v}) \tag{11.104}$$

Following the approach adopted in earlier chapters the FP associated with the IBVP, that is the problem (11.97), (11.98), can now be interpreted as an IVP for a second order ordinary differential equation in $\mathbf{L}_2(\mathbf{R}^3)$. Specifically we introduce the operator

$$A_0 : \mathbf{L}_2(\mathbf{R}^3) \to \mathbf{L}_2(\mathbf{R}^3) \tag{11.105}$$

$$A_0 \mathbf{u} = -\Delta^* \mathbf{u}, \quad \mathbf{u} \in D(A_0)$$

$$D(A_0) = \{\mathbf{u} \in \mathbf{L}_2(\mathbf{R}^3) : \Delta^* \mathbf{u} \in \mathbf{L}_2(\mathbf{R}^3)\} \equiv \mathbf{L}_2(\Delta^*, \mathbf{R}^3)$$

This then yields the IVP

$$\left\{ \frac{d^2}{dt^2} + A_0 \right\} \mathbf{u}(t) = \mathbf{f}(t) \tag{11.106}$$

$$\mathbf{u}(0) = \varphi, \quad \mathbf{u}_t(0) = \psi \tag{11.107}$$

where we understand

$$\mathbf{u} \equiv \mathbf{u}(\cdot, \cdot) : t \to \mathbf{u}(\cdot, t) =: \mathbf{u}(t)$$

It is readily seen that A_0 is positive, self-adjoint on $\mathbf{L}_2(\Delta^*, \mathbf{R}^3)$ provided λ and μ are strictly positive. Questions of the existence and uniqueness of solution to the IVP (11.106), (11.107) can be settled by using the limiting absorption principle (see Commentary). The required solution $\mathbf{u}$ can then be obtained in the form

$$\mathbf{u}(x, t) = \mathrm{Re}\{\mathbf{v}(x, t)\}$$
$$\mathbf{v}(x, t) = \{\exp(-itA^{1/2})\}\mathbf{h}(x)$$

where

$$\mathbf{h}(x) = \varphi(x) + iA^{-1/2}\psi(x)$$

when $\mathbf{f} \equiv 0$ otherwise $\mathbf{h}$ has to have an additional integral term in its definition.

Since A is self-adjoint then the spectral theorem can be used to interpret terms such as $A^{1/2}$. Generalised eigenfunction expansions can then be established to provide a basis for constructive methods.

As we have seen in the acoustic and electromagnetic cases a problem such as (11.106), (11.107) can be reduced to an equivalent first order system. The required reduction is obtained in the now familiar manner and yields the IVP

$$\Psi_t(t) + i\mathbf{G}\Psi(t) = \mathbf{F}(t), \quad \Psi(0) = \Psi_0 \tag{11.108}$$

where

$$\Psi(t) = \langle \mathbf{u}, \mathbf{u}_t \rangle(t), \quad \Psi(0) = \Psi_0 = \langle \varphi, \psi_t \rangle$$
$$\mathbf{F}(t) = \langle 0, \mathbf{f} \rangle(t)$$
$$i\mathbf{G} = \begin{bmatrix} 0 & -I \\ A & 0 \end{bmatrix}$$

Proceeding formally a simple integrating factor technique indicates that

$$\Psi(t) = \{\exp(-it\mathbf{G})\}\Psi_0 + \int_0^t \{\exp(-i(t-\eta))\mathbf{G}\}\mathbf{F}(\eta)d\eta \tag{11.109}$$

For this approach to be meaningful we must be able to show that the problem (11.108) is well-posed and that the relation (11.109) is well defined. As we have seen in the previous chapters all this can be settled by showing that $i\mathbf{G}$ is self-adjoint; Stone's theorem will then indicate that $\exp\{-it\mathbf{G}\}$ is a well-defined semigroup and the result of Chapter 5 will then show (11.108) is a well-posed problem.

To show that $i\mathbf{G}$ is self-adjoint we consider $i\mathbf{G}$ to be defined on an "energy space" $\mathbf{H}_E$ which is a Hilbert space with inner product

$$(\Psi, \Phi)_E := (\nabla \times \psi_1, \nabla \times \varphi_1)_{\mathbf{L}_2} + (\psi_2, \varphi_2)_{\Delta^*}$$

where

$$\Psi = \langle \psi_1, \psi_2 \rangle, \quad \Phi = \langle \varphi_1, \varphi_2 \rangle$$

A straightforward calculation establishes that $i\mathbf{G}$ is symmetric. The full proof that $i\mathbf{G}$ is self-adjoint then follows as in Chapters 5 and 6 and [5].

The required solution $\mathbf{u}(x, t)$ of the IVP (11.97) (11.98) is then meaningfully defined by the first component of the solution Ψ of the IVP (11.108). For example, in the case when $\mathbf{f} \equiv \mathbf{0}$ then we obtain the solution form

$$\mathbf{u}(x, t) = (\cos tA^{1/2})\varphi(x) + A^{-1/2}(\sin tA^{1/2})\psi(x) \tag{11.110}$$

The self-adjointness of A ensures that the spectral theorem is available for the interpretation of such terms as $A^{1/2}$. Consequently (11.110) is well defined and computable. As in the acoustic and electromagnetic cases the computation of the elastic wave field, $\mathbf{u}(x, t)$, is carried out using results of generalised eigenfunction expansion theorems. These theorems, once established, are essentially of two types, one to cater for longitudinal wave phenomena, the other to accommodate shear wave phenomena.

When we deal with perturbed problems, for example with target scattering problems, then the IBVP (11.97) to (11.99) has to be investigated. We remark that the accommodation of the effects of boundary conditions can possibly cause technical difficulties when the self-adjointness of associated operators has to be established. Nevertheless, we have now arrived at the stage when we have managed to give elastic wave scattering problems the same symbolic form that we investigated when dealing with acoustic and electromagnetic wave scattering problems. Consequently, we are now in the position of being able to follow step by step the procedures we have outlined in Chapters 5, 6 and 7 for acoustic and electromagnetic problems. However, although this is a straightforward matter it can be a lengthy process. It has been worked through by a number of authors and the details are to be found in the references cited in the Commentary.

References and Further Reading

[1] A.C. Aitken: *Determinants and Matrices*, Oliver and Boyd, Edinburgh, 1951.

[2] P.D. Lax and R.S. Phillips: *Scattering Theory*, Academic Press, New York, 1967.

[3] R. Leis: *Initial Boundary Value Problems of Mathematical Physics*, John Wiley & Sons, Chichester, 1986.

[4] P.M. Morse and H. Feshbach: *Methods of Theoretical Physics*, McGraw-Hill, New York, 1953.

[5] G.F. Roach: *An Introduction to Linear and Nonlinear Scattering Theory*, Pitman Monographs and Surveys in Pure and Applied Mathematics, Vol. 78, Longman, Essex, 1995.

[6] D.E. Rutherford: *Vector Methods*, Oliver and Boyd, Edinburgh, 1951.

[7] B. Spain: *Tensor Calculus*, Oliver and Boyd, Edinburgh, 1956.

[8] C.H. Wilcox: *Scattering Theory for the d'Alembert Equation in Exterior Domains*, Lecture Notes in Mathematics, No. 442, Springer, Berlin, 1975.

12

Commentary

12.1 Introduction

From the outset it has been emphasised that this book is an introductory text intended for the use of those wishing to begin studying the scattering of waves by time-dependent perturbations. For this reason we offer in this chapter some additional remarks on the material that has been presented in previous chapters. The main intentions are, on the one hand, to give some indications of the work that either has been or is being done to cater for more general situations than those considered here and, on the other hand, to suggest further reading directions. Whilst it is recognised that it is impossible to give credit to all sources, nevertheless, those cited in the extended Bibliography provided here will, in turn, give additional references.

12.2 Remarks on Previous Chapters

Chapter 1:

As its title suggests this chapter is purely introductory. The various aspects of scattering theory which will be needed in later chapters are illustrated here in an entirely formal manner. Most of the notions which are introduced would seem to have appeared initially in the theory of quantum mechanical scattering theory. In this connection see [6], [22], [79], [84], [57], [87]. In a series of papers Wilcox and his collaborators showed how many of the techniques used in the study of quantum mechanical scattering could be extended to deal with wave problems in classical physics. The foundations for this work are fully discussed in [131]. The Wilcox approach to wave scattering problems relies on the availability of suitable generalised eigenfunction expansion theorems [110], [43] and offers good prospects for developing constructive methods of solution. This approach is distinct from that adopted by Lax and Phillips [57]; a reconciliation of the two approaches is given in [56], [130]. Although mainly concerned with quantum mechanical scattering aspects the texts [12], [17] are worth bearing in mind when developing detailed analyses of wave scattering problems.

Chapter 2:

Wave motion on strings is developed in many texts. We would particularly mention [11], [117], [126]. The approach to solutions of the wave equation by considering an equivalent first order system is discussed from the standpoint of semigroup theory in [38], [72]. An investigation of solutions to the wave equation represented in the form (2.58) is detailed in [60], [131], [91]. The method of comparing solutions to equations that have the typical form given in (2.58) parallels that used so successfully in quantum scattering theory. In this connection we would particularly mention the texts [6], [12], [17]. A comprehensive account of waves on strings can be found in [58]. The discussion of a scattering problem on a semi-infinite string follows the treatment found in [50].

Chapter 3: [98], [113], [122]

In this chapter a number of mathematical facts which are used frequently in this monograph are gathered together. The material is included mainly for the newcomer to the area of scattering theory who might possibly not have had the training in mathematical analysis which present-day mathematics students receive. There are a number of fine texts available which provide a thorough development of the various topics introduced in this chapter, see [42], [46], [48], [52], [64], [135]. This being said, particular attention should be paid to the following topics. The notion of completeness is crucial for many of the arguments used in developing scattering theories. A good account of this concept can be found in [52] whilst fine illustrations of its use in practical situations can be found in [76], [117]. Distribution theory is comprehensively developed in [35], [48], [63], [64], [109]. The newcomer to this area should be encouraged to become familiar with distribution theory developed in $\mathbf{R}^n$, $n > 1$, and especially with the notion of the n-dimensional Dirac delta [90]. The theory of linear operators on Hilbert spaces is comprehensively treated in [4], [42].

Chapter 4:

Hilbert spaces provide generalisations of the familiar notions of algebra and geometry whch are used in a Euclidean space setting. In a Hilbert space setting we deal with the functions themselves rather than with their numerical value as would be the case in a Euclidean space setting. Representing a given problem in a Hilbert space therefore yields an abstact setting of that given problem. One of the main advantages in working in an abstract setting is that many quite difficult problems in classical analysis, such as, for example, obtaining existence and uniqueness results, can often be resolved more readily in the chosen absract setting.

There are many excellent texts dealing with Hilbert spaces, a number of which have already been mentioned. Whilst Hilbert spaces can form a study in themelves a good starting point for those who have applications in mind can be found in any of [42], [52], [46], [48], [64]. These will, if required, also lead to more general treatments of Hilbert space than given in this chapter.

Chapter 5:

The topic of spectral decomposition is now highly developed and has many far reaching applications. Comprehensive accounts of spectral decompositon methods, from a number of different standpoints, are available. In this connection we would particularly mention [4], [12], [24], [27], [33], [42], [48], [52], [66], [84], [87], [88], [106], [135], [94], amongst which will be found something to suit most tastes. The material in this chapter is quite standard. In the presentation given here the development has been mainly concerned with bounded linear operators simply for ease of presentation. Similar results can be obtained for unbounded operators but in this case more care is required when dealing with domains of operators. This aspect is discussed in detail in [42] and [88], the latter giving a fine account of reducing subspaces.

A comprehensive account of measure theory with applications to scattering theory very much in mind can be found in [84]. The account of spectral decomposition of Hilbert spaces given in [84] can usefully be augmented by the associated material given in [48] and [135].

A number of the techniques which have been used successfully when dealing with quantum mechanical scattering problems can be adjusted to deal with wave scattering problems [5], [38], [39], [48], [72], [74], [83], [88], [90], [91], [114], [119], [120], [123]. Essentially this amounts to replacing the partial differential equation associated with an IBVP for the wave equation by an equivalent IVP for an ordinary differential equation defined in a suitable energy space (H, say), which will have solutions that are H-valued functions of t. Indications were given of how existence and uniqueness results for solutions of the IVP could be obtained using results from semigroup theory. Suggestions were also given as to how constructive methods could be developed using results from the abstract theory of integral equations.

One of the earliest accounts of semigroup theory can be found in [116]. A fine introduction to the subject is given in [72]. More advanced modern texts which will be useful are [120], [38], [13], [14], [83].

Chapter 6: [1], [6], [11], [28], [29], [38], [48], [43], [57], [60], [83], [84], [87], [91], [107], [120], [131], [130]

The modelling of acoustic wave phenomena is introduced and discussed in [44] and [111]. The material presented here is based very much on the work of [131]. Some of the more important results and concepts are gathered together here in order that the newcomer to the area should gain familiarity, as quickly as possible, with the various strategies involved when developing scattering theories. Many of the results are simply stated as their proof is usually quite lengthy. These various proofs are given in the literature cited here and it is felt that they can be read more profitably once the overall strategy of the subject has been appreciated. In this connection see, for instance, [60], [61], [92], [93], [94]. A considerable amount of work has been done, in a quantum mechanics setting, on the existence, uniqueness and completeness of wave operators: see [6], [12], [17], [79], [87]. A comprehensive and unified account is given in [84]. A detailed account of how these various notions can be adapted to cater for a target scattering problem is to be found in [131] and the references cited there. A full account of radiation conditions and incoming and outgoing waves can be found in [28], [29], [125] whilst a treatment of generalised eigenfunctions theorems and completeness in this context can be found in [11], [6], [76]. The various types of solutions which can be obtained for the partial differential equations which arise when analysing scattering processes are introduced and discussed in texts dealing with the modern theories of partial differential equations; see for example [27], [35], [60], [134]. The notion of solutions with finite energy has been used extensively [131]. Methods for determining and discussing such solutions can be readily developed by reducing IBVPs for wave equations to equivalent IVPs for an ordinary differential equation and then using semigroup theory [38], [120], [72], [91].

Central to the development of the generalised eigenfunction expansion theorems required when dealing with acoustic scattering problems is an appreciation of the Helmholtz equation and its properties. A comprehensive treatment of boundary value problems for the Helmholtz equation, using integral equation methods is given in [51].

The limiting absorption principle was introduced in [28]. It has been applied to problems involving a variety of different differential expressions, boundary conditions and domains. Recently, in conjunction with the related limiting amplitude principle [29] it has been successfully used to analyse scattering problems in domains involving unbounded interfaces [92], [93] [94], [95].

Chapter 7:

An understanding of how a given incident wave evolves throughout a medium is the central problem when developing a scattering theory. In this chapter explicit attention has been paid to the nature of the signal transmitted [91] and the forms that it can adopt at large distances from the transmitter and scatterers. This then enables the notions of signal and echo wave forms to be introduced [131]. An alternative method to that adopted in this chapter for obtaining such quantities relies on the properties of retarded potentials. This is outlined in Appendix A 12.5 of this chapter.

Chapter 8:

The material presented in this chapter has been greatly influenced by that in [85], [132], [133]. These references provide a good starting point for those wishing to begin working in this particular area.

Chapter 9:

In this chapter Floquet theory plays a central rôle. A fine development of this theory can be found in [25], [26]. An analysis of the propagation of linear acoustic waves in an infinite string with periodic characteristics is discussed in detail in [75] and [128].

Chapter 10:

The inverse scattering problem has been investigated by many authors. In this connection particular reference can be made to the books of [2], [70], [71], [59],

[19], [20]. As might be expected there are also many papers in journals devoted to this area. Typical examples of works which cover many aspects of this general area can be found in [9], [10], [15], [16], [81], [18], [30], [31], [32], [80], [96], [97].

Chapter 11:

As has been pointed out, working through the details for wave propagation in the electromagnetic and the elastic cases whilst, potentially, might be considered a straightforward matter is nevertheless an extremely lengthy process. This fact is well illustrated in [60], [62], [8], [44] and in [68], [69] respectively. The details given in these references will provide a starting point for studying electromagnetic wave sand elastic waves in a similar manner to that outlined here for acoustic wave problems.

12.3 Appendices

This section is included to provide an easy and convenient reference to some of the more technical concepts which have been referred to but not developed in this monograph. The presentation is brief and is made, almost entirely, in spaces of one dimension. More details can be found in standard books on mathematical analysis (for example [52], [98]).

A12.1 Limits and Continuity

If B is an interval in $\mathbf{R}$ then we write $B \subseteq \mathbf{R}$.

If $B = (a, b) = \{x \in \mathbf{R}: a < x < b\}$ then B is said to be an **open** interval.

If $B = [a, b] = \{x \in \mathbf{R}: a \leq x \leq b\}$ then B is said to be a **closed** interval which we denote by B. It follows that

$$\bar{B} = B \cup \{a, b\} = B \cup \partial B$$

where ∂B denotes the boundary of B.

The notions of half closed intervals such as $[a, b)$ and $(a, b]$ can be introduced similarly.

A real-valued function f defined on $\mathbf{R}$, which symbolically is characterised by writing f: $\mathbf{R} \to \mathbf{R}$, is said to have a **limit** L as $x \to a \in \mathbf{R}$ if for any real number $\varepsilon > 0$, no matter how small, there is a $\delta > 0$ such that

$$|f(x) - L| < \varepsilon \quad \text{whenever} \quad 0 < |x - a| < \delta$$

and we write in this case

$$\lim_{x \to a} f(x) = L$$

An equivalent definition in terms of sequences is available. Specifically, if $\{x_n\}$ is a sequence such that $x_n \to a$ as $n \to \infty$ then it follows that $f(x_n) \to L$.

In order that a function f: $\mathbf{R} \to \mathbf{R}$ should have a limit at $x = a$ then the function must be defined on each side of $x = a$ but not necessarily at $x = a$. For example the function f defined by

$$f(x) = \frac{\sin x}{x}$$

is not defined at $x = 0$ since there it assumes the form 0/0. However, on expanding the numerator in series form, we see that it has the value unity at $x = 0$.

It is also possible to introduce the notion of one-sided limits. A function f: $\mathbf{R} \to \mathbf{R}$ is said to have a **limit** L **from the right** at $x = a$ as $x \to a$ if for any number $\varepsilon > 0$ there exists a number $\delta > 0$ such that

$$|f(x) - L| < \varepsilon \quad \text{whenever} \quad 0 < x - a < \delta$$

When this is the case we replace L by $f(x^+)$.

Similarly, a function f: $\mathbf{R} \to \mathbf{R}$ is said to have a **limit** L **from the left** at $x = a$ as $x \to a$ if for any number $\varepsilon > 0$ there exists a number $\delta > 0$ such that

$$|f(x) - L| < \varepsilon \quad \text{whenever} \quad 0 < a - x < \delta$$

When this is the case we replace L by $f(x^-)$.

Clearly, if both the limits from right and left exist and if $f(x^+) = f(x^-)$ then $\lim_{x \to a} f(x)$ exists and equals $f(x^+) = f(x^-)$.

A function f: $\mathbf{R} \to \mathbf{R}$ is **continuous** at a point $x = a \in \mathbf{R}$ if $\lim_{x \to a} f(x)$ exists and equals $f(a)$. In particular the function has to be defined at $x = a$. For example the function f defined by $f(x) = (\sin x)/x$ and $f(0) = 1$ is continuous at every point. A function is said to be **continuous in an interval** $a \leq x \leq b$ if it is continuous at each point in the interval.

The following are important results in applications.

Theorem: IntermediateValue Theorem. *If f: $\mathbf{R} \to \mathbf{R}$ is continuous in a finite (i.e. bounded) interval $[a, b]$ and if $f(a) < p < f(b)$ then there exists at least one point $x = c \in [a, b]$ such that $f(c) = p$.*

Theorem: Concerning Maxima. *If f: $\mathbf{R} \to \mathbf{R}$ is continuous in a finite closed interval $[a, b]$ then it has a maximum in that interval. That is, there exist a point $m \in [a, b]$ such that $f(x) \leq f(m)$ for all $x \in [a, b]$.*

Applying this result to the function $(-f)$ indicates that f also has a minimum.

Theorem: *Let f: $\mathbf{R} \to \mathbf{R}$ and assume*

(i+) *f is continuous on a finite closed interval $[a, b]$*

(ii) $f(x) \geqslant 0$, $x \in [a, b]$

(iii) $\int_a^b f(\eta)d\eta = 0$

then $f(x) \equiv 0$, $x \in [a, b]$.

A function f: $\mathbf{R} \to \mathbf{R}$ has a **jump discontinuity** if both one-sided limits $f(x^+)$ and $f(x^-)$ exist. The number $(f(x^+) - f(x^-))$ is the value of the jump.

A function f: $\mathbf{R} \rightarrow \mathbf{R}$ is **piecewise continuous** on a finite closed interval $[a, b]$ if there are a finite number of point

$$a = a_0 \leq a_1 \leq a_2 \leq \ldots \leq a_n = b$$

such that f is continuous on each sub-interval (a_{j-1}, a_j), $j = 1, 2, \ldots, n$ and all the one side limits $f(x^-)$ for $1 \leq j \leq n$ and $f(x^+)$ for $0 \leq j \leq n$ exist. Such a function has jumps at a finite number of points but otherwise is continuous.

It can be shown [98] that every piecewise continuous function is integrable. Similar results can be shown to hold in $\mathbf{R}^n$ [98], [88].

A12.2 Differentiability

A function f: $\mathbf{R} \rightarrow \mathbf{R}$ is **differentiable** at a point $x \in [a, b]$ if

$$\lim_{x \to a} \frac{|f(x) - f(a)|}{(x - a)}$$

exists. The value of the limit is denoted by either $f'(a)$ or (df/dx)(a).

A function f: $\mathbf{R} \rightarrow \mathbf{R}$ is differentiable in the open interval (b, c) if it is differentiable at each point of the interval.

When we work in $n > 1$ dimensions some rather more sensitive notation is required. We illustrate this in the following paragraphs.

A12.3 The Function Classes $C^m(B)$, $C^m(\overline{B})$

Let $B \subset \mathbf{R}^n$, n ≥ 1 and let f: $\mathbf{R}^n \rightarrow \mathbf{R}^n$. We shall assume

(i) $\alpha = (\alpha_1, \alpha_2, \ldots, \alpha_n)$ is a vector with non-negative integer components α_j, $j = 1, 2, \ldots, n$

(ii) $|\alpha| = \Sigma_{j=1}^{n} \alpha_j$

(iii) $\mathbf{x}^\alpha = x_1^{\alpha_1} \mathrm{x}_2^{\alpha_2} \ldots x_n^{\alpha_n}$.

We shall denote by $D^\alpha f(x)$ the derivative of order $|\alpha|$ of the function f by

$$D^\alpha f(x) = \frac{\partial^{|\alpha|} f(x_1, x_2, \ldots, x_n)}{\partial_{x_1}^{\alpha_1} \partial_{x_2}^{\alpha_2} \ldots \partial_{x_n}^{\alpha_n}}$$

$$= D_1^{\alpha_1} D_2^{\alpha_2} \ldots D_n^{\alpha_n}, \quad D_j^{\alpha_m} = \frac{\partial^{\alpha_m}}{\partial x_j^{\alpha_m}}, \quad D_j = \frac{\partial}{\partial x_j}, \quad j = 1, 2, \ldots, n$$

with

$$D^0 f(x) = f(x)$$

A set of (complex-valued) functions f on $B \subset \mathbf{R}^n$ which are continuous together with their derivatives $D^\alpha f(x)$, $|\alpha| \leq p$ where $1 \leq p < \infty$ form a **class of functions** denoted $C^p(B)$.

Functions $f \in C^p(B)$ for which all the derivatives $D^\alpha f(x)$ with $|\alpha| \leq p$ allow continuous extension into the closure, $\bar{B}$, form a class of functions $C^p(\bar{B})$. We shall also write

$$C^\infty(B) = \bigcap_{p \geq 0} C^p(B), \quad C^\infty(\bar{B}) = \bigcap_{p \geq 0} C^p(\bar{B})$$

These classes of functions are linear sets. Furthermore, if, for example, we endow the class $C(B)$ with a norm by setting

$$\|f\| := \max_{x \in B} |f(x)|$$

then $C(B)$ is converted into a normed linear space. Similarly we can convert the class $C(\bar{B})$.

A set of functions $f \in M \subset C(B)$ is said to be **equicontinuous** on B if for any $\varepsilon > 0$ there exists a number $\delta(\varepsilon)$ such that for all $f \in M$ the inequality $|f(x_1) - f(x_2)| \leq \varepsilon$ holds whenever $|x_1 - x_2| < \delta(\varepsilon)$ where $x_1, x_2 \in B$.

A function $f \in C(B)$ is said to be **Hölder continuous** on B if there are numbers $c > 0$ and $0 < \alpha \leq 1$ such that for all $x_1, x_2 \in B$ the inequality

$$|f(x_1) - f(x_2)| \leq c|x_1 - x_2|^\alpha$$

holds. In the case when $\alpha = 1$ the function $f \in C(B)$ is said to be Lipschitz continuous on B.

A12.4 Sobolev Spaces

Let $\Omega \subseteq \mathbf{R}^n$ be an open set. The Sobolev space $W_p^m(\Omega)$, $m \in N$ and $0 \leq p \leq \alpha$ is the space of all distributions which together with their distributional derivatives of order less than or equal to m are associated with functions that are pth power integrable on Ω, that is they are elements of $L_p(\Omega)$. Such a collection is a Banach space with respect to the norm $\|\cdot\|_{m,p}$ defined by

$$\|f\|_{m,p} := \left\{ \int_\Omega \sum_{0 \leq |m| \leq m} \left| \frac{\partial^{|m|} f}{\partial x_1^{m_1} \partial x_2^{m_2} \ldots \partial x_n^{m_n}} \right|^p \right\}^{1/p}$$

where $|m| = m_1 + m_2 + \ldots + m_n$ and $m_j, j = 1, 2, \ldots, n$ are positive integers.

If $p = 2$ then $W_2^m(\Omega)$ is a Hilbert space usually denoted $H^2(\Omega)$.

Of particular interest in this monograph are the Hilbert spaces $H^1(\Omega)$ and $H^2(\Omega)$. These are defined as follows

$$H^2(\Omega) := \left\{ f \in L_2(\Omega) : \frac{\partial f(x)}{\partial x_j} \in L_2(\Omega), j = 1, 2, \ldots, n \right\}$$

with inner product

$$(f, g)_{1,2} = (f, g)_1 = (f, g) + (\nabla f, \nabla g)$$

and norm

$$\|f\|_1^2 = \|f\|^2 + \|\nabla f\|^2$$

where $(\cdot, \cdot)$ and $\|\cdot\|^2$ denote the usual $L_2(\Omega)$ inner product and norm respectively.

Similarly

$$H_2(\Omega) = \left\{ f \in L_2(\Omega) : \frac{\partial f(x)}{\partial x_j} \in L_2(\Omega), \frac{\partial^2 f(x)}{\partial x_j \partial x_k} \in L_2(\Omega), j, k = 1, 2, \ldots, n \right\}$$

with inner product and norm

$$(f, g)_{2,2} = (f, g) + (\nabla f, \nabla g) + (D^\alpha f, D^\alpha g)$$

$$\|f\|_2^2 = \|f\|^2 + \|\nabla f\|^2 + \|D^\alpha f\|^2$$

where α is a multi-index such that $|\alpha| = \alpha_1 + \alpha_2$.

An important result in the general theory of Sobolev spaces [1] is the celebrated **Sobolev lemma** which states that an element $f \in H^{[n/2]+1+k}(\Omega)$, where n is the dimension of Ω and $[n/2]$ means rounded down to an integer, is also such that $f \in C^k(\Omega)$. Thus as the dimension n of Ω increases higher order Sobolev spaces must be taken in order to guarantee the continuity of the elements (functions) which they contain.

For more details of the properties of Sobolev spaces, especially concerning boundary values (traces) and imbedding theorems see [1].

A12.5 Retarded Potentials

Let f and g be locally integrable functions on $\mathbf{R}^n$ and assume that the function h defined by

$$h(x) = \int_{\mathbf{R}^n} |g(y) f(x - y)| dy$$

is also locally integrable on $\mathbf{R}^n$. The function $f * g$ defined by

$$(f*g)(x) = \int_{\mathbf{R}^n} f(y)g(x-y)dy$$
$$= \int_{\mathbf{R}^n} g(y)f(x-y)dy = (g*f)(x)$$

is called the **convolution** of the functions f and g.

Let $L(D)$ be a differential expression with constant coefficients $a_\alpha(x) \equiv a_\alpha$ which has the typical form

$$L(D) = \sum_{|a_\alpha|=0}^{m} a_\alpha D^\alpha$$

where α is a multi-index (see Appendix A12.3).

A generalised function $\varepsilon \in D'$ which satisfies in $\mathbf{R}^n$ the equation

$$L(D)\varepsilon(x) = \delta(x)$$

is called a **fundamental solution** of the differential expression. It should be noted that in general a fundamental solution is not unique.

Using a fundamental solution, $\varepsilon(x)$, of the differential expression $L(D)$ it is possible to construct a solution, $u(x)$, of the equation

$$L(D)u(x) = f(x)$$

where f is an arbitrary function. In this connection the following theorem is a central result.

Theorem ([35], [88]). *Let $f \in D'$ be such that the convolution $\varepsilon * f$ exists in D'. Then the solution $u(x)$ exists in D' and is given by*

$$u(x) = (\varepsilon * f)(x)$$

Moreover, this solution is unique in the class of generalised functions in D' for which a convolution with ε exists.

The above notions are very useful when we come to deal with wave equation problems. In the particular case of acoustic waves the wave equation has the form

$$\Box_a u(x, t) := \{\partial_t^2 - a^2\Delta\}u(x, t) = f(x, t)$$

where $\Box_a$ is referred to as the d'Alembert expression (operator).

A fundamental solution of the acoustic wave equation in $\mathbf{R}^n$ is denoted by $\varepsilon_n(x, t)$ and satisfies

$$\Box_a \mathcal{E}_n(x, t) = \delta(x, t)$$

Fourier transform techniques provide solutions of this equation in the form ([35], [88], [112])

$$\mathcal{E}_1(x,t) = \frac{1}{2a}\theta(at - |x|)$$

$$\mathcal{E}_2(x,t) = \frac{\theta(at - |x|)}{2\pi a\sqrt{a^2t^2 - |x|^2}}$$

$$\mathcal{E}_3(x,t) = \frac{\theta(t)}{2\pi a}\delta(a^2t^2 - |x|^2)$$

where here θ denotes the Heaviside unit function defined by

$$\theta(x) = 1 \text{ for } x \geqslant 0, \quad \theta(x) = 0 \text{ for } x < 0$$

The generalised function V_n defined by

$$V_n(x, t) = \mathcal{E}_n(x, t) * f(x, t)$$

where $\mathcal{E}_n(x, t)$ is a fundamental solution of the d'Alembert equation and f is a generalised function on $\mathbf{R}^{n+1}$ which vanishes on the half space $t < 0$ and is called a **retarded potential** with density f. (With a slight abuse of notation we might write $f(x, t) \in D'(\mathbf{R}^{n+1})$.)

The retarded potential V_n will, according to the above theorem, satisfy the equation

$$\Box_a V_n(x, t) = f(x, t)$$

It can be shown [35], [112] that if f is a locally integrable function on $\mathbf{R}^{n+1}$ then V_n is a locally integrable function on $\mathbf{R}^{n+1}$ and assumes the following particular forms

$$V_1(x,t) - \frac{1}{2a}\int_0^t \int_{x-a(t-\tau)}^{x+a(t-\tau)} f(\xi, \tau)d\xi d\tau$$

$$V_2(x,t) = \frac{1}{2\pi a}\int_0^t \int_{S(x;a(t-\tau))}^{x+a(t+\tau)} \frac{f(\xi, \tau)}{\left\{a^2(t-\tau)^2 - |x-\xi|^2\right\}^{1/2}} d\xi d\tau$$

$$V_3(x,t) = \frac{1}{4\pi a^2}\int_{U(x;at)} \frac{f(\xi, t - |x-\xi|/a)}{|x-\xi|} d\xi$$

where

$$U(x;\, at) = \text{ball centre } x \text{ radius } at$$

$$S(x,\, at) = \text{surface of } U(x;\, at)$$

As an illustration of the use of retarded potentials recall [90] that the generalised Cauchy problem for the wave equation is to determine $u \in D'(\mathbf{R}^{n+1})$ which is zero for $t < 0$ and which satisfies

$$\Box_a u(x, t) := \{\partial_t^2 - a^2\Delta\}u(x, t) = f(x, t), \quad u(x, 0) = \varphi(x), \quad u_t(x, 0) = \psi(x)$$

where $f \in D'(\mathbf{R}^{n+1})$ with data $\varphi \in D'(\mathbf{R}^n)$ and $\psi \in D'(\mathbf{R}^n)$. It can be shown[134], [90] that this generalised Cauchy problem has a unique solution of the form

$$u(x, t) = V_n(x, t) + V_n^{(0)}(x, t) + V_n^{(1)}(x, t)$$

where

$$V_n(x, t) = \varepsilon_n(x, t)*f(x, t),\ V_n^{(0)}(x, t) = \varepsilon_n(x, t)*\psi(x),\ V_n^{(1)}(x, t) = (\varepsilon_n(x, t))_t*\psi(x)$$

This leads to the following classical solutions of the Cauchy problem for the wave equation [90].

n = 3 (Kirchhoff's formula)

$$u(x,t) = \frac{1}{4\pi a^2}\int_{U(x;at)} \frac{f(\xi, t - |x-\xi|a)}{|x-\xi|}d\xi + \frac{1}{4\pi a^2 t}\int_{S(x;at)} \psi(\xi)dS + \frac{1}{4\pi a^2}\frac{\partial}{\partial_t}\left\{\frac{1}{t}\int_{S(x;at)} \varphi(\xi)dS\right\}$$

n = 2 (Poisson's formula)

$$u(x,t) = \frac{1}{2\pi a}\int_0^t \int_{U(x;a(t-\tau))} \frac{f(\xi,\tau)}{(a^2(t-\tau)^2 - |x-\xi|^2)^{1/2}}d\xi d\tau + \frac{1}{2\pi a}\int_{U(x;at)} \frac{\psi(\xi)}{(a^2t^2 - |x-\xi|^2)^{1/2}}d\xi + \frac{1}{2\pi a}\frac{\partial}{\partial t}\int_{U(x;at)} \frac{\varphi(\xi)}{(a^2t^2 - |x-\xi|^2)^{1/2}}d\xi$$

n = 1 (d'Alembert's formula)

$$u(x,t) = \frac{1}{2a}\int_0^t \int_{(x-a(t-\tau))}^{(x+a(t-\tau))} f(\xi,\tau)d\xi d\tau + \frac{1}{2a}\int_{(x-at)}^{(x+at)} \psi(\xi)d\xi + \frac{1}{2a}\int_{(x-at)}^{(x+at)} \varphi(\xi)d\xi + \frac{1}{2}\{\varphi(x+at) + \varphi(x-at)\}$$

A12.6 An Illustration of the Use of Stone's Formula

For the purpose of illustration we consider an operator which occurs frequently in scattering problems. Specifically, let $A: H \supset D(A) \to H = L_2(\mathbf{R})$ be defined by

$$Au = u_{xx},\ u \in D(A)$$
$$D(A) = \{u \in H\colon u_{xx} \in H \text{ and } u(0) = 0\}$$

To use Stone's formula we must compute the resolvents, $R(t \pm i\varepsilon)$, of A. To this end, recalling the definition of A, we consider the boundary value problem

$$(A - \lambda I)v(x) = f(x), \quad x \in (0, \infty), \quad v(0) = 0 \tag{A12.1}$$

This is an ordinary differential epuation which has a solution given by

$$v(x) = (A - \lambda I)^{-1} f(x) = \int_0^\infty G(x, y) f(y) dy \tag{A12.2}$$

where $G(x, y)$, which is the Green's function for the problem (A12.1), is readily found to have the form [90]

$$G(x,y) = \begin{cases} \dfrac{(\exp i\sqrt{x})\sin(\sqrt{\lambda} y)}{\sqrt{\lambda}}, & 0 \le y \le x \\ \dfrac{(\exp i\sqrt{y})\sin(\sqrt{\lambda} x)}{\sqrt{\lambda}}, & 0 \le x \le y \end{cases}$$

We now define

$$\lambda_+ := t + i\varepsilon = Re^{i\theta}, \quad R = \sqrt{t^2 + \varepsilon^2}, \quad \theta = \tan^{-1}(\varepsilon / t)$$
$$\lambda_- := t - i\varepsilon = Re^{-i\theta}$$

and choose

$$\sqrt{\lambda_+} = R^{1/2} e^{i\theta/2}, \quad \sqrt{\lambda_-} = R^{1/2} e^{i(\pi - \theta/2)}$$

We then see that as $\varepsilon \downarrow 0$

$$\sqrt{\lambda_+} \to +\sqrt{t}, \quad \sqrt{\lambda_-} \to -\sqrt{t}$$

Therefore

$$R(t + i\varepsilon) - R(t - i\varepsilon) = (A - \lambda_+ I)^{-1} - (A - \lambda_- I)^{-1}$$

and on first writing (A12.2) out in full and collecting terms we obtain

$$\lim_{\varepsilon \downarrow 0} \{R(t + i\varepsilon) - R(t - i\varepsilon)\} f(x) = \frac{2i \sin(\sqrt{t} x)}{\sqrt{t}} \int_0^\infty f(y) \sin(\sqrt{t} y dy)$$

For convenience at this stage assume $f, g \in C_0[a, b]$, $a < b < \infty$. Stone's formula now reads

$$([E_b - E_a]f, g) = \frac{1}{2\pi i}\int_a^b \int_0^\infty \frac{2i\sin(\sqrt{t}x)}{\sqrt{t}} \int_0^\infty f(y)\sin(\sqrt{t}y)dy\overline{g(x)}\,dxdt$$

If we now set $s = \sqrt{t}$ and introduce

$$\tilde{f}(s) = \left(\frac{2}{\pi}\right)^{1/2} \int_0^\infty f(y)\sin(sy)dy$$

then

$$([E_b - E_a]f, g) = \int_{\sqrt{a}}^{\sqrt{b}} \tilde{f}(s)\overline{\tilde{g}(s)}ds$$

It can be shown [48] that $\sigma(A) \subset (0, \infty)$. Consequently, $E_a \to \Theta$ as $a \to 0$. Hence

$$(E_\lambda f, g) = \int_0^{\sqrt{\lambda}} \tilde{f}(s)\overline{\tilde{g}(s)}ds \tag{A12.3}$$

which in turn implies (write $g(s)$ in full and interchange the order of integration)

$$(E_\lambda f)(x) = \left(\frac{2}{\pi}\right)^{1/2} \int_0^{\sqrt{\lambda}} \tilde{f}(s)\sin(sx)ds =: \int_0^{\sqrt{\lambda}} \tilde{f}(s)\theta_x(s)ds$$

where $\theta_s(s)$ has been introduced for ease of presentation.

The spectral theorem indicates that for $f \in D(A)$

$$(Af)(x) = \int_0^\infty \lambda dE_\lambda f(x)$$

In the present case we obtain from (A12.3)

$$\begin{aligned} dE_\lambda f(x) &= \frac{d}{d\lambda}\left(\int_0^{\sqrt{\lambda}} \tilde{f}(s)\theta_x(s)ds\right)d\lambda \\ &= \frac{1}{2\sqrt{\lambda}}\tilde{f}(\sqrt{\lambda})\theta_x(\sqrt{\lambda})d\lambda \\ &= f(s)\theta_x(s)ds, \quad s = \sqrt{\lambda} \end{aligned}$$

and hence

$$\begin{aligned} (Af)(x) &= \int_0^\infty \lambda dE_\lambda f(x) = \int_0^\infty s^2 \tilde{f}(s)\theta_x(s)ds \\ &= \left(\frac{2}{\pi}\right)^{1/2} \int_0^\infty s^2 \tilde{f}(s)\sin(sx)ds \end{aligned} \tag{A21.4}$$

which we notice involves the Fourier sine transform of f. This is to be expected, bearing in mind results of a classical analysis of the boundary value problem (A12.1).

To see how this material can be used in the analysis of wave motions consider, as an example, the transverse vibrations of a string occupying the region $\Omega \subset \mathbf{R}^+$.

The associated wave motions are governed, typically, by an initial boundary value problem of the form

$$\{\partial_t^2 + A\}u(x, t) = 0, \ (x, t) \in \Omega \times \mathbf{R}^+ \tag{A12.5}$$

$$u(x, 0) = u_o(x), \ u_t(x, 0) = u_1(x), \ u(0, t) = 0 \tag{A12.6}$$

where $A: H \to H =: L_2(\Omega)$ is defined by

$$Au(x, t) = -\partial_x^2 u(x, t), \quad u \in D(A)$$

$$D(A) = \{u \in H: -\partial_x^2 u \in H \text{ s.t. } u(0, t) = 0\}$$

We have seen that a solution of this initial value problem can be written in the form

$$u(x,t) = (\cos(tA^{1/2}))u_0(x) + A^{-1/2}(\sin(tA^{1/2}))u_1(x) \tag{A12.7}$$

The spectral theorem and the above use of Stone's formula enales us to write (A12.7) in the form

$$\begin{aligned} u(x,t) &= \int_0^\infty \cos\left(\sqrt{\lambda}t\right) dE_\lambda u_0(x) + \int_0^\infty \frac{\sin\left(\sqrt{\lambda}t\right)}{\sqrt{\lambda}} dE_\lambda u_1(x) \\ &= \int_0^\infty \cos(st\tilde{u}_0)(s)\theta_s(x)ds + \int_0^\infty \frac{\sin(st)}{s} dE_\lambda \tilde{u}_1(s)\theta_s(x)ds \end{aligned} \tag{A12.8}$$

Once the initial conditions have been given explicitly then their Fourier sine transforms $\tilde{u}_0$ NS $\tilde{u}_1$, respectively, can be calculated as indicated above. Hence a completely determined representation of the solution to (A12.5), (A12.6) can be obtained from (A12.8) [60]. For example, consider the particular case when

$$u_0(x) = 0, \quad u_1(x) = 2\left\{\frac{\sin x}{x^2} - \frac{\cos x}{x}\right\} \tag{A12.9}$$

It then follows that

$$\tilde{u}_1(s) = 2s\left(\frac{2}{\pi}\right)^{1/2} \int_0^\infty \frac{\sin x \cos sx}{x} dx =: 2s\left(\frac{2}{\pi}\right)^{1/2} g(s)$$

To compute $g(s)$ take an arbitrary $\varphi \in C_0^\infty(\mathbf{R}^+)$ and consider

$$\begin{aligned} (g, \varphi') &= \lim_{R\to\infty} \int_0^\infty \int_0^\infty \frac{\sin x \cos(sx)}{x} \varphi'(s) ds dx \\ &= \left(\frac{\pi}{2}\right)^{1/2} \int_0^\infty \sin(x\tilde{\varphi}(x)) dx \\ &= \left(\frac{\pi}{2}\right)\varphi(1) \end{aligned} \tag{A12.10}$$

This last result follows from the fact that the Fourier sine transform is its own inverse. (Set $s = 1$ in the definition of the Fourier sine transform given above.)

Furthermore,

$$(g, \varphi') = (-1)(g', \varphi) = \frac{\pi}{2}(\varphi, \delta(s-1))$$

which implies

$$g'(s) = -\frac{\pi}{2}\delta(s-1) \tag{A12.11}$$

Integrating (A12.11) we obtain

$$g(s) - g(0) = -\frac{\pi}{2}\int_0^s \delta(\xi - 1)d\xi = -\frac{\pi}{2}H(s-1)$$

Since $g(0) = \frac{\pi}{2}$ we can re-write this in the form

$$g(s) = \frac{\pi}{2}\{1 - H(s-1)\} = \frac{\pi}{2}H(1-s)$$

and obtain

$$\tilde{u}_1(s) = 2s\left(\frac{\pi}{2}\right)^{1/2} H(1-s)$$

Consequently, in this particular case (A12.8) reduces to

$$u(x,t) = 2\int_0^1 \sin(st)\sin(sx)ds = \int_0^1 \{\cos(t-x)s - \cos(t+x)s\}ds$$
$$= \frac{\sin(t-x)}{t-x} - \frac{\sin(t+x)}{t+x}$$

which nicely displays the travelling wave components.

References and Further Reading

[1] R. Adams: *Sobolev Spaces*, Academic Press, New York, 1975.

[2] Z.S. Agranovich and V.A. Marchenko: *The Inverse Problem of Scattering Theory*, Gordon and Breach, New York, 1963.

[3] A.C. Aitken: *Determinants and Matrices*, Oliver and Boyd, Edinburgh, 1951.

[4] N.I. Akheizer and L.M. Glazman: *Theory of Linear Operators in Hilbert Space*, Pitman-Longman, London, 1981.

[5] H. Amann: *Ordinary Differential Equations, An Introduction to Nonlinear Analysis*, W. de Gruyter, Berlin, 1990.

[6] W.O. Amrein, J.M. Jauch and K.B. Sinha: *Scattering Theory in Quantum Mechanics*, Lecture Notes and Supplements in Physics, Benjamin, Reading, 1977.

[7] J. Arsac: *Fourier Transforms and Theory of Distributions*, Prentice Hall, New York, 1966.

[8] C. Athanasiadis, G.F. Roach and I.G. Stratis: A time domain analysis of wave motions in chiral materials, *Math. Nachr.* 250, 2003, 3–16.

[9] G.N. Balanis: The plasma inverse problem, *Jour. Math. Physics* 13(7), 1972, 1001–1005.

[10] G.N. Balanis, Inverse scattering: Determination of inhomogeneities in sound speed, *Jour. Math. Physics* 23(12), 1982, 2562–2568.

[11] G.R. Baldock and T. Bridgeman: *Mathematical Theory of Wave Motion*, Ellis Horwood, Chichester, 1981.

[12] H. Baumgärtel and M. Wollenberg: *Mathematical Scattering Theory*, Operator Theory: Advances and Applications, Birkhaüser-Verlag, Stuttgart, 1983.

[13] A. Belleni-Morante: *Applied Semigroups and Evolution Equations*, Clarendon Press, Oxford, 1979.

[14] A. Belleni-Morante and A.C. McBride: *Applied Nonlinear Semigroups*, Mathematical Methods in Practice, **3**, Wiley, Chichester, 1998.

[15] A. Belleni-Morante and G.F. Roach: A mathematical model for gamma ray transport in the cardiac region, *Jour. Math. Anal Appl.*, **244**, 2000, 498–514.

[16] A. Belleni-Morante and G.F. Roach: Gamma ray transport in the cardiac region: an inverse problem, *Jour. Math. Anal Appl.* **269**, 2002, 200–2115.

[17] A.M. Berthier: *Spectral Theory and Operators for the Schrödinger Equation*, Pitman Research Notes in Mathematics No. 71, Pitman, London, 1982.

[18] R. Burridge: The Gel'fand-Levitan, the Marchenko and the Gopinath-Sondhi integral equations of the inverse scattering theory regarded in the context of inverse impulse-response problems, *Wave Motion* **2**, 1980, 305–323.

[19] K. Chadan and P. Sabatier: *Inverse Problems in Quantum Scattering Theory*, Springer-Verlag, New York, 1977.

[20] D. Colton and R. Kress: *Inverse Acoustic and Electromagnetic Scacttering Theory*, Applied Mathematical Sciences No. 93, Springer-Verlag, Berlin, 1991.

[21] R. Courant and D. Hilbert: *Methods of Mathematical Physics Vol. II*, Wiley Interscience, New York, 1962.

[22] H.L. Cyon, R.G. Froese, W. Kirsh and B. Simon: *Schrödinger Operators with Applications to Quantum Mechanics and Global Geometry*, Texts and Monographs in Physics, Springer-Verlag, Berlin, 1981.

[23] J.W. Dettman: *Mathematical Methods in Physics and Engineering*, McGraw-Hill, New York, 1962.

[24] N. Dunford and J.T. Schwartz: *Linear Operators*, Vol. 1–3, Wiley Interscience, New York, 1958.

[25] M.S.P. Eastham: *Theory of Ordinary Differential Equations*, Van Nostrand, London, 1970.

[26] M.S.P. Eastham: *The Spectral Theory of Periodic Differential Equations*, Scottish Academic Press, Edinburgh, 1973.

[27] D.E. Edmunds and W.D. Evans: *Spectral Theory and Differential Operators*, Clarendon Press, Oxford, 1987.

[28] D.M. Eidus: The principle of limiting absorption, *Math. Sb.*, **57**(99), 1962 and *AMS Transl.*, **47**(2), 1965, 157–191.

[29] D.M. Eidus: The principle of limiting amplitude, *Uspekhi Mat. Nauk.* **24**(3), 1969, 91–156 and *Russ. Math. Surv.* **24**(3), 1969, 97–167.

[30] L.D. Faddeev: On the relation between S-matrix and potential for the one-dimensional Schrödinger operator, *Dokl. Akad. Nauk SSSR Ser. Mat. Fiz.* 121(1), 1958, 63–66 (in Russian).

[31] L.D. Faddeev: The inverse problem in the quantum theory of scattering, *Uspekhi Mat. Nauk* 14, 1959, 57–119 (in Russian); English translation: *Jour. Math. Physics* 4, 1963, 72–104.

[32] J. Fawcett: On the stability of inverse scattering problems, *Wave Motion* 6, 1984, 489–499.

[33] K.O. Friedrichs: *Spectral Theory of Operators in Hilbert Space*, Springer-Verlag, Berlin, 1973.
[34] I.M. Gel'fand and B.M. Levitan: On the determination of a differential equation from its spectral function, *Izv. Akad. Nauk SSSR Ser. Mat.* 15, 1951, 309–360 (in Russian); English translation: *Amer. Math. Soc. Transl.* 1, 1955, 253–304.
[35] M. Gelfand and G.E. Shilov: *Generalised Functions*, Academic Press, New York, 1964.
[36] J.A. Goldstein: Semigroups and second order differential equations, *J. Functional Anal.* **4**, 1969, 50–70.
[37] J.A. Goldstein: Abstract evolution equations, *Trans. American Math. Soc.* **141**, 1969, 159–185.
[38] J.A. Goldstein: *Semigroups of Linear Operators and Applications*, Oxford University Press, Oxford, 1986.
[39] E. Goursat: Cours d'analyse Mathématique, Paris, 1927.
[40] S.H. Gray: Inverse scattering for the reflectivity function, *Jour. Math. Physics* 24(5), 1983, 1148–1151.
[41] K.E. Gustafson: *An Introduction to Partial Differential Equations and Hilbert Space Methods*, John Wiley and Sons, New York, 1980.
[42] G. Helmberg: *Introduction to Spectral Theory in Hilbert Space*, Elsevier, New York, 1969.
[43] T. Ikebe: Eigenfunction expansions associated with the Schrodinger Operators and their application to scattering theory. *Arch. Rat. Mech. Anal.* **5**, 1960, 2–33.
[44] D.S. Jones: *Acoustic and Electromagnetic Waves*, Clarendon Press, Oxford, 1986.
[45] T.F. Jordan: *Linear Operators for Quantum Mechanics*, John Wiley & Sons, New York, 1969.
[46] L.V. Kantorovich and G.P. Akilov: *Functional Analysis in Normed Spaces*, Pergamon, Oxford, 1964.
[47] T. Kato: On linear differential equations in Banach spaces, *Comm. Pure Appl. Math.* **9**, 1956, 479–486.
[48] T. Kato: *Perturbation Theory for Linear Operators*, Springer, New York, 1966.
[49] I. Kay: The inverse scattering problem when the reflection coefficient is a rational function, *Comm. Pure Appl. Math.* 13, 1960, 371–393.
[50] R.E. Kleinman and R.B. Mack: Scattering by linearly vibrating objects, *IEEE Trans. Antennas & Propagation* AP-27 (3), 1979, 344–352.
[51] R.E. Kleinman and G.F. Roach: Boundary integral equations for the three dimensional Helmholtz equation, *SIAM Reviews*, **16**(2), 1974, 214–236.
[52] E. Kreyszig: *Introductory Functional Analysis with Applications*, Wiley, Chichester, 1978.
[53] R.J. Krueger: An inverse problem for a dissipative hyperbolic equation with discontinuous coefficients, *Quart. Appl. Math.* 34(2), 1976, 129–147.
[54] A. Kufner and J. Kadelec: *Fourier Series*, Illiffe, London, 1971.
[55] S.T. Kuroda: On the existence and the unitary property of the scattering operator, *Nuovo Cimento*, **12**, 1959, 431–454.
[56] J.A. LaVita, J.R. Schulenberg and C.H. Wilcox: The scattering theory of Lax and Phillips and propagation problems of classical physics, *Applic. Anal.* **3**, 1973, 57–77.
[57] P.D. Lax and R.S. Phillips: *Scattering Theory*, Academic Press, New York, 1967.
[58] H. Levin: *Unidirectional Wave Motions*, North Holland, Amsterdam, 1978.
[59] B.M. Levitan: *Inverse Sturm-Liouville Problems*, VNU Science Press, Utrecht, 1987.
[60] R. Leis: *Initial Boundary Value Problems of Mathematical Physics*, John Wiley & Sons, Chichester, 1986.

[61] R. Leis and G.F. Roach: A transmission problem for the plate equation, *Proc. Roy. Soc. Edinburgh* **99A**, 1985, 285–312.

[62] A.E. Lifschitz: *Magnetohydrodynamics and Spectral Theory*, Kluwer, Dordrecht, 1988.

[63] J. Lighthill: *Introduction to Fourier Analysis and Generalised Functions*, Cambridge University Press, Cambridge, 1958.

[64] B.V. Limaye: *Functional Analysis*, Wiley Eastern, New Dehli, 1981.

[65] W. Littman: Fourier transforms of surface-carried measures and differentiability of surface averages, *Bull. Amer. Math. Soc.* **69**, 1963, 766–770.

[66] E.R. Lorch: *Spectral Theory*, Oxford University Press, Oxford, 1962.

[67] M. Matsumura: Uniform estimates of elementary solutions of first order systems of partial differential equations, *Publ. RIMS, Kyoto Univ.* **6**, 1970, 293–305.

[68] M. Mabrouk and Z. Helali: The scattering theory of C. Wilcox in elasticity, *Math. Meth. in Appl. Sci.* **25**, 2002, 997–1044.

[69] M. Mabrouk and Z. Helali: The elastic echo problem, *Math. Meth. in Appl. Sci.* **26** , 2003, 119–150.

[70] V.A. Marchenko: Some problems in the theory of one-dimensional linear differential operators of second order. I. *Trudy Mosk. Ob.* 1, 1952, 327–420 (in Russian).

[71] V.A. Marchenko: The construction of the potential energy from the phases of the scattered wave, *Dokl. Akad. Nauk SSSR* 104, 1955, 695–698 (in Russian).

[72] A.C. McBride: *Semigroups of Linear Operators; an Introduction*, Pitman Research Notes in Mathematics No156, Pitman, London, 1987.

[73] I.V. Melnikova and A. Filinkov: *Abstract Cauchy Problems: Three Approaches*, Monographs and Surveys in Pure and Applied Mathematics, Vol. 120, Chapman and Hall, London, 2001.

[74] S.G. Mikhlin: *Integral Equations and their Applications to Certain Problems in Mechanics, Physics and Technology*, Pergamon Press, Oxford, 1957.

[75] K. Morgenröther and P.Werner: Resonances and standing waves, *Math. Meth. Appl. Sci.* **9**, 1987, 105–126.

[76] P.M. Morse and H. Feshbach: *Methods of Theoretical Physics* Vol 1,2 McGraw-Hill, New York, 1953.

[77] N.F. Mott and M.S.W. Massey: *The Theory of Atomic Collisions*, Oxford University Press, Oxford, 1949.

[78] A.W. Naylor and G.R. Sell: *Linear Operator Theory in Engineering and Science*, Holt Rinehart and Winston, New York, 1971.

[79] R.G. Newton: *Scattering Theory of Waves and Particles*, McGraw-Hill, New York, 1966.

[80] R.G. Newton: Inverse scattering, *J. Math. Phys.*, **23** , 1982, 594–604.

[81] L.P. Niznik: Inverse problem of nonstationary scattering, *SSSDokl. Akad. Nauk* R.196(5), 1971, 1016–1019 (in Russian).

[82] A. Olek: *Inverse Scattering Problems for Moving, Penetrable Bodies*, PhD Thesis, University of Strathclyde, Glasgow, 1997.

[83] A. Pazy: *Semigroups of Linear Operators and Applications to Partial Differential Equations*, Springer, New York, 1983.

[84] D.B. Pearson: *Quantum Scattering and Spectral Theory*, Academic Press, London, 1988.

[85] C.L. Perkeris: Theory of propagation of explosive sound in shallow water, *Geo. Soc. Am. Memoir* 27, 1948.

[86] E. Prugovecki: *Quantum Mechanics in Hilbert Space*, Academic Press, New York, 1981.

[87] M. Reed and B. Simon: *Methods of Mathematical Physics, Vols 1–4*, Academic Press, New York, 1972–1979.

[88] F. Riesz and B. Sz-Nagy: *Functional Analysis*, Ungar, New York, 1981.
[89] M.A. Rincon and I. Shih Lin: Existence and uniqueness of solutions of elastic string with moving ends, *Math. Meth. Appl Sci.*, **27**, 2004.
[90] G.F. Roach: *Greens Functions* (2nd Edn), Cambridge Univ. Press, London, 1970/1982.
[91] G.F. Roach: *An Introduction to Linear and Nonlinear Scattering Theory*, Pitman Monographs and Surveys in Pure and Applied Mathematics, Vol. 78, Longman, Essex, 1995.
[92] G.F. Roach and B. Zhang: A transmission problem for the reduced wave equation in inhomogeneous media with an infinite interface, *Proc. Roy. Soc. Lond.* **A436**, 1992, 121–140.
[93] G.F. Roach and B. Zhang: The limiting amplitude principle for wave propagation with two unbounded media, *Proc. Camb. Phil. Soc.* **112**, 1993, 207–223.
[94] G.F. Roach and B. Zhang: Spectral representations and scattering theory for the wave equation with two unbounded media, *Proc. Camb. Phil. Soc.* **113**, 1993, 423–447.
[95] G.F. Roach and B. Zhang: On Sommerfeld radiation conditions for wave propagation with two unbounded media, *Proc. Roy. Soc.* Edinburgh, 1992, 149–161.
[96] J. Rose, M. Cheney and B. De Facio: Three dimensional inverse plasma and variable velocity wave equations, *J. Math. Phys.*, **26**, 1985, 2803–2813.
[97] J. Rose, M. Cheney and B. De Facio: Determination of the wave field from scattering data, *Phys. Rev. Lett.* **57**, 1986, 783–786.
[98] W. Rudin: *Principles of Mathematical Analysis*, (3rd Ed), McGraw-Hill, New York, 1976.
[99] D.E. Rutherford: *Vector Methods*, Oliver and Boyd, Edinburgh, 1951.
[100] Y. Saito: An inverse problem in potential theory and the inverse scattering problem. *J. Math. Kyoto Univ.* **22**, 1982, 307–321.
[101] Y. Saito: Some properties of the scattering amplitude and the inverse scattering problem, *Osaka J. of Math.* **19**(8), 1982, 527–547.
[102] Y. Saito: An asymptotic behaviour of the S-matrix and the inverse scattering problem, *J. Math. Phys.* **25**(10), 1984, 3105–3111.
[103] Y. Saito: Inverse scattering for the plasma wave equation starting with large-t data, *J. Phys. A: Math. Gen.* **21**, 1988, 1623–1631.
[104] E. Sanchez-Palencia: *Non-homogeneous Media and Vibration Theory*, Lecture Notes in Physics Vol. 127, Springer-Verlag, Berlin, 1980.
[105] J. Sanchez Hubert and E. Sanchez-Palencia: *Vibration and Coupling of Continuous Systems: Asymptotic Methods*, Springer-Verlag, Berlin, 1989.
[106] M. Schechter: *Spectra of Partial Differential Operators*, North Holland, Amsterdam, 1971.
[107] E.J.P. Schmidt: On scattering by time-dependent potentials, *Indiana Univ. Math. Jour.* 24(10) 1975, 925–934.
[108] J.R. Schulenberger and C.H. Wilcox: Eigenfunction expansions and scattering theory for wave propagation problems of classical physics, *Arch. Rat. Mech. Anal.* 46, 1972, 280–320.
[109] L. Schwartz: *Mathematics for the Physical Sciences*, Hermann, Paris, 1966.
[110] N.A. Shenk: Eigenfunction expansions and scattering theory for the wave equation in an exterior domain, *Arch. Rational Mechanics and Anal.*, **21**, 1966, 120–1506.
[111] E. Skudrzyk: *The Foundations of Acoustics*, Springer-Verlag, New York, 1977.
[112] V.I. Smirnov: *Course of Higher Mathematics*, Pergamon, New York, 1965.
[113] I.N. Sneddon: *Fourier Transforms*, McGraw-Hill, New York, 1951.
[114] P.E. Sobolevski: Equations of parabolic type in a Banach space, *Amer. Math. Soc. Transl.* **49**, 1996, 1–62.
[115] B. Spain: *Tensor Calculus*, Oliver and Boyd, Edinburgh, 1956.

[116] M.H. Stone: *Linear Transformations in Hilbert Space and their Appliations to Analysis, Amer. Math. Soc. Coll. Publ.* **15**, Providence, RI, 1932.

[117] W.A. Strauss: *Partial Differential Equations: An Introduction*, Wiley, New York, 1992.

[118] W.W. Symes: Inverse boundary value problems and a theorem of Gel'fand and Levitan, *Jour. Math. Anal. Applic.*, 71, 1979, 379–402.

[119] H. Tanabe: On the equation of evolution in a Banach space, *Osaka Math. J.* **12**, 1960, 363–376.

[120] H. Tanabe: *Evolution Equations*, Pitman Monographs and Studies in Mathematics, Vol. 6, Pitman, London, 1979.

[121] J.R. Taylor: *Scattering Theory: The Quantum Theory of Non-Relativistic Collisions*, University of Boulder, Colorado.

[122] E.C. Titchmarsh: *Introduction to the Theory of Fourier Integrals*, Oxford Univ. Press, 1937.

[123] F. Tricomi: *Integral Equations*, Interscience, New York, 1957.

[124] B. Vainberg: *Asymptotic Methods in Equations of Mathematical Physics*, Gordon and Breach, New York, 1989.

[125] V.S. Vladymirov: *Equations of Mathematical Physics*, Marcel Dekker, New York, 1971.

[126] H.F. Weinberger: *A First Course in Partial Differential Equations*, Blaisdell, Waltham, MA, 1965.

[127] P. Werner: Regularity properties of the Laplace operator with respect to electric and magnetic boundary conditions, *J. Math. Anal. Applic.* **87**, 1982, 560–602.

[128] P. Werner: Resonances in periodic media, *Math. Methods in Appl. Sciences*, 14, 1991, 227–263.

[129] V.H. Weston: On the inverse problem for a hyperbolic dispersive partial differential equation, *Jour. Math. Physics* 13(12), 1972, 1952–1956.

[130] C.H. Wilcox: Scattering states and wave operators in the abstract theory of scattering, *Jour. Functional Anal.* 12, 1973, 257–274.

[131] C.H. Wilcox: *Scattering Theory for the d'Alembert Equation in Exterior Domains*, Lecture Notes in Mathematics, No.442, Springer, Berlin, 1975.

[132] C.H. Wilcox: Transient Electromagnetic Wave Propagation in a Dielectric Waveguide, Proc. Conf. on the Mathematical Theory of Electromagnetism, Instituto Nazionale di Alto Mathematica, Rome, 1974.

[133] C.H. Wilcox: Spectral analysis of the Perkeris operator in the theory of acoustic wave propagation in shallow water, *Arch. Rat. Mech. Anal.* 60, 259–300, 1976.

[134] J. Wloka: *Partial Differential Equations*, Cambridge University Press, Cambridge, 1987.

[135] K. Yosida: *Functional Analysis*, Springer-Verlag, Berlin, 1971.

Index

Printed in the United States
113163LV00002B/223-246/P